Comparative ecology

Comparative ecology

Y. ITÔ, D.SC.

Associate Professor, Faculty of Agriculture, Nagoya University

EDITED AND TRANSLATED BY
JIRO KIKKAWA, D.SC.

Professor, Department of Zoology, University of Queensland

CAMBRIDGE UNIVERSITY PRESS

Cambridge
London New York New Rochelle
Melbourne Sydney

Published by the Press Syndicate of the University of Cambridge
The Pitt Building, Trumpington Street, Cambridge CB2 1RP
32 East 57th Street, New York, N.Y. 10022, U.S.A.
296 Beaconsfield Parade, Middle Park, Melbourne 3206, Australia

Hikaku Seitaigaku (*Comparative Ecology*, 2nd edition) by Yosiaki Itô
Originally published in Japanese by Iwanami Shoten, Publishers, Tokyo, 1978

English translation first published 1980

Photoset and printed in Malta by Interprint Ltd

British Library Cataloguing in Publication Data

Ito, Yosiaki
Comparative ecology.
1. Ecology
I. Title
574.5 QH541 79-41581
ISBN 0 521 22977 4 hard covers
ISBN 0 521 29845 8 paperback

CONTENTS

PREFACE TO THE ENGLISH EDITION

This is an edited translation of the second Japanese edition (1978) of Dr Yosiaki Itô's *Comparative Ecology* [*Hikaku Seitaigaku*], Tokyo: Iwanami. In translating it I have followed Dr Itô's interpretations, views and theories faithfully, though in factual detail this edition is different in places from the Japanese edition. I have consulted the author on numerous points which I thought might create difficulty in understanding if translated without further information or deliberation. I must thank Dr Itô for his time and patience in listening to my questions and discussing his theories at length. In some cases it meant checking of the original work that the author had cited in the Japanese edition and rephrasing of quotations. In other cases the author provided additional materials or new diagrams to replace those in the Japanese edition. His help with the tables and figures was indeed indispensable as was his advice on the phonetics of Japanese proper nouns.

In editing the text the author's style of presentation was preserved as far as possible. Most footnotes in the Japanese edition were incorporated into the text where they were appropriate. Where it was found that the flow of discussion would be interrupted I have used square brackets to indicate the insertion of notes. In a few cases in which the information contained was applicable only to Japanese readers the footnote was omitted with the author's consent.

The translation was made possible by a leave granted to me by the University of Queensland, the accommodation provided by the Laboratory of Applied Entomology and Nematology, Nagoya University, and the Primate Research Institute, Kyoto University, and above all by the encouragement received from the publisher and many of our colleagues in Japan and English-speaking countries. I am especially grateful to Alan Crowden, Martin Walters and Peter Silver of Cambridge University Press for their advice and various arrangements. Iwanami, the Japanese publisher, made the original drawings available, and many authors and publishers who granted permission to use their published materials in the Japanese edition gave permission to reproduce them in the English edition.

J. K.

PREFACE TO THE SECOND JAPANESE EDITION

It is eighteen years since the first edition of this book was published. Until recently the book has been reprinted, though in small numbers, at a rate of once every one or two years. This is very gratifying to the author of a natural science book.

As in other fields of science the progress of field zoology in the past twenty years has been explosive. For the first edition only a few papers giving life tables of field populations were available, but today the number of such papers far exceeds one hundred. I tabulated all known social structures of primates in the first edition because only nine species were involved. Today there are more than fifty species of primates for which there is at least one reported study of social structure. Furthermore, we now know social relationships of such animals as the crocodile, lion and spotted hyaena, which we did not dream of learning then. Of course I have examined the validity of my earlier theory in the light of these new findings. However, the new facts fortunately did not alter the fundamental point of my thesis; even though many modifications were necessary as a result, they seemed to re-inforce my theory.

Another incident which occurred in the ensuing period was the proposal of an *r-K* strategy theory by MacArthur and Wilson about ten years ago. This theory now dominates the American ecological scene. In revising my book I naturally had to determine my attitude towards this theory. I came to the conclusion that the acceptable side of MacArthur–Wilson theory was close to what I had asserted in my book, regretably only in Japanese, and that my original idea appeared rather closer to truth than theirs in other respects. Thus I have decided to re-publish the book in a completely revised form.

In the first edition I cited most of the published work on animal population dynamics and social behaviour, but to compile such a bibliographic monograph would be not only almost impossible today for an individual but it would also make the book unnecessarily bulky. In revising the book I have given up citing all relevant literature, instead I have referred only to some important work. Of course subjectivity cannot be avoided in selecting the literature, but I do not think I have been biased too much in this. A large weakness resulted rather more from the shortage of literature on the extraordinarily diverse life histories of animals in the tropical rain forest, from which I could not find enough examples for the present discussion. When research is advanced in this region the book will have to be revised extensively. I would be satisfied if this book would be found to play a small predictive role for that occasion.

Since MacArthur and Wilson there has been a proliferation of mathematical papers on the evolution of life history strategies, but I have omitted most of them. The purpose of this book is not to present mathematical discussions but to provide a new generation of researchers with as wide a range of materials as possible to examine the validity of my theory with this sort of approach, which could not be achieved in the first edition.

A few remarks are in order concerning the use of scientific names in the book. It is nearly impossible to find from ever-changing nomenclature the most appropriate name for every species mentioned in a book such as this, which deals with the entire animal kingdom. Thus I followed the authors cited, as a rule, in the use of scientific names (including the taxa above the genus), though some of them might be of doubtful status. I apologize for any possibility that the same species might be given two different names. [The scientific names have been checked and updated as far as possible in the English edition.]

The assistance of many people was necessary, as it was with the first edition, to complete this edition while I was working in the remote island of Okinawa as an applied entomologist involved in practical problems of pest control. The following people willingly provided me with copies of the literature needed: Kazuki Miyashita, Kazuo Nakamura, Takeshi Yushima, Syoiti Tanaka, Tadao Matsumoto, Hiroshi Hasegawa, Akira Suzuki, Eiji Kuno, Yūiti Ono, Seiji Goshima and Yoshimi Hirose. Professor S. F. Sakagami kindly spared a whole day to give me private tutoring for writing Chapter 5 and also critically read the manuscript of that chapter for me. Professor Masao Kawai gave me many useful comments on the draft of Chapters 4 and 6. As a result these chapters have been improved considerably but since not all advice has been followed the ultimate responsibility for the contents must rest with me alone. Professor Syun'iti Iwao (Chapter 1), Professor Sinzō Masaki and Dr Yasuyuki Sirota (Chapters 1–4) and members of Kyushu University particularly a group of graduate students from the Laboratory of Biological Control (Chapters 1–3) listened to my talks and debated the contents for me. Dr Eric Pianka of the University of Texas and Professor Gordon Orians of the University of Washington each kindly spent nearly a day listening to my ideas and commenting on them. Equally invaluable was the expert advice on various groups of animals received from the following people: Professor Fujio Yasuda (fishes), Dr Nobuhiko Mizuno (freshwater fishes). Dr Shigemitsu Shokita (freshwater shrimps), Dr Takao Yamaguchi (crabs), Dr Masafumi Matsui (frogs), Mr Hiroshi Hasegawa (birds) and Professor Hideo Obara (mammals). I thank them all most sincerely. Finally, I wish to express my gratitude to Dr Minoru Murai who came to Okinawa with me five years ago, bringing with him a large library of useful literature, and who has discussed the ideas with me endlessly since; to those members of Okinawa Agricultural Experimental Station, who have been involved in the designated experiments on sugar cane pests and fruit-flies; to Miss Midori Hoshino for helping with the reference list and index; and to Mr Hideo Arai of Iwanami-shoten, who has given much encouragement throughout the period of the preparation of this book since the publication of the first edition.

February 1978 Yosiaki Itô at Naha-shi, Okinawa

1

Evolution of reproductive rates

Introduction

The relationship between reproduction and death in a species determines the position of the species in the natural world. It appears to be Darwin (1859) who first paid due attention to the importance of interactions between these two contradictory phenomena. He discussed various factors affecting mortality in the third chapter 'Struggle for Existence' of *The Origin of Species* ... He himself investigated the survival-rate of seedlings in a prescribed area by marking the plants. Darwin also considered the reproductive rate as a factor in the maintenance of a population, but his idea was later popularized and misconstrued. As a result, the simple idea has spread that the variations in the reproductive rate are due to the fact that 'animals reproduce enough to match the number of deaths'.

Wunder's (1934) review on the number of eggs produced by the reptiles showed that unlike fishes (generally laying more than 10 000) or amphibians (often laying more than 1000) the reptiles seldom lay more than 100 eggs at a time. This suggests that the number of eggs laid decreases with evolutionary advancement. When examined in detail, however, one finds a great deal of variation even in one group of animals. Among the fishes, for example, the sunfish lays as many as several hundred million eggs while the stickleback produces only a few eggs at a time. Among the frogs an egg mass of *Rana nigromaculata* contains about 1000 eggs but the number of eggs laid by *Flectonotus pygmaeus* ranges from four to seven, which is smaller than the clutch size of many birds or the litter size of rats. Accepting that 'the number of births matches the number of deaths', the question is: what evolutionary forces are responsible for the direction of changes in the reproductive capacity?

Is a mere 'economic' viewpoint sufficient to explain the reduction of reproductive capacity through evolution? As Tokuda (1957) has pointed out, the reproductive capacity of animals is a product of 'the effort for survival' in given conditions whether its evolution is towards high fecundity or low fecundity.

Theories on the evolution of reproductive rates

The reproductive rate is the number of eggs or young produced by a female in her lifetime. Because this is not always known the number produced at a time is often referred to as the reproductive rate for convenience. [Cole (1954) discussed the importance in population dynamics of the distinction between those organisms that die after one reproductive season (e.g. annual plants, many insects, salmon) and those that reproduce more than once (e.g. many perennial plants, birds, many mammals). He called the former semelparity and the latter iteroparity. According to his calculations for an annual species the maximum benefit that a species gains in intrinsic population growth by changing from semelparity to iteroparity would be equivalent to adding one individual to the average clutch size (or litter size) without considering density-dependent mortality. Thus in animals with a large reproductive capacity this effect is very limited (see also Bryant, 1971).]

Lack (1954a) found the following ecological tendencies in the reproductive rates of animals.

1. The reproductive rate, like other characteristics of animals, is a product of natural selection and tends to produce within each species that number of eggs which results in the maximum number of surviving offspring.
2. In animals which do not feed their young, the number of eggs laid is probably the maximum that the parent can produce. In many species clutch size can be modified within limits to suit the particular feeding conditions. For example, in an environment in which food is scarce and predation is rare, animals tend to lay larger but fewer eggs while under conditions of abundant food supply and high predation pressure animals tend to lay smaller but many eggs.
3. In animals which feed their young, the number of eggs laid is inversely correlated with the degree of parental care and clutch size is limited by the number of young that the parents can feed. Parental care is developed particularly in an environment where food is scarce.

As I shall discuss later the above tendencies of reproductive rates appear to hold for organisms in general and we refer to these as Lack's laws.

Lack's laws imply that the reproductive rate is no longer accepted merely as 'balancing mortality' but as a positive adaptation by organisms through their 'effort for survival', whether the evolution is towards greater or smaller reproductive capacity. Specially important in Lack's arguments is the understanding that animals have always faced the choice between the two contrasting evolutionary strategies of high fecundity/no parental care and low fecundity/parental care. I have stressed this point of Lack's in the first edition (1959) of this book.

It is widely accepted that among species of similar adult weights those laying large eggs have low fecundity while those laying small eggs have high fecundity; this fact reflects the limited energy source for the production of eggs. If animals try to utilize limited energy in rearing young, this effort is incompatible with high fecundity. This allocation of energy was examined by Cody (1966) using 'the principle of allocation'. [The principle of allocation in the adaptation of phenotypes is given by a generalized fitness curve in Levins (1968).]

Cody (1966) examined the data on clutch size and attempted an expansion of Lack's theory. According to Cody the evolutionary strategies of birds in a particular environment are composed of (1) clutch size, (2) escape from predators, and (3) the ability to succeed in intraspecific competition. Cody expressed these adaptive functions in a three-

Fig. 1.1 Cody's (1966) scheme showing changes of clutch size from temperate to tropical latitudes for mainland open-nesting passerines ($P' \rightarrow P$), predation-free species ($P' \rightarrow P''$), hole-nesters ($H' \rightarrow H$) and island-nesting species ($Q \rightarrow Q'$). In the tropics the clutch size is reduced in spite of increased competition and predation pressure.

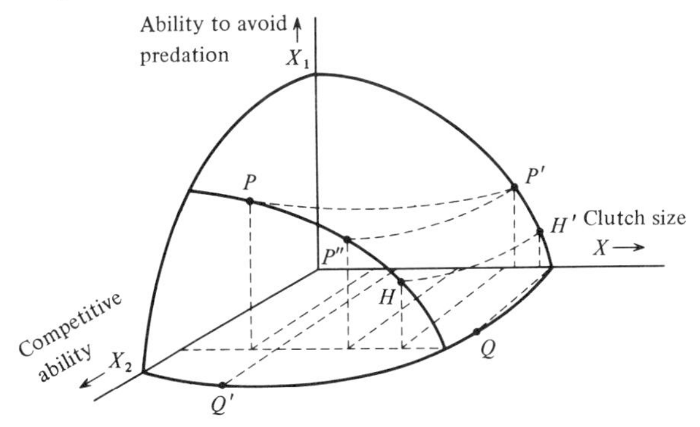

dimensional diagram (Fig. 1.1). In this diagram the convex surface of something like one-eighth of a sphere represents the phenotypes of species. In the temperate region where the climatic conditions are severe and, in the case of the migratory birds, mortality is high, the density is kept low and competition is avoided. Where there is no intraspecific competition, open-nesting passerine birds (under a moderately high risk of predation) would take the phenotype P' with a large clutch size (X axis). On the other hand, even in the same genus the tropical species under favourable and stable climatic conditions may achieve a high density resulting in severe competition. To be successful in competition the young must be more developed at hatching or fledging than in low-density populations, and by the principle of allocation the clutch size must be reduced ($P' \rightarrow P''$). In fact, Lack (1968) in his wide search of the literature on this topic found that the tropical species have smaller clutch sizes than the temperate counterparts among the passerines, the tree-nesting non-passerines and ground-nesters.

Cody argues that the severity of competition is in some way related to climatic stability; the more stable the environment the higher the population density and the severer the competition. In equatorial Africa Moreau (1944) found the clutch size of rain forest birds to be smaller than that of (seasonal) rain-green forest birds. The oceanic birds nesting on cliffs of islands also have small clutch sizes (normally one or two eggs) and this fact is considered to reflect the climatic stability of the ocean.

Cody's hypothesis is good in incorporating the idea of energy economics but differs partially from Lack's. It is not totally acceptable: Fig. 1.1 shows a decrease in clutch size with increasing predation pressure, but this is contrary to the fact as will be discussed later. Cody himself has presented data showing that the clutch size of island birds (considered to have low predation pressure) is smaller than that of related continental species though under low predation pressure the clutch size is supposed to increase.

Safriel (1975) has examined clutch sizes of birds whose young feed themselves (e.g. Galliformes, Anseriformes), and has arrived at the following equation:

$$\text{clutch size} \propto \frac{(\text{protection}) \times (\text{density of food})}{\text{no. of items taken by one young})}$$

This equation matches Lack's laws, but Safriel's data are limited in coverage.

Dobzhansky in 1950 (cited by MacArthur & Wilson, 1967) argued that natural selection in the tropics was fundamentally different from that in the temperate region; in the temperate region most of the deaths would be unrelated to particular genotypes and independent of population density whereas in the tropics most deaths would occur favouring the genotypes with greater competitive ability. In this view natural selection would favour those with high fecundity and fast growth in the temperate region and those with low fecundity and slow growth in the tropics.

MacArthur (1962) further advanced this theory. In the tropics (here the tropics means humid tropics where rain forest is the climax) the climate is stable and unpredictable natural disasters, such as blizzards, cyclones and drought, are rare, so that the animal populations have reached a level close to the carrying capacity of the environment – near the saturation density K in the logistic model of population growth. If so, the species that could utilize the carrying capacity more effectively – the species that could reach a higher level of K in the same environment (Fig. 1.2) – would be at an advantage. On the other hand, in the unstable environment where unpredictable disasters occur frequently (temperate–subarctic regions, the arid zone, the lower parts of large rivers and the marine zone susceptible to sudden changes of ocean currents), the high reproductive capacity to compensate for the catastrophes would be advantageous.

One of the measures of the reproductive capacity is r, the intrinsic rate of natural increase (an instantaneous rate of increase during the period of geometric growth of the population, see Fig. 1.2). Species living in unstable environments would have greater values of r than those living in stable environments. The expression is: $r \simeq \ln R_0/T$, where R_0 is the mean number of female young produced per female and T is the mean period of

Fig. 1.2. *A*, Geometric growth curve of a population $dN/dt = rN$; *B*, logistic growth curve of a population $dN/dt = rN(1-N/K)$. The diagram on the right gives increase of *A* in logarithmic scale.

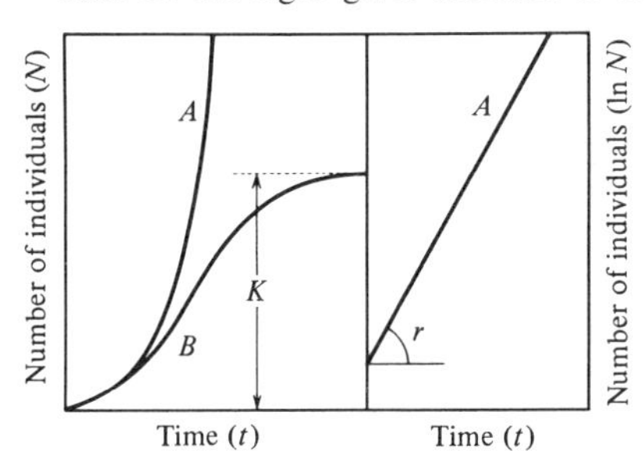

one generation. Thus r changes proportionally with the logarithm of the number of young and is inversely proportional to the length of time taken from birth to reproduction. Species with a large value for r produce a large number of young or breed early (the generation time is short) or both. [Of the two parameters (the number of young and the generation time) the latter is more important than the former (Lewontin, 1965; and others). Thus the 'large r' does not necessarily mean 'high fecundity'. Nevertheless, many papers being published on reproductive strategies, mainly in North America, have generalized the expression of high-fecundity species as r-strategists. I have followed a similar line of argument in this book but have not neglected the dual character of r. It will become clear in the text that this duality is the cause of contradiction in the theory of r-K strategies.]

Species living in stable environments, such as tropical forests, do not require a large r but instead show adaptation to win the competition with other species. In animals this competitive ability generally takes the form of mobility and, in the case of plants, the ability to grow fast and tall using the nutrients stored in the seed. Such eggs or seeds are generally large, and this parallels with the degree of parental protection. In the principle of allocation this is incompatible with a large r.

MacArthur & Wilson (1967) called the selection that favours larger r the r-selection and one that favours increased competitive ability the K-selection. The r-selected species became known as r-strategists and the K-selected species as K-strategists. The characteristics of r-selection and K-selection summarized by Pianka (1970) are reproduced in Table 1.1.

The viewpoint reached by MacArthur & Wilson (1967) is that species have always faced the crossroad of two contrasting evolutionary pathways and have survived by selecting one or the other strategy and that what determines the strategy to be adopted has been the interrelation of two contrasting factors – birth and death. The recognition of these two points is extremely valuable and generally conforms with Lack's viewpoint. In recent years the concept of r-K selection has been used widely to explain evolutionary strategies of organisms and has been discussed in relation to the problems of species becoming weeds or pests (see p. 31). However, the view of MacArthur & Wilson is different from Lack's in some respects. They consider that competition is severe in the region of a rich biota and evolution here has favoured low fecundity. The risk of predation may also be high in such a region. Lack's theory indicates that in such an environment high fecundity rather than low fecundity would have evolved; low fecundity would have evolved in the region of low

Table 1.1. *Some of the correlates of r- and K-selection*

	r-selection	*K*-selection
Climate	Variable and/or unpredictable: uncertain	Fairly constant and/or predictable: more certain
Mortality	Often catastrophic, non-directed, density-independent	More directed, density-dependent
Survivorship curve[a]	Often Type C (III) (Deevey, 1947)	Usually Type A (I) and B (II) (Deevey, 1947)
Population size	Variable in time, non-equilibrium; usually well below carrying capacity of environment; unsaturated communities or portions thereof; ecological vacuums; recolonization each year	Fairly constant in time, equilibrium; at or near carrying capacity of the environment; saturated communities; no recolonization necessary
Intra- and interspecific competition	Variable, often lax	Usually keen
Relative abundance	Often does not fit MacArthur's broken stick model (King, 1964)	Frequently fits the MacArthur model (King, 1964)
Selection favours	1. Rapid development 2. High r_{max}[b] 3. Early reproduction 4. Small body size 5. Semelparity: single reproduction	1. Slower development, greater competitive ability 2. Lower resource thresholds 3. Delayed reproduction 4. Larger body size 5. Iteroparity: repeated reproductions
Length of life	Short, usually less than one year	Longer, usually more than one year
Leads to	Productivity	Efficiency

After Pianka (1970); MacArthur & Wilson first proposed the two strategies as a solution to the problem of island colonization but Pianka extended them to all organisms.
[a]See Fig. 2.1.
[b]High intrinsic rate of increase is used synonymously with high reproductive capacity. See text for further explanation.

species diversity such as mountain streams. I shall discuss this problem fully after surveying the facts (p. 38). For the moment we shall examine the validity of Lack's laws.

Egg numbers of fishes

Table 1.2 summarizes the number and the type of eggs laid (or contained in one fish – usually exuded in one spawning act) by fishes. Amongst the teleosts the sardine, mackerel, yellowtail, etc. lay pelagic eggs in tens of thousands to hundreds of thousands. In the sunfish the number reaches 200 million. These eggs are extremely small (mostly less than 0.5 mm in diameter) compared with the size of parent. In tri-

Table 1.2. *Clutch (brood) size of Japanese fishes*

Common name	Scientific name	Clutch (litter) size	Type of egg
1. Gummy shark	*Mustelus manazo*	10 ind.	Ovoviviparous
2. Blue shark	*Prionace glauca*	60 ind.	Ovoviviparous
3. Sawshark	*Pristiophorus japonicus*	12 ind.	Ovoviviparous
4. Sea wall ray	*Raja kenojei*	4–5	Laid in case
5. Red skate	*Dasyatis akajei*	10 ind.	Ovoviviparous
6. Green sturgeon	*Acipenser medirostris*	8×10^5–21×10^5	Adhesive
7. Sardine	*Sardinops melanosticta*	5×10^4–8×10^4	Pelagic
8. Pacific herring	*Clupea pallasi*	30 000[a]	Adhesive
9. Glassfish	*Salangichthys microdon*	600–800[a]	Adhesive
10. Chum salmon	*Oncorhynchus keta*	$2 \times 10^3 - 3 \times 10^3$	Demersal
11. Kokanee salmon[b]	*O. nerka* var. *adonis*	100–800	Demersal
12. Cherry salmon[b]	*O. masou* var. *ishikawae*	200[a]	Demersal
13. Ayu[b]	*Plecoglossus altivelis*	10 000	Adhesive
14. Carp[b]	*Cyprinus carpio*	4×10^5–6×10^5	Adhesive
15. Crucian carp[b]	*Carassius auratus*	1×10^5–2×10^5	Adhesive
16. Willow gudgeon[b]	*Gnathopogon caerulescens*	2×10^3–8×10^3	Demersal
17. Oily gudgeon[b]	*Sarcocheilichthys variegatus*	1×10^3–2×10^3	Demersal in shells
18. Japanese dace[b]	*Tribolodon hakonenses*	10×10^3–15×10^3	Demersal
19. Common minnow[b]	*Zacco platypus*	1200[a]	Demersal
20. Medaka[b]	*Orizias latipes*	500–800	Adhesive
21. Loach[b]	*Misgurnus anguillicaudatus*	*ca* 10 000	Adhesive
22. Japanese eel[b]	*Anguilla japonica*	5×10^6	In deep sea
23. Japanese tubesnout[b]	*Aulichthys japonicus*	*ca* 200	Laid inside body of sea squirt
24. Stickleback[b]	*Gasterosteus aculeatus*	*ca* 50	Care of eggs and young
25. Japanese mackerel	*Scomber japonicus*	4×10^5	Pelagic
26. Yellowtail	*Seriola quinqueradiata*	18×10^5[a]	Pelagic
27. Jewfish	*Stereolepis ischinagi*	24×10^6	Pelagic
28. Sailfish sandfish	*Arctoscopus japonicus*	270–2000	Adhesive
29. Sea chub	*Ditrema temmincki*	10–24 ind.	Ovoviviparous, care of young
30. Ocean sunfish	*Mola mola*	2×10^8	Pelagic
31. Oblong rockfish	*Sebastichthys mitsukurii*	6000[a]	Ovoviviparous
32. Japanese sculpin[b]	*Cottus hilgendorfi* (large egg type)	*ca* 300[a]	Adhesive, care of eggs
33. Lizard goby[b]	*Rhinogobius flumineus*	100[a]	Adhesive, care of eggs
34. Bastard halibut	*Paralichthys olivaceus*	45×10^4	Pelagic
35. Dusky sole	*Lepidopsetta mochigarei*	51×10^4–55×10^4	Pelagic & adhesive
36. Pacific cod	*Gadus macrocephalus*	2×10^6–3×10^6	Pelagic
37. Weasel fish	*Brotula multibarbata*	2–7	Eggs covered with gelatin

Mainly from Okada (1955).
[a] Data were taken from Suehiro (1942) and Miyadi *et al.* (1976).
[b] Freshwater fish.

chodontid fishes whose eggs adhere to algae and other objects in the sea relatively small numbers of eggs are laid, but in the sole, cod and herring whose semi-adhesive eggs are attached to floating weeds the number laid is large. A typical adhesive egg is about 1 mm in diameter. The freshwater fishes and anadromous fishes, such as salmon and trout, lay fewer eggs and the egg size is large in proportion to the body length of adults. Many freshwater fishes spawn on the river bed and the eggs are attached to rocks or covered with sand or gravel. In many river fishes the protection of eggs, and the production of fry that can swim, is advantageous even with a reduction in the number of eggs laid, because streams and small oligotrophic lakes are poor in food organisms, such as plankton and small plankton feeders, and compared with the marine environment contain fewer predators. Among freshwater fishes carp and crucian carp lay comparatively large numbers of eggs, but it is worth noting that their habitat is eutrophic. These habitat relations of egg numbers are shown in Fig. 1.3 (left) based on the data from Table 1.2.

The numbers of eggs laid are naturally influenced by the body length

Fig. 1.3. The relation between the habitat and the number of eggs (fry) laid by fish. Open circles, species which protect eggs or which are ovoviviparous; solid circles, species which do not protect eggs; crosses, ovoviviparous elasmobranchs (sharks and rays). S, sea; F. freshwater; L, lakes and lower parts of rivers; U, upper parts of rivers. Species numbers refer to Table 1.2. Arrows on the bottom indicate the direction from spawning areas to maturing areas. Average body lengths and egg numbers were obtained from handbooks etc.

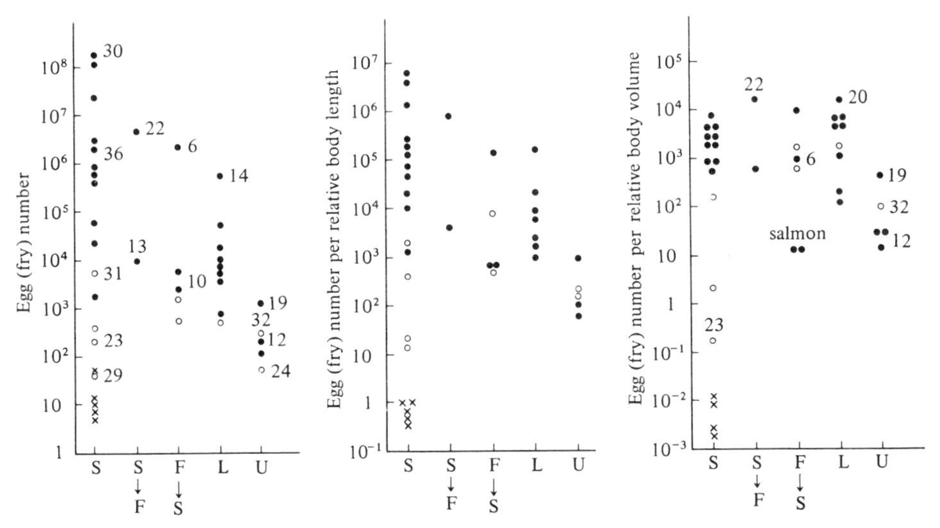

and weight of the parents. The diameter of eggs has a minimum value and it is impossible for carp to have several million eggs like tuna. However, the egg number per unit body weight is not necessarily the right measure for the comparison of species strategies. The colonization of the middle of the ocean was made possible by becoming large enough to lay an enormous number of eggs. Thus it is meaningless to compare the ratio of egg number to body weight between the tuna and the minnow. Fig. 1.3 (centre and right) also shows the ratios of egg number to body length (in cm $\times 10^3$) and to body volume (cm$^3 \times 10^3$), but in the latter the value for the medaka *Orizias latipes* (a minnow-like fish) becomes the greatest. For the comparison of reproductive strategies the ratio of egg number to body length may be the most appropriate. In any case such comparison is very coarse, yet the tendency for marine fishes to lay more or river fishes (specially in the upper waters) to lay fewer eggs is clear.

Those that lay few eggs show parental protection: in the weasel fish *Brotula multibarbata* eggs are protected by gelatin, in the tubesnout *Aulichthys japonicus* eggs are laid in the body cavity of sea squirts and the sea chub *Ditrema temmincki* is ovoviviparous. Also, the male stays near the eggs in the sculpin *Cottus pollux* which lays 200–300 and in the sticklebacks (e.g. *Gasterosteus aculeatus* = *Pygosteus sinensis*) which lay less than 100. The male stickleback constructs a nest, fans eggs (for aeration) and protects young. The redfish and stingfish (*Sebastes*) which live in the sea, relatively rich in nutrients, are ovoviviparous and produce young in thousands. This number is still exceptionally small among the marine fishes. In general, parental protection is developed in the oligotrophic environment. Among the paradise fishes (Anabantidae) males of *Macropodus opercularis* and *Betta splendens* (Siamese fighting fish) make floating nests with bubbles, in which 100–500 and 20–300 eggs respectively, are laid whereas *Anabas testudineus* and *Helostoma rudolphi* which do not make nests lay 500–2000 and 1000–3000 respectively. The egg numbers of deep-sea fishes and fishes of the polar regions are known to be small (see Chapter 2).

The ancestors of fishes probably had high fecundity, but in view of the fact that early fishes evolved in the freshwater or brackish water [the earliest teleost is known from the freshwater deposit (Colbert, 1969)], the fecundity of palaeozoic fishes may be considered to have been lower than that of modern oceanic fishes. Fishes may then be divided into two evolutionary groups, one invading the eutrophic but risky ocean with increased numbers of eggs and the other invading oligotrophic streams with large but few eggs. This was a result of the choice of strategy that the fishes made as they faced the situation at different times. The tuna might be an extreme example of fishes that took the former course of evolution while the salmonid fishes might be a group which took the latter course of evolution.

Sharks and rays are special cases. Although they appeared at about the same time as the teleosts, their prosperity preceded the great expansion of teleosts by about 100 million years in the Devonian–Pennsylvanian period. There is no question that this group is a survivor of the ancient animals. Their low fecundity is of course related to ovoviviparity in many cases. As their copulatory organ, claspers, is also known from early representatives the history of their large egg size and ovoviviparity is probably very old. The combination of their predatory habits and reproductive characteristics might be the cause for the survival of elasmobranchs over a period of 200 million years. Predatory life requires that the young are mobile or that the young metamorphose from non-predatory forms. In other words, sharks might be considered a group which exploited an evolutionary pathway in the sea totally different from the course taken by fishes like tuna and bonito. There are some large

Fig. 1.4. Numbers of eggs laid by marine invertebrates (after Thorson, 1950). *a*, species viviparous or with protection of young; *b*, species with limited protection of young, large eggs, no planktonic life; *c*, species with small eggs, long planktonic life.

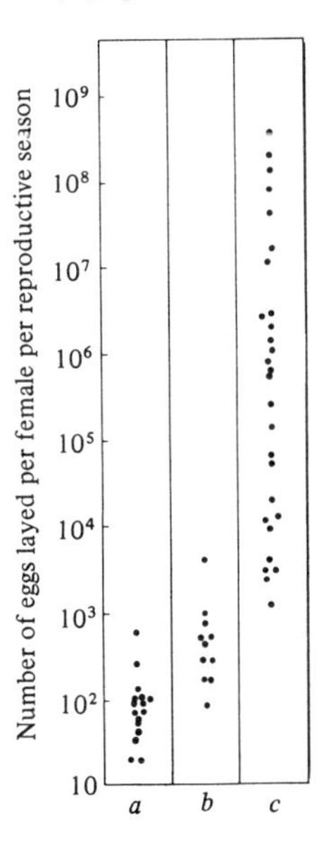

sharks which feed on plankton (e.g. *Cetorhinus maximus*, *Rhincodon typus*). Although we do not have enough knowledge to solve this problem it is probable that they specialized into plankton feeding later, just as baleen whales evolved after toothed whales had appeared.

Egg numbers of marine invertebrates and amphibians

What has been found in fishes also applies to marine invertebrates. Thorson (1950) has shown that the species whose young go through planktonic life have high fecundity whereas those whose young have no planktonic stage lay relatively few eggs (Fig. 1.4). Needless to say, the

Table 1.3. *The number of eggs laid by representative Japanese marine invertebrates and related species living in freshwater or on land*

	Species	Egg number	Diameter of egg (mm)
Bivalves:	*Anadara subcrenata*	$25 \times 10^5 - 30 \times 10^5$	0.05
	Mytilus edulis	5×10^6–12×10^6	0.07
	Corbicula sandai[a]	65×10^4	0.13
	Ostrea futamiensis	7×10^4	—
Gastropods:	*Lymnaea japonica*[a]	*ca* 500	—
	Semisulcospira bensoni[a]	100–180	See below[d]
	Euhadra quaesita[b]	Several hundreds	3.0
	Bradybaena sieboldiana[b]	*ca* 100	2.0
	Achatina fulica[b]	200	5.0
Cephalopods:	*Octopus vulgaris*	$10 \times 10^4 - 15 \times 10^4$	2.3×0.9
	Sepia esculenta	500–2000	–
	Sepiella maindroni	*ca* 1000	—
Crustaceans:	*Panulirus japonicus*	4×10^5–5×10^5	0.5
	Penaeus japonicus	*ca* 7×10^5	0.3
	Acetes japonicus	*ca* 3000	0.1
	Neomysis intermedia	3–20[c]	–
	Palaemon nipponensis[a]	500–3500	0.7
	Leander paucidens[a]	40–250	1.0
	Cambarus clarki[a]	250–600	—
	Cambaroides japonicus[a]	50	2.6
	Paralithodes camtschatica	*ca* 3×10^5	0.8
	Sesarma intermedea	2×10^4–6×10^4	—

Data source: papers published in Japanese fisheries journals, mostly from Proceedings of Japanese Fisheries Society and Japanese Journal of Scientific Fisheries.
[a]Freshwater species. [b]Terrestrial species. [c]Care of eggs.
[d]An ovoviviparous species which produces large young with a shell diameter up to 1 mm.

former suffer high mortality of young while the latter lose little. While the number of eggs laid by free-living invertebrates on land and in freshwater reaches several hundreds to several thousands, most marine invertebrates, including phylogenetically different molluscs, arthropods and others, discharge several thousand to several million eggs. In the species with high fecundity eggs are small and the young spend several weeks to several months as plankton feeding on phytoplankton. In the low-fecundity species eggs are large and rich in yolk, and the young spend only a few hours to a few days as plankton. Among the latter, there are also many which complete metamorphosis during the short planktonic life, using their own yolk as the main source of nutrients (lecithotrophic). As shown in Fig. 1.4, in the ovoviviparous species or the species which protect eggs, the number of eggs laid is small (usually fewer than 1000). The phenomenon described above may be seen in familiar animals. Table 1.3 lists the numbers of eggs laid by representative Japanese marine invertebrates and their relatives in freshwater and on land. Taking molluscs for example, marine bivalves often lay several million eggs but the size of each egg is very small. Although the freshwater bivalve *Corbicula sandai* living in eutrophic lakes lays as many as several hundred thousands, other freshwater molluscs, such as the pond snail *Lymnaea japonica*, lay only about 500 and land snails even fewer. These eggs are very large relative to the length of the adult shells. There are, as expected, differences in the developmental stage at the time of hatching, between the marine and the terrestrial snails. In the former the planktonic young are microscopic and look totally different from adults whereas in the latter the eggs hatch into small shells similar to adults in morphology.

The situation is similar in the Crustacea. The marine crustaceans lay tens or hundreds of thousands while the freshwater species lay hundreds to thousands. The freshwater crayfish *Cambaroides japonicus* living in oligotrophic lakes of Hokkaido lays the least number of eggs but the egg size is the largest among the crustaceans listed in the table.

Table 1.4 lists the number of eggs laid by amphibians. It is noteworthy that among the Japanese species *Polypedates buergeri* living in running water lays relatively few eggs. [Recently a second running-water species in Japan has been discovered (Matsui, 1975, 1976). This is a toad, *Bufo torrenticola*, which resembles *B. bufo japonicus*, but its young are attached to rocks in the torrent. Unfortunately its egg number is not known but the egg diameter reaches 2.7 mm compared with 2.4 mm (average) in *B. bufo japonicus* (the egg volume is 1.5 times that of the latter).] Among the foreign species only those showing special parental

Table 1.4. *Nature and development of eggs and the number laid by some frogs and toads; non-Japanese species are selected from those with parental care*

Family name	Common name	Scientific name	Egg characteristics	No. of eggs laid
Discoglossidae	Painted frog	*Discoglossus pictus*	Eggs scattered in water	300–1000
	Midwife toad	*Alytes obstetricans*	Males carry egg mass on hind legs till hatching	20–60
Pipidae	Surinam toad	*Pipa pipa*	Eggs hatch in dorsal sponge-like skin of female and larvae live in it	60–70
Pelobatidae	Asiatic horned frog	*Megophrys longipes*	Large eggs in soil. Metamorphosis takes place in egg	12
Hylidae	Japanese tree frog	*Hyla arborea japonica*[a]	Egg mass laid in water	500
	Marsupial frog	*Flectonotus marsupiata*	Eggs hatch on back of female	*ca* 50
	Pygmy marsupial frog	*F. pygmaeus*	Larvae nourished from dorsal blood vessels of female and metamorphosis takes place on female's back	4–7

Bufonidae	Japanese common toad	*Bufo bufo japonicus*[a]	Egg mass laid in water	5000–14 000
Rhinodermatidae	Darwin's toad	*Rhinoderma darwini*	Larval stage passes in vocal pouch of male	20–30
Leptodactylidae	Tiny Cuban frog	*Sminthillus limbatus*	Large eggs in soil. Metamorphosis takes place in egg	1
Ranidae	Japanese pond frog	*Rana nigromaculata*[a]	Egg mass laid in water	840–1260
	Japanese rugose frog	*R. rugosa*[a]	Several small egg masses laid in water	900–1700
Rhacophoridae	Japanese stream frog	*Polypedates buergeri*[a]	Egg mass laid in clear flowing water	580
	Japanese tree frog	*Rhacophorus arboreus*[a]	Eggs laid in spittle made by female on tree and larvae fall into water	400–600
	Terrestrial Rhacophorid frog	*Rh. microtympanum*[a]	Large eggs in soil. Metamorphosis takes place in egg	20

Based on Tokuda (1957), Uchida (1976) and M Matsui (personal communication).
[a]Japanese species.

care are listed. Note that they lay only very small numbers of eggs. It is not known in what environment these frogs have evolved (they seem to have evolved parental care independently in different phylogenetic groups; see p. 220.

Clutch size of birds

Table 1.5 lists the clutch size of some Japanese birds. Birds may be divided into two groups according to the mode of early development and parental feeding: nidicolous and nidifugous birds. [Similarly, 'altricial' and 'precocial' (see Kuroda, 1967, for Japanese translation) distinguish the hatchlings that cannot walk or swim and those that can. They generally correspond to 'nidicolous' and 'nidifugous' respectively, but the distinction is not always clear. For example, in gulls the chick can walk and hide by itself soon after hatching but is fed by its parents.] The chicks of the nidicolous birds are typically not mobile when they hatch and remain in the nest for a period to be fed by their parents. They include passerines, raptors, herons and egrets. The chicks of the nidifugous birds can walk or swim when they hatch (in an advanced stage of development) and feed themselves as a rule though they are protected and may be led to the feeding ground by their parents. They include gallinaceous birds, ducks and geese. This life-form differentiation is not necessarily uniform within the same phylogenetic group. Within the Charadriiformes the gulls are nidicolous but waders are nidifugous, whereas within the Columbiformes pigeons and doves are all nidicolous but the sand grouse *Syrrhaptes paradoxus* is nidifugous. The distinction between these two types is ecologically very important, and is indicated in both Tables 1.5 and 1.6.

The nidifugous birds (mostly Galliformes and Anseriformes) have the largest clutch size among the various groups shown in Table 1.5. Clutch size is smallest in the oceanic birds nesting on cliffs, such as the gannet and albatrosses, followed by the raptors (mostly tree-nesting species). Among the ducks the mandarin duck, *Aix galericulata*, nests in tree holes but has a large clutch size like other members of the family. The nesting trees of this species are located near water and the chicks when they hatch plunge into the water from their nesting hole and can swim. They are thus nidifugous and their tree-nesting habit may have evolved comparatively recently. Clutches of intermediate size are found among the passerines with a large number of species (mostly tree-nesting), woodpeckers and herons.

Table 1.6 gives examples of clutch size in non-Japanese species, mainly

Table 1.5. *Clutch size of some Japanese birds*

Common name	Scientific name	Clutch size	Size of egg (mm)
Passeriformes			
Tree sparrow[a]	*Passer montanus saturatus*	4–8	19.6 × 14.6
Meadow bunting[a]	*Emberiza cioides ciopsis*	3–6	20.7 × 16
Skylark[a]	*Alauda arvensis japonica*	3–5	22 × 16.5
Japanese white-eye[a]	*Zosterops japonica japonica*	4–5	16.3 × 12.7
Eurasian nuthatch[a]	*Sitta europaea*	5–8	19 × 14
Great tit[a]	*Parus major minor*	7–12	17 × 13.3
Varied tit[a]	*P. varius*	5–8	18.9 × 14.4
Brown-eared bulbul[a]	*Hypsipetes amaurotis*	4–5	30 × 20
Bush warbler[a]	*Cettia diphone cantans*	4–6	18.2 × 14
Japanese robin[a]	*Erithacus akahige*	5	21.5 × 16.1
Swallow[a]	*Hirundo rustica gutturalis*	3–7	20 × 14
Caprimulgiformes			
Jungle nightjar[a]	*Caprimulgus indicus jotaka*	2	32.6 × 23
Piciformes			
Japanese green woodpecker[a]	*Picus awokera*	7–8	25 × 17
Great spotted woodpecker[a]	*Dendrocopus major hondoensis*	6–7	25 × 16.9
Strigiformes			
Long-eared owl[a,b]	*Asio otus otus*	3	33 × 34
Ural Owl[a,b]	*Strix uralensis hondoensis*	2–3	50 × 40
Falconiformes			
Golden eagle[a,b]	*Aquila chrysaëtos japonica*	1–3	76.8 × 59
Hodgson's hawk-eagle[a,b]	*Spizaëtus nipalensis*	1–2	70 × 55
Goshawk[a,b]	*Accipiter gentilis fujiyamae*	2–3	51.7 × 39.3
Ciconiiformes			
Grey heron[a]	*Ardea cinera*	4	60.7 × 42.8
Little egret[a]	*Egretta garzetta*	3–5	45.7 × 33
Anseriformes			
Mallard[c]	*Anas platyrhynchos*	8–12	58 × 40.5
Spotbill duck[c]	*A. poecilorhyncha zonorhyncha*	10–12	56 × 43
Teal[c]	*A. crecca*	8–12	46 × 33.5
Mandarin duck[c]	*Aix galericulata*	10–13	54 × 40
Pelecaniformes			
Brown booby[a,d]	*Sula leucogaster*	2	60 × 40
Procellariiformes			
Tristram's storm petrel[a,d]	*Oceanodroma tristrami*	1	42 × 29
Streaked shearwater[a,d]	*Calonectris leucomelas*	1	68 × 45
Short-tailed albatross[a,d]	*Diomedea albatrus*	1	115 × 70

table 1.5. *continued*

Common name	Scientific name	Clutch size	Size of egg (mm)
Gruiformes			
Moorhen[c]	*Gallinula chloropus indica*	5–10	41 × 29
Galliformes			
Ptarmigan[c]	*Lagopus mutus japonicus*	5–12	47 × 33
Japanese quail[c]	*Coturnix coturnix japonica*	6–10	30 × 23.5
Bamboo partridge[c]	*Bambusicola th. thoracica*	7–8	20.5 × 15.5
Green pheasant[c]	*Phasianus colchicus tohkaidi*	9–12	44 × 34
Copper pheasant[c]	*P. soemmerringii*	7–8	47.7 × 35

Mainly from Shimomura (1938) and Kumagai (1952).
[a]Nidicolous birds. [b]Predatory birds. [c]Nidifugous birds. [d]Oceanic birds.

Table 1.6. *Clutch size of birds studied outside Japan*

Common name	Scientific name	Clutch size	Habitat
Ratites			
Ostrich[a]	*Struthio camelus*	12–15	Grassland, terrestrial
Emu[a]	*Dromaius novae-hollandiae*	7–8	Savanna, terrestrial
Cassowary[a]	*Casuarius casuarius*	3–6	Forest, terrestrial
Brown kiwi[a]	*Apteryx australis*	1–2	Forest, terrestrial[c]
Galliformes			
Micronesian megapode[a]	*Megapodius laperouse*	(6–8)	Terrestrial[d]
Red grouse[a]	*Lagopus l. scoticus*	6–8	Bush, terrestrial
Gruiformes			
Limpkin[b]	*Aramus guarauna*	6	Near water surface
Charadriiformes			
Gulls[b]	*Larus* spp.	2–3	Colonial breeding
Anseriformes			
Indian whistling duck[a]	*Dendrocygna javanica*	10	Usually nest on land
Whooper swan[a]	*Cygnus cygnus*	5–6	Waterside, nest on land
White-fronted goose[a]	*Anser albifrons*	5	Waterside, nest on land
Mallard[a]	*Anas platyrhynchos*	11	Waterside, nest on land
Wood duck[a]	*Aix sponsa*	12	Nest in tree[e]
Circoniiformes			
Flamingos[b]	*Phoenicopterus* spp.	1	Colonial nests on land
Falconiformes			
California condor[b]	*Gymnogyps californianus*	1	Rocky shelf
Black kite[b]	*Milvus migrans*	2	Open forest
African vulture[b]	*Pseudogyps africans*	1	Rocky shelf, top of tree

table 1.6. *continued*

Common name	Scientific name	Clutch size	Habitat
Sphenisciformes			
Emperor penguin[b]	*Aptenodytes forsteri*	1	Antarctic, colonial
Yellow-eyed penguin[b]	*Megadyptes antipodes*	2	Terrestrial, colonial
Columbiformes			
Wood pigeon[b]	*Columba palumbus*	2[f]	Arboreal
Sand grouse[a]	*Syrrhaptes paradoxus*	2–4	Terrestrial
Coraciiformes			
Hornbills[b]	*Bycanistes* spp.	1–2	Hole in wood[g]
Apodiformes			
Swift[b]	*Apus apus*	2–3	Rocky shelf, tower
Hummingbirds[b]	*Trochilidae*	2	Arboreal
Passeriformes			
European robin[b]	*Erithacus rubecula*	4–7	Arboreal
Great tit (Europe)[b]	*Parus major*	8–12	Arboreal

Data are taken mainly from the appendix in Lack (1968) and supplemented from Kuroda (1962).
[a]Nidifugous birds.
[b]Nidicolous birds.
[c]Hatches in an advanced stage.
[d]Hatches in an advanced stage. Eggs are laid in a mound 2–8 days apart.
[e]Breeding habit same as *Aix galericulata*; eggs are laid in a hollow of a tree overhanging a lake or river; the hatched young fall into the water and can swim immediately.
[f]The number of broods is greater than other non-passerine birds.
[g]The female during incubation is plastered in and fed through a hole by the male.

taken from Lack (1968). Emphasis is given to those groups not represented in Table 1.5. The ostrich and the emu are typical nidifugous species nesting on the ground and their clutch size is very large; the rhea is the same. There are reports of scores of eggs in a nest but in this group (particularly in the ostrich) more than one female may lay in the same nest. Among the ratites the cassowary which lives in the forest has a somewhat smaller clutch size than other members of this group. The small clutch size of the kiwi is probably due to the special ecological characteristics of this primitive bird; the incubation period of more than 70 days is the longest among all birds and the large young hatch into adult plumage.

Gulls have a clutch of 2–3 eggs, which is in the small clutch-size group though not as small as that of the shearwaters and other oceanic birds.

In general clutch size is small among the birds that nest in trees, on cliffs of islands and on the ice or snow. Penguins are nidicolous but nest on the ground. In Antarctica where the climate is severe but predators are absent, some species (e.g. emperor penguin, king penguin) lay a single egg while others produce 1–2 eggs. The yellow-eyed penguin breeding in southern New Zealand has a normal clutch of two eggs like many other species living in the southern temperate region.

Our examples of fishes and some invertebrate groups examined earlier appear to support Lack's notion that large egg size, and small egg number have evolved in the environment where both food and predators are scarce whereas small egg size and large egg number have evolved in the opposite conditions. What about in birds? The forest environment is rich in organic energy compared with grassland, yet the clutch size of forest birds is smaller than that of grassland birds, which is contrary to the above notion. Sea-birds living in a food-rich environment also have a small clutch size. On the other hand, the fact that small clutch size has evolved in birds living in severe conditions of exposed rocks or snow fields does not lend support to the theory of r-K selection which states that fecundity is low where competition is severe. I shall elaborate on this problem later (p. 39), but my conclusion is as follows. The question is not the absolute amount of food in the environment but the *procurability* of food by the young, whether they are hatchlings of nidifugous birds or fledglings of nidicolous birds. Similarly the question of predation is concerned not only with absolute number of predators in the area but also with the accessibility of prey. From this viewpoint arboreal life makes it difficult for young birds to find food and for predators to approach young birds. Cliff-living on oceanic islands has the same effect. In raptors prey catching requires the power of skilled locomotion, and prolonged protection of the young is necessary by parents. Fledged young of eagles are fed by parents for a long time (Cody, 1971).

Among arboreal nidicolous birds other than the raptors, small clutch size is the rule for pigeons (1–2 eggs), hummingbirds (mostly 2 eggs) and hornbills (1–2 eggs). Hummingbirds take nectar in flight and young birds must be reared till the stage when they acquire the strong flight power necessary for this specialized life. In this sense they resemble raptors and benefit from their low fecundity. But how did they reduce the risk of predation? It may be that their nest made on twigs (sometimes on a leaf!) is not approachable by predators. Obviously many field studies are necessary to unravel the evolutionary pathways that this specialized group of birds has taken. The male hornbill feeds the incubating female through the window of the nesting hole which she has otherwise sealed.

Apart from the question of how this habit has evolved, the low fecundity may be related to it. In the case of pigeons the low fecundity may be related to their special habit of feeding young with 'milk' secreted from their crop. The small clutch size is compensated for by having more than one clutch in the same breeding season; according to Murton *et al.* (1974) the wood pigeon has a mean number of 2.5 clutches per season.

Among the ground-nesting nidifugous birds in the forest habitat the low-fecundity species lay relatively large eggs in relation to their body weight (Fig. 1.5). The largest egg of the living birds is that of the ostrich, but its weight (1.5 kg) is only 1/60 of the parent's (100 kg). In the forest-dwelling kiwi (2.5 kg) the egg weighs 450–500 g, which is about 1/5 of the body-weight of the parent. The next heaviest relative weight of eggs is seen among the mound-builders (Megapodidae). *Megapodius laperouse* (see Table 1.6) lays a 'clutch' of 6–8 large eggs, 2–8 days apart, and the chicks do not hatch at the same time. Thus its clutch size is not comparable to the clutch size of other birds. In the case of the malleefowl *Leipoa ocellata* the egg weight (200 g) is about 1/7 of the adult weight (1.4 kg). The incubation period of mound-builders is about 50 days, which is longest after the ritites. The mound-builders are so called because of their mound-building habit with fallen leaves. It is the fermentation heat of leaf litter in the mound that hatches the eggs (Kuroda, 1962, for references). The chicks hatch at an advanced stage of

Fig. 1.5. Relationship between the body weight of adult females and the egg weight as a percentage of body weight in birds (from Lack, 1968).

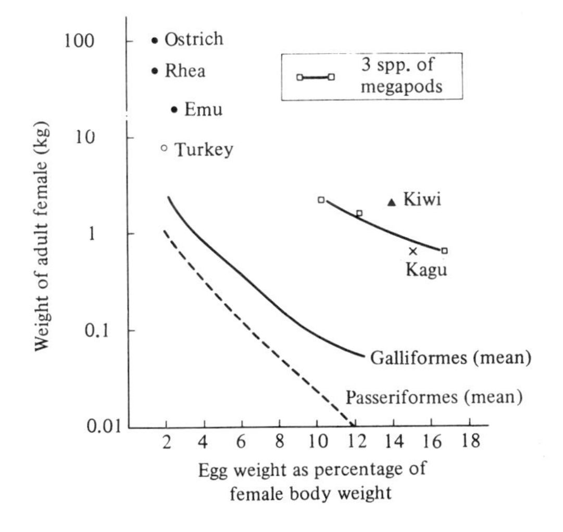

development and can fly. Following the mound-builders, the kagu *Rhinochetos jubatus* (not in Table 1.6) has the longest incubation period. It nests on the ground in forest habitat and has a clutch of one. As can be seen in Fig. 1.5, in which the lines of body weight/egg weight ratios have been drawn for the gallinaceous birds and passerine birds from Lack's (1968) data, the three groups discussed above have much greater egg weight/body weight ratios than even the gallinaceous birds which have relatively large eggs and hatch at more advanced stages than the passerines

Litter size of mammals

Bourlière (1964) gives the litter size of many mammals from which representative species are listed in Table 1.7. The hedgehog and moles (also shrews) – the living members of the insectivores that have evolved earliest among the mammals – have a large litter in general. By contrast almost all bats produce a litter of one or rarely two. Among the primates, apart from the primitive Prosimii (particularly Tupaiidae, closest to Insectivora) and Callithricidae (marmosets *Callithrix* and tamarins *Saguinus*) considered to be the most primitive of the Anthropoidea (Platyrrhini of the New World and Catarrhini of the Old World), the litter size is one (not counting the twins which appear occasionally in the chimpanzee and man). Even in the Prosimii, Daubentoniidae, Indriidae and most members of Lemuridae produce a litter of one. Among other mammals the Cetacea is one group, all members of which have a litter of one. The carnivores generally have large litters from several to more than ten whereas the ungulates other than the pigs produce one (usually large species) or two young at a time.

On the whole the terrestrial mammals, with the exception of the ungulates, tend to have high fecundity whereas the arboreal, cliff-living and cave-dwelling groups tend to have low fecundity. This fact seems to be related also to the procurability of food by the young. In both monkeys and bats young must be reared until they are well developed and able to jump from branch to branch or fly in the air. It is interesting to note that the arboreal sloth produces only one young while the terrestrial armadillo in the same order, Edentata, has a large litter. In Australia where the marsupials radiated, Dasyuridae corresponding to the eutherian carnivores and insectivores, have high fecundity while the arboreal koala, *Phascolarctos cinereus*, has a litter of one, showing parallel phenomena to the eutherians.

The low fecundity of large ungulates and cetaceans may be related to the size that these animals have attained, which permits them to escape from predators other than large carnivores. Also worth noting is the fact

Table 1.7. *Litter size of mammals*

Common name	Scientific name	Litter size
European hedgehog	*Erinaceus europaeus*	4–6
European mole	*Talpa europa*	1–7
Bats	Chiroptera	1(2)
Black lemur	*Lemur macaco*	1–2
Spider monkey	*Ateles fusciceps*	1
Japanese macaque	*Macaca fuscata*	1
Chimpanzee	*Pan troglodytes*	1
Sloth	*Bradypus griseus*	1
Texas armadillo	*Dasypus novemcinctus*	4–5
Euopean hare	*Lepus europaeus*	1–4
Eastern chipmunk	*Tamias striatus*	2–8
American beaver	*Castor canadensis*	1–6
Boreal red-backed vole	*Clethrionomys gapperi*	3–8
House mouse	*Mus musculus*	4–7
Whales and dolphins	Cetacea	1
Brown bear	*Ursus arctos*	1–3
Striped skunk	*Mephitis mephitis*	4–7
Coyote	*Canis latrans*	5–10
European red fox	*Vulpes vulpes*	3–7
Lion	*Panthera leo*	2–6
Indian elephant	*Elephas maximus*	1(2)
Mountain zebra	*Equus zebra*	1
Wild boar	*Sus scrofa*	3–12
Moose	*Alces alces*	2
Giraffe	*Giraffa camelopardalis*	1–2
Bighorn sheep	*Ovis canadensis*	1

Mainly from Bourlière (1964).
Note: Numerals in parentheses show values seen at low frequencies. 'Twins' rarely seen are not included in this table.

that the large ungulates are born at a well advanced stage (young can walk soon after birth).

Geist (1971) expressed an interesting view on the strategy of survival in large ungulates. The moose *Alces alces* can increase rapidly under protection and disperse, whereas the bighorn sheep *Ovis dalli* and *O. canadensis* do not increase their number and seldom disperse. He attributed this difference to the fact that the bighorn sheep live in relatively stable montane grassland and eat grass whereas the moose lives in temporary habitats of deciduous forest and scrub vegetation that develop within the boreal coniferous forest biome. He claims that, as the habitat of the moose undergoes repeated changes of rapid expansion and slow contraction, the moose has evolved an ability for rapid increase (compared with other large ungulates) and great power for migration. It is interesting to note that among the horse-sized ungulates the moose is unique in having a normal litter of two.

Egg numbers of insects (mainly Hymenoptera)

As the insects consist of an enormous number of species and have differentiated into a great many modes of life it is difficult to survey the relationship in the same way as treated above. Fortunately the fecundity of one highly evolved group – the Hymenoptera – which contains ants and bees with a long evolutionary series of sociality has been surveyed by Iwata (1955a–1960, 1964, 1966a, b) and Iwata & Sakagami (1966). Their data are used to construct Table 1.8 in which the number of ovarioles with the number of mature eggs in them and the egg length as a measure of egg size are given for selected species. The number of eggs laid is not represented by either of the first two parameters as, for example, the number of mature eggs is reduced in species which lay over a long period. In social hymenopterans (*Polistes yokohamae* and *Bombus diversus* in the table) the eggs mature gradually and both the number of ovarioles and the number of mature eggs at any time are small in spite of the fact that they lay a large number of eggs in a lifetime. In non-social groups, laying occurs in a short space of time and these measurements may be considered to indicate proportionately the number laid. Strictly speaking the Symphyta with free-living larvae and many parasitic wasps do not lay all mature eggs while hunting wasps and bees being in the early evolutionary stages of social development lay two to ten times the number of mature eggs found at one time.

Among the groups listed in Table 1.8 the Symphyta is of early evolutionary origin and its members lay eggs in leaves and woody tissues of plants. Their larvae are mobile and find their own food. Apart from the wood wasp *Orussus japonicus* which parasitizes buprestid beetles, showing the evolutionary trace of parasitism in this group, the number of eggs laid by the Symphyta is more than a few score (there are no species in the Hymenoptera which scatter eggs or attach them on the surface of the host plant). Attention should be drawn to the relatively large size of eggs in *Orussus japonicus*. The Parasitica, such as the ichneumonid flies, belonging to the suborder Apocrita, are parasitic (strictly speaking they are parasitoids) and have a large number of mature eggs but *Polysphincta tuberculata*, a species occupying an intermediate position in the social evolution of this group, contains fewer mature eggs. In this species the female lays in a paralysed spider in a hole and the larva feeds on the immobilized spider. In other parasitoids the host is usually mobile.

In the Aculeata the eggs mature gradually and the number is not proportional to the number laid, but the data in the table indicate that wasps belonging to the Scolioidea (e.g. *Tiphia popilliavora*) and Sphecoidea (e.g. *Larra amplipennis*), which do not make nests, have large egg numbers while others which construct nests and carry prey to the nests (the habit-type) (containing I in Table 1.8) have much reduced numbers of eggs. Among the families in which the actual number of eggs laid is known, the egg number of Bethylidae, Scoliidae, Tiphiidae and Chrysididae ranges between 50 and 100, that of nest-building solitary wasps and bees (Pompilidae, most of Sphecidae) is usually up to thirty and that of subsocial wasps and bees is less than ten. With decrease in the number of eggs laid the longevity of adults increases, their habits become complex and more parental care is given to the larvae. At the same time the egg becomes large, one-third of the adult's body length in *Bembicinus japonicus* and as much as one-half of the body length in *Xylocopa appendiculata.* The largest eggs measured by Iwata & Sakagami (1966) are those of *Xylocopa latipes* and *X. auripennis* of Thailand, the egg length being 16.5 mm in both species. In its ratio to the adult's body length it is said to be the largest among all metazoans, and in its absolute length excluding the appendages it is the largest known among all living insects.

With further increase in sociality the number of eggs laid increases again. In *Polistes* the egg becomes medium-sized and the number is in the order of tens. In ants and honeybees (*Apis*) which have reached a peak of social evolution in another direction, one queen lays several hundred to several thousand eggs. However, as most other females become workers and do not take part in reproduction, we should perhaps consider here the number of eggs laid per genetically determined female in the group.

Iwata (1966a, b) also examined the egg size of the Curculionoidea and Scarabaeoidea. In his results the correlates of fecundity are not as clear as in the Hymenoptera. The eggs of the Attelabidae, beetles that construct a special 'leaf-roll' (swing-cot) from leaves to protect and provide for the larvae, are relatively large compared with the eggs of other snout beetles (Curculionoidea). In the case of the Scarabaeoidea, out of the seventeen species for which the ratio of egg length to forewing length of adults is available, sixteen species ranged from 0.08 to 0.29 in this ratio

Table 1.8. *Numbers of ovarioles and mature eggs in the ovaries of hymenopteran insects*

Family	Scientific name	Habit-type[a]	Total no. of ova-rioles	No. of mature eggs[b]	Egg length (mm)	Egg index[c]
Symphyta						
Siricidae	*Urocerus japonicus*	Surface feeder	153	1207	1.2	0.18
Diprionidae	*Neodiprion sertifer*		46	88	0.9	0.38
Tenthredinidae	*Tenthredo analis*		16	22	1.7	0.49
Orussidae	*Orussus japonicus*	Parasitoid	8	16	2.7	1.10
Parasitica						
Pteromalidae	*Pteromalus puparum*	Parasitoid	6	125	—	—
Braconidae	*Apanteles glomeratus*		4	400	—	—
Cynipidae	*Trichagalma serrata*	Gall wasp	45	300	—	—
Ichneumonidae	*Habronyx insidiator*		63	130	0.5	0.11
Ichneumonidae	*Trogus mactator*	Parasitoid	27	60	1.1	0.31
Ichneumonidae	*Diplazon laetatorius*		28	28	0.6	0.43
Ichneumonidae	*Polysphincta tuberculata*	PO	6	17	0.8	0.44
Aculeata						
Bethylidae	*Goniozus japonicus*	PO	6	26	0.6	—
Tiphiidae	*Tiphia popilliavora*	PO	6	20	1.2[d]	0.67
Sphecidae	*Larra amplipennis*	PO	8	21	2.1	0.54
Chrysididae	*Chrysis ignita*	Parasitoid	10	21	1.5	0.48
Anthophoridae	*Nomada japonica*	Labour parasite	6	22	1.4	0.39
Megachilidae	*Coelioxys japonica*		6	6	—	—
Pompilidae	*Batozonelus maculifrons*	PTIOC	6	2	4.1	0.96
Sphecidae	*Sphex nigellus*	PTOPTC	6	2	2.8	0.60

Eumenidae	*Eumenes micado*	IOPTC	6	1	3.5	0.81
Sphecidae	*Psenulus lukricus*	IPTOPTC	6	1	1.3	0.90
Halictidae	*Nomia punctulata*	IPTOC	6	1	3.7	0.98
Megachilidae	*Megachile kobensis*	IPTOC	6	1	3.4	0.81
Eumenidae	*Ancistrocerus fukaianus*	IOPTC	6	0	4.4	0.88
Sphecidae	*Bembicinus japonicus*	IOPTC	6	0	3.3	1.01
Anthophoridae (Xylocopinae)	*Xylocopa appendiculata*	IPOC	8	0	12.5	1.38
Vespidae (Polistinae)	*Polistes yokohamae*	O (queen)	6	6	3.7	0.69
Apidae (Bombinae)	*Bombus diversus*	O (queen)	8	5	3.4	0.44

Data from Iwata (1955a–1964) and Iwata & Sakagami (1966).

[a]For habit-type, see Chapter 5, p. 264.

[b]Counted under a microscope. They do not correspond with the number of eggs laid in the species in which eggs mature gradually (e.g. Aculeata).

[c]Egg index = length of egg/length of mesosoma.

[d]Substituted with the value for *T. bicarinata*.

whereas in the seventeenth species, *Geotrupes laevistriatus*, which protects young, this ratio reached 0.44. In this species there are two pairs of ovarioles compared with a minimum of five pairs in other scarabaeid beetles examined, in addition the number of mature eggs in the ovarioles was very small. The number of eggs laid is thought to be between one-tenth and one-hundredth of the number produced by other scarabaeid beetles.

One of the coleopterans that shows well-developed parental care is the dung beetle *Copris acutidens*. According to M. Yajima (personal communication, 1977) the female of this species lays one egg in each of three or four dung pellets she has rolled and remains there for about two months. The diameter of the egg in this species reaches about 2 mm. Since the female normally lives for one year the total number of eggs laid by one female in her lifetime may only be four.

Seed numbers of plants

Can we see the above tendencies in plants? It is generally known that plants with large seeds are more advanced in phylogeny and produce fewer seeds per plant than the lower groups. Although there are some exceptions, if we compare the numbers of spores produced by fungi and ferns and the numbers of seeds produced by the gymnosperms, dicotyledons and monocotyledons the rule is that the higher the plants the larger the seeds and the smaller the number of seeds produced. However, here too it is necessary to consider the habitat relations in some detail.

Salisbury (1942) examined the weight of seeds for many species of plants occurring in different habitats and presented a table of seed-weight distribution (Table 1.9). The purpose of his investigation was to show that the size and number of seeds were the weapons in the struggle for existence among plants. Like Cody, Salisbury argued that for a plant of a given leaf area and therefore having a certain potential for photosynthesis the credit balance of food production might be expended on seed formation in one of two ways: the species could either utilize the available food material for the production of a large number of small seeds, each with just enough food to give the seedlings a start in life, or it could produce comparatively few seeds each with a copious food reserve, thus enabling the offspring to survive for a longer period before they became self-supporting. The former is the *r*-strategist type, advantageous in scattering the offspring widely into all suitable places. In this case the intraspecific competition is reduced. The latter strategy is advantageous

Table 1.9. *Distribution of mean seed weights of* 187 *species of plants. Figures show the numbers of species in the weight class intervals which are arranged in descending order from* 24 (8–16 *g*) *to* 3 (0.00381–0.00763 *mg*). *The modes of frequency distribution are indicated by underlines.*

Weight class	Open habitat	Short turf	Meadow	Scrub	Woodland herbs	Shrubs	Trees
24							1
23							1
22							2
21						1	—
20						1	—
19						—	2
18						1	3
17						2	3
16					4	5	1
15	1	1		3	3	2	3
14	3	3	3	4	7	5	1
13	2	1	—	9	10	4	1
12	7	7	—	9	4	1	—
11	10	10	4	8	3	2	1
10	15	10	—	—	4	—	—
9	17	4	1	6	—	—	—
8	18	9	—	2	1	—	1
7	8	1	—	9	—	—	—
6	4	1					
5	6	2					
4	6	1					
3	1	1					
Total no. of species	98	51	8	50	36	24	20

From Salisbury (1942).

in competition with other species, for the embryo is supplied with sufficient nutrients in the seed to grow until it becomes independent. Thus large seeds are said to be advantageous in species-rich, dense communities. The seedlings in such communities must grow fast and taller than the surrounding vegetation in order to obtain light, and to be successful they would have to depend on the food reserve carried in the seed. From the data in Table 1.9 the mean weight of seeds for ninety-eight species growing in open habitats may be found 1.315 mg and that for twenty-two species of Gramineae growing in short turf 2.124 mg

(0.928 mg for those growing in meadow). The mean weight of seeds for thirty-two species growing in scrub is 4.438 mg while that for twenty-seven species of woodland herb is 13.686 mg. The mean weight of fruits and seeds for twenty-nine species of tree is 653 mg. From these data Salisbury concluded that 'in general the larger the supply of food material provided by the parent plant in the seed, or other propagule, the more advanced the phase of succession that the species can normally occupy'. There are some exceptions to this rule. One of them is that in some of the open habitats, such as coastal sand or gravel land, seeds cannot establish themselves unless they are large enough to contain sufficient food reserve, permitting growth of roots down to the low water table. This is exemplified by the coconuts. Another is that orchids in forest habitats produce thousands of small seeds. In this case growth is made possible by the mycorrhizal habit in germination.

The above rule also holds for species within the same genus. The genus *Euphorbia* is a typical example of this (Salisbury, 1942):

1. *E. lathyris* (seed: 26 mg): annual or biennial, growing in woodland but also invading cultivated fields. This seed is the heaviest in the genus.
2. *E. amygdaloides* (seed: 4.05 mg) and *E. hiberna*: perennial, growing in woodland. The seed is large.
3. *E. esula* (seed: 3.5 mg): growing in shrubland.
4. *E. portlandica* (seed: 1.6 mg): perennial, growing in open habitats such as sand dunes.
5. *E. peplus* (seed: 0.497 mg), *E. exgua* (seed: 0.51 mg), *E. helioscopia* (seed: 2.49 mg), etc.: annual, growing in dunes, grassland or cultivated fields. The seeds are mostly small.

In *Trifolium* the mean seed weight of nineteen species growing in relatively closed habitats of woodland, tall grassland and pasture was 1.7 mg ($\sigma = 2.0$) whereas the mean for fourtees species growing in dunes and wasteland was 0.59 mg ($\sigma = 0.8$).

Thus in sessile plants we can recognize two opposed directions of evolution, one scattering a large number of small seeds and the other producing a small number of large seeds. The fact that seeds tend to be larger in the vegetation appearing in more advanced stages of succession may provide a proof for the hypothesis of r-K selection. In other words, the r-strategy is advantageous for colonists that colonize open habitats where the environment is unstable with little competition either within and between species. The K-strategy, on the other hand, is advantageous in the highly competitive environment of late successional stages.

Gadgil & Solbrig (1972) collected species of the genus *Solidago* from various communities and investigated the ratio of the weight of the reproductive organ to the dry weight of the plant above ground. They considered that the weight of the reproductive organ was related to the number of seeds rather than the size of the seed. This is probably true for *Solidago* though it may not hold for other groups. They found the relationships shown in Fig. 1.6. The ratio of reproductive organ to the above-ground weight is highest in the populations growing on dry sites in which environmental factors fluctuate (communities consisting of herbs and grasses in early successional stages) and it is lowest for hardwood sites of advanced communities consisting mainly of trees and shrubs. From these data and the observation reported on three varieties of *Taraxacum officinale*, Gadgil & Solbrig concluded that the theory of *r*-*K* strategies holds in these cases. *Taraxacum officinale* invaded Japan where it attained the status of a weed plant in urban areas. The indigenous species, *T. albidum*, which produces a small number of large seeds remains in cultivated fields. Here, the former may be considered an *r*-strategist and the latter a *K*-strategist.

Reproductive strategies of weeds, pests and parasitoids

Although certain criticism will be given later, the theory of *r*-*K* strategies is appealing in that, as already mentioned, animals are con-

Fig. 1.6. Relationship between reproductive effort (ratio of dry weight of reproductive tissue to total aerial tissue) and the dry weight of total aerial tissue in various species populations of *Solidago* (from Gadgil & Solbrig, 1972). Each enclosed area is represented by the individuals of a single population on a dry site (enclosed in dotted curve), a wet site (with horizontal shading) or a hardwood site (vertical shading).

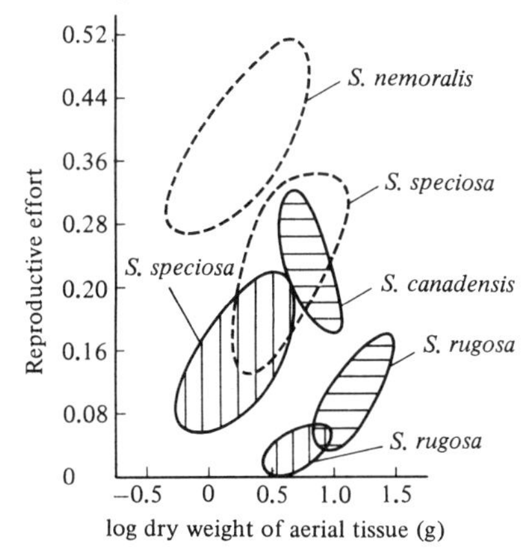

sidered to select between the two opposing strategies. It has given much stimulus to American ecologists and resulted in many publications in the 1970s.

Here I shall provide some discussions from the standpoint of this theory.

First of all, the question is what sorts of species can become a weed or a pest when the choice between the two strategies is given to the organisms. As Lack (1954b) himself considered this choice and discussed the variation of reproductive rates among closely related species and within the same species, the proponents of *r*-*K* selection have asserted that this theory not only applies to organisms in general (higher categories) but also among closely related species for which there have always been two choices.

The cultivated field is an ecosystem which is interrupted by harvesting, ploughing and other agricultural activities which today include spraying of chemicals. The species becoming weeds or pests should have (1) a high reproductive potential to be able to compensate for the loss caused by harvesting, ploughing, etc. and (2) the strong dispersal power to be able to colonize a newly created field. Thus one may predict that most weeds and pests are *r*-strategists.

Kawano (1975) listed sixteen items characteristic of weeds. Of these the main items concerned with the reproductive strategy are as follows:

1. The weeds consist largely of annual herbs though perennial grasses and herbs appear occasionally. The woody plants seldom become weeds.
2. After germination seedlings grow rapidly.
3. The period of vegetative growth is very short, if any, and reproductive growth takes over quickly.
4. Throughout the growth period the reproductive growth – seed production – is continuous.
5. If the environmental conditions are suitable an extremely large number of seeds will be produced.
6. In the case of perennial plants very conspicuous vegetative reproduction occurs; the nodes of stems and roots are readily separated and their re-growth can produce new plants.
7. Almost all weeds show phototropism.

The problem of the reproductive rate is complex as there are many plants in which vegetative reproduction is common, but as far as sexual reproduction is concerned the weeds in well-managed, cultivated fields and in the vacant land of city areas tend to be annuals and disperse small seeds. In Japan they include *Digitaria abscendens*, *Setaria viridis*, *Erigeron canadensis* and *Ambrosia elatior*. Strictly speaking the first colonizers of bare grounds are carried there by birds and have slightly larger seeds (e.g. *Ambrosia elatior*, *Polygonum persicaria*) than the colonizers that follow immediately in the early stages of succession (e.g. *Erigeron annus*, *E. canadensis*). However, in the case of cultivated fields partial weeding encourages the spread of perennial *K*-strategists such as those with large seeds (e.g. legumes), rhizomes or runners. In Japanese grassland the succession culminating in dwarf bamboos proceeds in the following order: *Erigeron* type→lawn-grass type→bracken type→*Miscanthus* type→*Pleioblastus* type→*Sasa* type (Hayashi & Numata, 1968). Recent Japanese studies may be seen in Kawano & Nagai (1975).

In the case of pest insects we can also see that they have been able to achieve the status of a pest by being more *r*-strategic than the closely related species that have not become pests.

Occupying a special position among the pests are the migratory pest insects. The planthopper *Nilaparvata lugens*, which recently became one of the greatest pests of the rice plant in Japan, seldom over-winters in Japan. The original stock of pest individuals which emerges in rice paddies every autumn has been shown to derive from the Chinese continent, and is carried on the front that crosses the East China Sea to Japan during the rainy season (Kisimoto, 1971). The irruption of this species in Japan seems to occur in the years when the migratory population is large. This species has no population regulation mechanisms so that if the number of arrivals in the rice paddy is great in the spring the irruption in the autumn is inevitable (Kuno, 1968; Kuno & Hokyo, 1970). When this occurs several hundred planthoppers congregate on a single plant of rice and kill the plant. When the plants die the planthoppers change their phase to the long-winged form and emigrate (Kisimoto, 1965). Of course the original migration phase is also this form but two subsequent generations or so consist mainly of the brachypterous form. On the other hand, the leafhopper *Nephotettix cincticeps*, which is also common in rice paddies, has a well developed regulatory mechanism and even when the density is high in spring it is stabilized to less than twenty per plant from summer to autumn (Kuno, 1968; Kuno & Hokyo, 1970). We may say that the former is a typical *r*-strategist and the latter a *K*-strategist (see p. 154).

There are many other pest species which damage crops after invading in large numbers from afar. The tobacco cutworm *Spodoptera litura* overwinters in small numbers in the southern part of Japan, whence the population spreads widely over central Japan from summer to autumn and attacks green vegetables and sweet potatoes. The armyworm *Mythimna separata* of the Chinese continent overwinters in Kuantung (Canton) Province and moves northwards from spring to summer in two or three generations. A large-scale marking experiment has shown that a part of the population reaches the northeastern part of Chilin Province and some of them even perform the 'return' migration (Li *et al.*, 1964). Other well-known examples include the milkweed bug *Oncopeltus fasciatus* which performs a long-distance migration from the south to the north of the American continent (Dingle, 1968), an African bug, *Dysdercus fasciatus*, and the cutworm *Spodoptera exempta* which emerge in equatorial Africa and migrate to Ethiopia and Sudan (Brown *et al.*, 1969).

According to Dingle (1974), who studied African species of the genus *Dysdercus*, *D. fasciatus* which always migrates has an intrinsic rate of increase (r) of 0.0939 per day (he used r_m to designate the maximum r but this is untenable; in this book r_m is replaced by r). This species eats fallen fruits of the baobab tree which occurs scattered in the savannah and bears numerous fruits. If the fallen fruits are abundant the wing muscles of the female *D. fasciatus* degenerate and the bugs continue breeding without migration. If, however, food is scarce the females migrate in search of other baobab trees. By contrast, the r for *D. nigrofasciatus*, which has several alternative host trees, is 0.878 per day and that for *D. superstitiosus*, which attacks grasses, is 0.0616 per day. These two species utilize the abundant food supply and do not migrate. Dingle concluded that the migratory species or the migratory generations, in the case of a species with alternation of sedentary and migratory generations, must be r-strategists.

Among the aphids there are many species which have apterous and alate (wingled) forms in adults. The female of the apterous form reproduces rapidly but viviparously as a rule, staying on the new buds of a plant. The young often increase in number and cover the entire buds in less than 20 days. As the conditions of the plant deteriorate some members of the population become alate and fly to new host plants. Many aphids change host plants seasonally and have different summer hosts and winter hosts. The change between host plants is made by the alate form. Kennedy & Stroyan (1959) considered that the existence of the two forms is a division of labour within the species. Although the new buds are highly nutritious they provide only a temporary habitat. Here in the temporary (unstable) environment the rapid 'vegetative' reproduction is advantageous. But the condition of the new leaves soon changes and the alate form capable of migration becomes advantageous. They claim that the aphids have established a unique position on the earth by having two co-existing forms within the same species. We may say that the division of labour is between the *r*-strategist of the apterous form and the *K*-strategist of the alate form.

There are many aphids which attack only wild plants, such as the chestnut tree aphid *Nippocallis kurikola* which has only the alate form. On the other hand, among the species famous for their pest status there are many that have lost their migratory habits. For example, the green peach aphid *Myzus persicae*, which usually selects Rosaceae plants as the winter host and cruciferous plants as the summer host, has a race which produces many generations exclusively in vegetables (Tanaka, 1967). The corn leaf aphid *Rhopalosiphum maidis* does not migrate and may be considered to have intensified its *r*-strategy as it has turned into a pest.

It is well known today that the migratory locusts which nearly darken the sky during migration are the gregarious phase of ordinary grasshoppers found in that region and that they are derived from the green or light-brown solitary phase under conditions of high density. Kennedy (1961) thought that this change between the two phases, providing a division of labour, enables a versatile response to an unstable habitat, such as that found in the delta regions of the Nile, the Niger and the Hwang Ho, where periods of flood and drought alternate. When the wet grassland shrivels as a result of prolonged drought and these gigantic deltas are exposed, the solitary-phase grasshoppers breed rapidly in the manner of an *r*-strategist (Ma, 1958; Waloff, 1966). Their high reproductive rate is assured when the density is low, but as the density increases the new phase emerges. The body colour darkens and the wing length increases, and after several generations the new phase looks like a different species. This gregarious phase migrates in tens or hundreds of thousands. The eggs laid by the individuals of the gregarious phase are large and the nymphs are also large and highly active. The number of eggs laid is somewhat lower than that of the solitary phase. As the gregarious phase requires a certain amount of flight before laying, its laying is retarded. These characteristics may be considered as those of the *K*-strategist (this idea is different from Dingle's view that the migratory individuals are of an *r*-type; Kennedy is critical of the theory of *r*-*K* selection). Thus the aphids, planthoppers and migratory locusts may be considered special organisms that employ two opposing strategies alternately.

Table 1.10. *Certain characteristics of four parasitoids that attack a gall midge,* Rhopalomyia californica

Species	r^a	Characteristics
Tetrastichus sp. (Larval parasitoid)	0.200	Highly restrained in parasitizing hosts already parasitized by other species. Larvae usually lose (die) in competition for host with larvae of other species
Platygaster californica (Egg parasitoid)	0.092	Larvae nearly always die in competition for host with larvae of other species
Torymus koebelei (Larval parasitoid)	0.090	No restraint in parasitizing hosts already parasitized by another species. Larvae nearly always win in competition for host with larvae of other species
Torymus baccharicidis (Larval parasitoid)	0.080	No restraint in parasitizing hosts parasitized by another species – except in case of *T. koebelei*. Larvae nearly always win in competition for host with larvae of other species

Modified from Table 3 in Force (1975).
[a]Determined by laboratory experiments (Force, 1970).

Recently, Force (1972, 1975) and Price (1973, 1975) considered the quality of natural enemies for the control of pest insects in terms of r-K strategies. Force (1975) found that, among the four parasitic wasps that parasitize the gall midge *Rhopalomyia californica*, the species with a large r loses in competition with others that parasitize the same host. The species with a small r is favoured in competition (Table 1.10). Multiple parasitism was found commonly in the field and here *Torymus* with the lowest reproductive rate was dominant. However, in the unstable stand where the plants were clipped continually, *Tetrastichus* was found in large proportions. Thus Force (1975) thought that many parasitoids became adapted to many long-established insects and K-strategists would survive in this process. Price (1973) has investigated the parasitoid fauna that attacks the Swaine jack-pine sawfly *Neodiprion swainei* in pine forests of various successional stages. He found that the number of parasitoid species increased with the progress of succession until it reached an equilibrium or decreased slightly in late stages and that in this process the proportion of larval parasitoids decreased and that of the pupal parasitoids (strictly speaking these oviposit on the larval host and adults emerge from the host's pupa) increased. If the host was parasitized by both a larval parasitoid and a pupal parasitoid the latter usually won the competition even when the former parasitized the host first. The number of eggs laid by the pupal parasitoid was usually small. Force thought that this fact would verify his prediction and made a proposal for the introduction of natural enemies; namely that r-strategists should be introduced against invading pests and K-strategists against those with several established parasitoids.

Price (1975) noted that the larval mortality of the parasitoid was proportional to the mortality of the parasitized host, and compared the reproductive rates of parasitoids in the hosts of various developmental stages. He considered that because many insects have a high larval mortality (see next chapter) the egg parasitoid must be more r-strategic than the larval parasitoid and that the larval parasitoid must be more r-strategic than the pupal parasitoid. [Price refers to the egg parasitoids as those that lay in host's eggs and emerge from the larvae of the host and the larval parasitoids as those that lay in larvae of the host and emerge mostly from the pupae of the host.] There is very little information available on the intrinsic rate of increase of parasitoids but there are many data, such as Iwata's mentioned earlier, on the number of eggs (ovarioles) in the ovaries of the female. In the parasitoids the number of ovarioles may be considered proportional, to some extent, to the intrinsic rate of increase. Fig. 1.7 was drawn from the numbers of ovarioles per

Fig. 1.7. Relationship between the number of ovarioles per ovary of ichneumonid parasitoids and the developmental stages of the attacked hosts (from Price, 1975). The oviposition sites include plant leaves (F), host eggs (E), young (Y), mid-instar (M) and old larvae (O) and pupae (P). The eggs laid on leaves are very small and they enter the digestive tract of the host when the leaf is eaten.

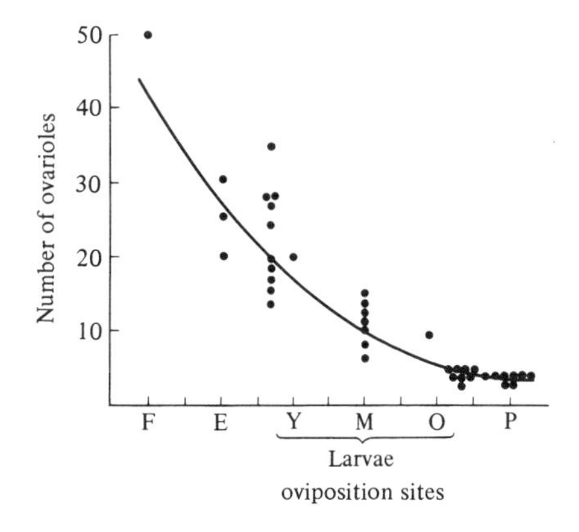

ovary of many ichneumonid wasps that parasitize various developmental stages of host insects (tachinid flies showed similar results). The figure shows that the earlier the developmental stage of the host parasitized the greater the number of eggs carried by the parasitoid. Thus early parasites are *r*-strategic in Price's interpretation.

A critique of *r*-*K* selection theory

As mentioned earlier the *r*-*K* selection theory of MacArthur & Wilson (1967) and the idea of Pianka (1970, 1974) who advanced this theory recognized that organisms always faced the evolutionary crossroad of two opposing strategies, high fecundity and low fecundity, and that they survived the struggle for existence by choosing one or the other. This understanding is in accord with the ideas of Lack (1954a, b) and Itô (1959). In the table constructed by Pianka (1970) – see Table 1.1 – the following points are supported by the facts presented in this or the next chapter: (1) *r*-selection operates in those environments in which the conditions fluctuate and these changes occur irregularly (unpredictably), (2) *K*-selection operates in those environments in which the conditions are stable or change predictably, (3) in many *r*-selected species major mortality processes are density-independent and the survivorship curve is of C type (see next chapter), (4) in *K*-selected species major mortality processes are density-dependent and the survivorship curve is of type A or B. However, the characteristics of the environment that induce *r*-selection or *K*-selection are not correctly interpreted. Take fish for example; we have seen that low fecundity and parental protection have developed in the rapids of the upper reaches of streams. In molluscs and crustaceans also large eggs and low fecundity are found in oligotrophic waters. In the freshwater environment the land-locked race of Pacific salmon (cherry salmon) *Oncorhynchus masou ishikawae*, the char *Salvelinus leucomaenis*, the Japanese sculpin *Cottus hilgendorfi* and the Japanese crayfish *Cambaroides japonicus* are *K*-strategists, yet their habitat is not endowed with a rich biota which would impose severe interspecific and intraspecific competition as Pianka might have expected. On the contrary, the biota is poor and predators are few and, in some cases, there are no closely related species in the habitat. By contrast the oceanic fish are typical high-fecundity animals, but they spawn in the surface zone of the ocean where the biota is rich and predators are numerous. This also applies to molluscs and large crustaceans of the marine environment.

Lack's theory and mine, though applicable to fishes and molluscs, had some difficulty in explaining the clutch size of birds and the litter size of mammals as touched on earlier. Although the forest habitat is richer in organic energy and species diversity than grassland, the clutch size of tree-living forest birds is smaller than that of grassland birds. Oceanic birds breeding on islands have a clutch of one but the surrounding sea is often rich in fishes on which they prey. I attempted to give a unified explanation for such phenomena by introducing the concept of 'relative oligotrophy' (Itô, 1959, p. 298; Itô, 1970). The question is not concerned with the total amount of organic energy in the environment but with the form it exists in, that is its availability for the young. Sea-birds reared on an island need nursing until their power of flight is sufficient to enable them to dive and catch their prey underwater; no matter how abundant the fish is in the surrounding sea it is not available till then. In the forest habitat young birds in trees cannot obtain their food until they can fly or at least hop from branch to branch; many fruits on the forest floor are not available to them. Thus arboreal species require parental care for a long period and this is incompatible with high fecundity according to the principle of allocation. The small clutch size of antarctic penguins is due to limited food availability for the young and not because of severe interspecific competition or the existence of many predators. It is easy to see that the antarctic is not an environment for K-selection.

From the above consideration it becomes clear that the contrasting relation of grassland (high-fecundity habitat) versus forest trees (low-fecundity habitat) is similar to the relation of downstream versus upstream for rivers, plains versus rocky mountains or temperate regions versus polar regions, though such relations appear different at first sight. However, the second habitat in each of the above contrasts has no relationship with the average amount of nutrition in the environment. What is common among these habitats is the difficulty of procuring food by the young (and also the inaccessibility of the young to predators). I would now like to abandon the use of the term 'relative oligotrophy' which might cause misinterpretation, and instead unify this relationship of habitats from the viewpoint of *the procurability of food by the young*.

One may also say that in each of the habitat contrasts given above the first is a two-dimensional and continuous habitat whereas the second is a three-dimensional and discontinuous habitat. This discontinuity contributes to the difficulty of procuring food by the young and also to the inaccessibility of the prey to predators. I should think, however, that the main factor in the evolution of low fecundity and parental care is the

Table 1.11. *Mean clutch size of passerine birds inhabiting tropical Africa and middle Europe (figures in parentheses indicate the number of species examined)*

Family or Subfamily	Tropical Africa	Middle Europe
Alaudidae	2.1 (6)	4.4 (3)
Hirundininae	2.9 (17)	5.0 (4)
Motacillidae	2.9 (10)	5.3 (7)
Laniinae	3.2 (6)	5.5 (4)
Muscicapinae	2.1 (36)	5.9 (4)
Sylviinae	2.5 (65)	5.8 (18)
Turdinae	2.5 (37)	5.3 (15)
Parinae	3.5 (3)	9.2 (6)
Emberizinae	2.5 (4)	5.0 (5)
Carduelinae	2.8 (10)	4.9 (9)
Passerinae	3.7 (17)	5.2 (3)
Sturninae	3.0 (13)	6.0 (1)
Oriolidae	2.3 (3)	4.3 (1)
Corvidae	4.4 (5)	5.2 (8)

From Lack (1968).

difficulty of gathering food rather than the degree of predation. First, the low degree of predation does not positively select for low fecundity though it may *permit* low fecundity. Secondly, when weedy plants invade areas of lava or cultivated fields the high-fecundity species first succeed in colonization. In this case natural enemies do not exist or are very few. Not only plants but also pest insects may be able to produce small eggs and high fecundity under little or no predation pressure.

However, Lack (1968) found among the passerine birds that tropical African species have a considerably smaller clutch size than European species of the same family or subfamily (Table 1.11). Lack points out that this also applies to the tree-living species of non-passerines (e.g. nightjars, pigeons, woodpeckers) and, though the data are very limited, to the terrestrial species. What does this mean? Should there not be more food in the tropics than in the temperate region when the forest habitat is compared between the two regions? Lack tried to explain this fact in terms of (1) day length and (2) migration: the further north one goes the longer the daylight in summer in which to feed young, and many northern birds are migrants and need to compensate for the bigger toll they suffer during migration. There are some opposing views concerning the first point and criticisms of the second as being teleological. [Royama (1969) put forward a model for the clutch size of the robin in relation to the heat metabolism of nestlings. In colder regions the heat radiation of the nestlings is greater and the large clutch can reduce this heat loss more than the small clutch. This can explain why the clutch size of the robin in the colder region of the interior of the continent is larger than that in the warmer region of the same latitude.]

I would emphasize that the high species diversity of animals in the tropics reflects low densities of special fruits and prey insects. In the temperate or boreal regions the forest contains many trees of the same species; in the boreal region there are areas in which no more than ten species of tree are found in several square kilometres of forest. In a typical tropical rain forest, on the other hand, species diversity is so high that one seldom encounters two trees of one species in a 100-metre walk. The number of individuals of a species is small; in the tropical rain forest of Paso, Malay Peninsula, 1174 tall trees of 276 species were counted in 10 ha but only seven species were represented by more than twenty individual trees in this area (Itô, 1975–76, p. 384). This would make it very difficult for certain animals to collect certain food items. Therefore, even in the same forest habitat, some animals will find tropical rain forest more difficult than temperate forests to search for food. Lack (1968) and Owen (1977) suggested similar explanations. The hypothesis of MacArthur and others that the lower fecundity in the tropics is due to the stronger interspecific competition cannot explain all cases of low fecundity in the unified way that I have given above.

What about the examples of higher plants (Table 1.9) given by Salisbury (1942)? It seems reasonable that competition for light made forest plants evolve in the direction of large seeds and low fecundity, but even here it is possible to unify our theory with the situation in animals by considering the evolution in relation to the procurability of sunlight as 'food' (how much growth is required in order to reach light), without invoking interspecific competition. In the sense that the sea is said to be eutrophic the lava field and flood plains are not eutrophic. They are oligotrophic. However, here young plants sprouting from small seeds can obtain sunlight. In effect, we can say for plants, as I have said for animals, that high fecundity has been selected for in the environment in which it is easy for the young plants to obtain 'food' (solar energy) and low fecundity and parental protection have been selected for in the environment in which it is difficult. [The Orchidaceae produces a large number of very small seeds. Apart from the mycorrhizal habit mentioned earlier, they occupy an exceptional position among the forest plants. This may be explained by the fact that many orchids live high on forest trees as epiphytes and can readily obtain solar energy.]

Let us review what I have discussed so far. Table 1.12 lists conditions

Table 1.12. *Conditions considered to be related to the evolution of high-fecundity and low-fecundity/parental protection strategies in major groups of organisms*

Group	High fecundity A	v.	Low fecundity B	Total amount of organic matter (H, high; L, low) A	B	Procurability of food (E, easy; D, difficult) A	B	Predators[a] (M, many; F, few) A	B	Competitors (M, many; F, few) A	B	Unpredictability of environment (P, predict.; U, unpredict.) A	B	Examples of low-fecundity strategies
Aquatic crustaceans	Sea	:	Inland waters	H	L	E	D	M	F	M	F	U	P	
	Rivers, lower parts	:	Upper waters	H	L	E	D	M	F	M	F	U	P	Land-locked atyid shrimps
	Eutrophic lakes	:	Oligotrophic lakes	H	L	E	D	M	F	M	F			*Cambaroides*
Fishes	Sea	:	Inland waters	H	L	E	D	M	F	M	F	U	P	
	Lakes, rivers lower parts	:	Upper waters	H	L	E	D	M	F	M	F	U	P	Land-locked sculpin
	Temperate seas	:	Arctic & antarctic seas			E	D	M	F	M	F			
	Shallow water	:	Deep sea			E	D	M	F	M	F	U	P	
Molluscs	Sea	:	Inland waters	H	L	E	D	M	F	M	F	U	P	*Lymnaea*
	Aquatic	:	Terrestrial			E	D			M	F			Land snails, slugs
Amphibians	Standing water	:	Running water			E	D	M	F	M	F			*Polypedates*
	Aquatic	:	Semi-terrestrial			E	D			M	F			Semi-terrestrial salamanders

Reptiles	Aquatic	:	Terrestrial			E	D							Desert lizards
Birds	Terrestrial	:	Arboreal	L	H	E	D	M	F[b]	F	M	U	P	
	Herbivorous, insectivorous	:	Carnivorous			E	D	M	F			U	P	Eagles, owls
	Continents; large islands	:	Cliffs, oceanic islands			E	D	M	F	M	F			Albatrosses
	Temperate forests	:	Tropical forests	L	H	E	D?[c]	F	M	F	M	U	P	
Small mammals	Terrestrial	:	Arboreal	L	H	E	D	M	F					Monkeys, sloth, koala
Higher plants[d]	Bare land, grasslands	:	Forests	L	H	E	D	F	M	F	M	U	P	
	Crop fields, towns	:	Natural vegetation	L	H	E	D	F	M	F	M	U	P	

[a]Number of species of predators is assumed to be proportional to species diversity of the community.
[b]Predators, if present, cannot approach prey easily.
[c]Because of great diversity it is considered difficult to collect fruits and insects of specific species.
[d]Procurability of light (=food for higher plants) is lower in dense forests than on bare land, fields of short grass, etc.

considered to be related to the strategy of low fecundity and parental protection in major groups of organisms. The only factor common to all pairs of high fecundity and low fecundity is the procurability of food by the young; other factors, such as the total organic matter, abundance of predators or competitors, cannot explain this contrast consistently. Thus the factors listed by Pianka do not provide a correct explanation. Further, the characteristic of the environment that parallels most closely to the procurability of food by the young is the structure of the environment. That is, the environment in which the young will not find it easy to obtain food may be said to be generally more *three-dimensional and discontinuous*. By contrast the environment in which food is easily procured by the young is more flat and continuous. It is this increased dimensionality that reduces the danger of attack by predators. Since such dimensionality and discontinuity of the habitat are difficult to define (for example, how to categorize the antarctic and arctic seas, deep seas), the procurability of food by the young seems to be a clearer indicator of fecundity.

When MacArthur & Wilson (1967) proposed the concept of *r*-*K* selection, they had the biota of islands in mind and put the primary emphasis on the strength of competition; fluctuations of the physical factors of the environment received only a brief mention. On the other hand, Pianka (1970) listed climatic stability as one of the criteria to distinguish between *r*- and *K*-selection (Table 1.1). Thus *K*-selection is said to occur where the climate is either stable or changes regularly as in the temperature changes between summer and winter (animals often avoid such changes by means of diapause or emigration), and *r*-selection where climatic factors fluctuate greatly or irregularly. This idea has since gained importance as the concept of unpredictability of climatic conditions. Recently, for example, Southwood (1977) considered this problem specially in relation to generation time. The habitats in which irregular climatic changes occur frequently, or temporary habitats, such as dry river-beds, are said to be *r*-selective.

As can be seen in Table 1.12, the amplitude of irregular fluctuations of the physical factors (here I refer not only to the climate but also to other factors including changes of an ocean current, river bottom, harvesting or pest control in cultivated fields) has a close association with the strategies of high fecundity versus low fecundity/parental protection. There is no reverse combination to be seen. As mentioned in the section on insect pests and weeds (p. 31) it is clear that this factor has played an important role in determining the evolution of reproductive rates.

Do we then consider the frequency or amplitude of irregular fluctuations in environmental factors as the most important criterion, instead of the procurability of food by the young, for separating high fecundity and low fecundity/parental protection? I do not think we should. In the irregularly fluctuating environment high fecundity is certainly advantageous because there is often high accidental mortality or even local extinction, but what would be the cause for not selecting for high fecundity in the stable environment? MacArthur, Pianka and many other American ecologists thought that in the stable environment the density of each species would increase and many species would co-exist, resulting in severe intraspecific and interspecific competition which would lead to K-selection. However, this does not always hold as we have seen in the case of the stream-dwelling fishes. It is difficult to explain consistently the evolution of low fecundity and parental protection as a strategy unless it is considered in the light of procurability of food by the young.

Here let us remind ourselves that r equals $\ln R_0/T$ (see p. 5) and thus the mean generation time (this may be taken as the time from birth to peak reproduction) influences r more than R_0. [Since $r=\ln(\Sigma l_x m_x)/T$, where l_x is the survival-rate of females and m_x is the age-specific birth-rate; r is influenced by three different factors, namely the survivorship curve, the number of young produced and the generation time. r does not simply mean high fecundity. One of the reasons that I do not like to use the expression r-selection is its multiple meaning and the fact that these three factors do not always change in the same combinations.] What is influenced most by the irregular fluctuation of environmental factors may be the mean generation time. Under the conditions of mass mortality due to irregular fluctuations of the physical factors, fast breeding and strong migratory habits as well as high fecundity become advantageous characters. There are groups like aphids which meet this situation of temporary habitat not by high fecundity but by shortening the generation time and increasing the migratory ability. But the irregularly fluctuating environment is usually (because it is non-structured) an environment in which young animals can obtain food easily. This may result in the coincidence of irregular fluctuations of the environment and high fecundity.

If we take only the stability of the environment both short and long life-spans can be selected for, but the stable environment is usually an environment in which young animals cannot easily obtain their food. This is because stability is usually related to the complexity of the structure of the environment (e.g. forest). This is the reason for the development of low fecundity in stable environments. Since the protection of young requires a certain degree of longevity there may be many opportunities for the long life-span to be selected for even if there are negative effects of longevity on r through an increased value of T in such an environment. In effect the fundamental factor encouraging high fecundity is the procurability of food by the young and the unstable nature of the environment would accelerate this tendency.

Let me give my conclusion: following Lack's original idea faithfully, I would propose the concept of the *high-fecundity strategist* versus the *low-fecundity/parental-protection strategist* and assert that the common factor among all organisms in determining the selection between these two types is the procurability of food by the young.

2

Strategies of survival

The survivorship curve

In Chapter 1, I have considered the ecological significance of birth from a comparative point of view. How, then, does its opposite character, death, occur? Plants and animals in nature seldom live out their physiological longevity. Thus their ecological longevity is much shorter than their physiological longevity. The longevity of animals in zoos listed in various books generally approaches the latter. In most animals death occurs not in one period but throughout the life-span of the species. For a comparative ecological consideration of death we must deal with age-specific mortality.

The life-table and the survivorship curve meet this purpose. An example of the life-table is given in Table 2.1. It shows how the young born at one time (or assumed to have been born at the same time) die in various age intervals, and usually includes the following terms: the number surviving (l_x), the number dying (d_x), the death-rate (q_x) and the life expectancy ($\dot{e}_x$). The life-table was contrived by Halley (1693) and developed with the life insurance business (the premium rate is determined by the value of $\dot{e}_x$), but it was probably Pearl (1922), one of the founders of population ecology, who first applied it to non-human animals. Application to wild animal populations was made much later, and Deevey's (1947) review became the starting point. However, because of the difficulty of measuring life-spans in the field there were only few life-tables in existence when the first edition of this book was published. It is since the 1960s that the life-tables of many animal species in the field have been constructed actively. In this chapter I shall use the survivorship curve which plots l_x against the age x as a criterion for comparison. The l_x curve cannot be drawn without the number of births,

Table 2.1. *Life-table of a natural population of the black-tailed deer; see text for explanation*

x (years)	x'	l_x	d_x	xd_x	q_x	$\dot{e}_x$
0–1	−100	1000	526	263	526	2.3
1–2	− 56	474	97	145	204	3.3
2–3	− 13	377	137	343	363	3.0
3–4	31	240	75	263	313	3.4
4–5	74	165	20	90	121	3.7
5–6	118	145	14	77	97	3.2
6–7	161	131	14	91	107	2.5
7–8	205	117	30	225	256	1.7
8–9	248	87	35	298	402	1.1
9–10	292	52	52	494	1000	0.5
10–	335	0	=1000	=2289		

Calculated from data of Taber & Dasmann (1957).
Mean longevity = 2289/1000 = 2.289 years.
q_x per thousand.

l_0. On the other hand, since q_x is the ratio of the number of deaths d_x (from l_x to l_x+1) to l_x, this can be used without the knowledge of l_0 which is difficult to estimate. Thus Caughley (1966) proposed the use of q_x for comparative ecological studies. Throughout this book, however, I shall use l_x; the reason for this will become clear in the following.

Pearl & Miner (1935) compared life-tables of various organisms by taking the percentage deviation, x', from the mean life-span (the mean life-span is taken as 0 and the time of birth as −100) and distinguished three basic types: type A (I) has low early mortality and high mortality near the end of the physiological longevity. Type C (III) has a marked

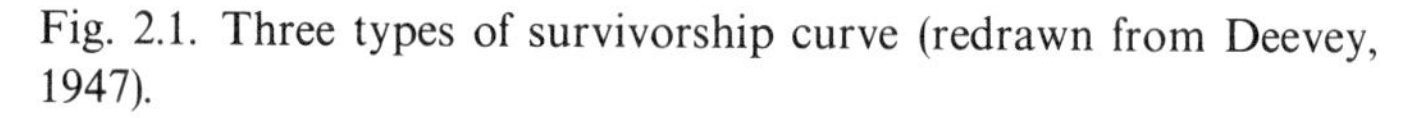

Fig. 2.1. Three types of survivorship curve (redrawn from Deevey, 1947).

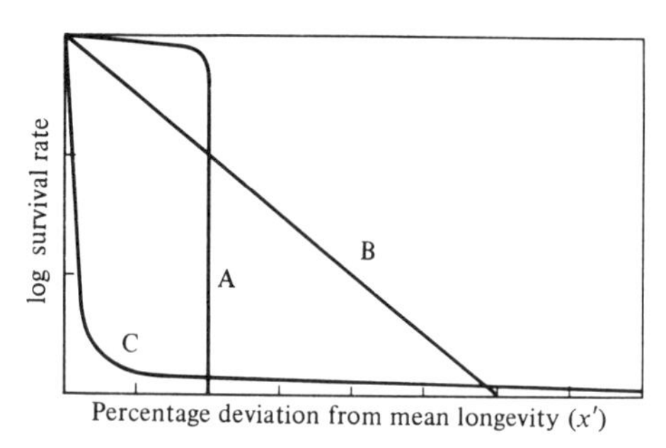

deviation from the mean life-span as a result of high early mortality, and type B (II) is intermediate.

Deevey (1947) expressed these three basic types by plotting the survival-rate along a logarithmic scale as in Fig. 2.1. According to him type A is characteristic of large mammals including man and a few invertebrates excluding the planktonic life stage. It means that accidental deaths are few and there is little difference between the mean longevity and the physiological longevity. Type B is the survival-type seen in many wild birds and is characterized by a constant death-rate (therefore survival-rate) over all age groups. Type C is the survival-type of those aquatic invertebrates and fishes that have even higher juvenile mortality, showing marked deviation from the mean life-span.

The adoption by Pearl and Deevey of percentage deviation from the mean life-span opened the way for the comparison of survivorship curves among animals of very different life-spans. The three types of Deevey's survivorship curve have been illustrated in most ecological textbooks. However, even the type C survivorship curve would have to turn downwards at the end of the curve when the last individual dies so long as the logarithmic scale is used for survival. Also, in the standardization of time used for the mean life-span it is difficult to compare those animals that live long after attaining maturity (starting reproduction) and those that die after one breeding season. In birds and large mammals the mean life-span, that is $x'=0$, lies far to the right of the point where breeding begins whereas in the insects that lay eggs at the end of their life-span the zero point lies to the left of the breeding age. For this reason the former gives an impression of having high mortality in early life though mortality before breeding is in fact lower than in the latter (Fig. 2.2). To overcome this difficulty I proposed the use of the period from birth to the commencement of breeding instead of the mean life-span as a unit of time, at the Population Ecology Symposium held at Sugadaira, Nagano Prefecture, in 1974. At this symposium Morisita independently proposed the same method and, for the comparison of species with very different mortality patterns during the pre-reproductive period, he further proposed adjusting the ordinate so that l_x is 1 at $x'=0$ (commencement of breeding). In this way the number of female offspring to be produced by a female in order to maintain the stability of the population is shown approximately by $l_{(-100)}$. [This value gives a slight underestimate for those females that breed more than once.] Unfortunately, it may not be easy to follow the graph when survivorship curves of several species are drawn on the same graph. In this book I shall use the pre-breeding period as a

Fig. 2.2. Methods of drawing survivorship curves (examples taken from Morisita, 1975): (*a*) Deevey's method, giving an impression that early mortality is lowest in the cabbage white and similar between the brown trout and the great tit; (*b*) taking the percentage deviation from the duration of the pre-reproductive period on the abscissa (commencement of reproduction = 0), the curve shows particularly high early mortality in trout, which conforms to expectations; (*c*) sliding the curves of (*b*) along the *y*-axis so that the number of individuals, l'_x, becomes 1 at $x = 0$ (assuming the sex ratio 1:1) and l'_{-100} is the number of female young required to replace one breeding female (this is more than 100 for the brown trout, about 40 for the cabbage white and 12 for the great tit).

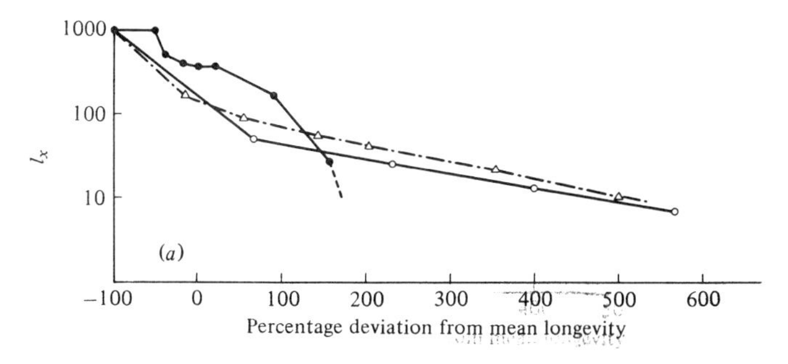

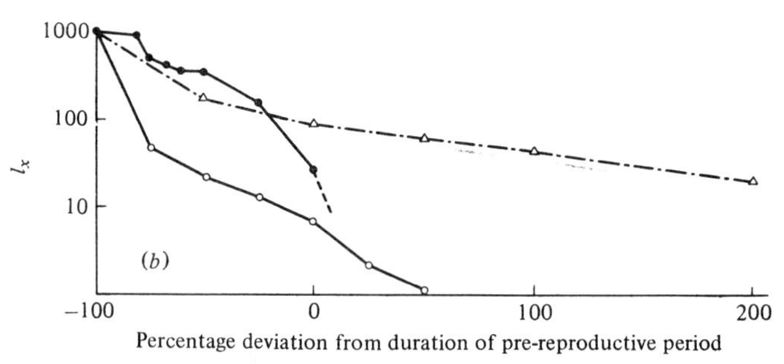

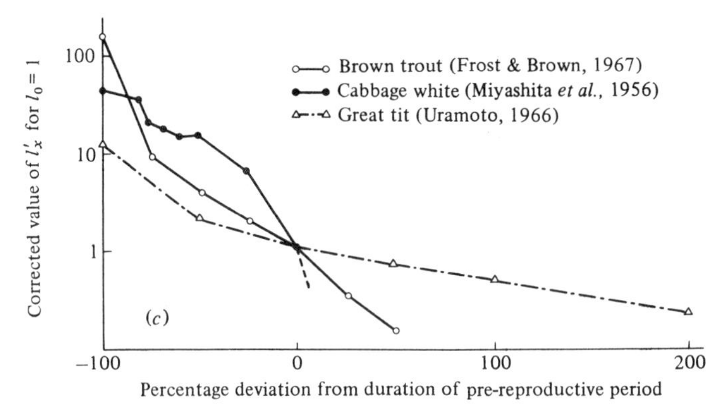

unit of time and express cumulative mortality of the number born as a logarithm along the ordinate. In so doing we may evaluate the pre-breeding mortality of each species. [There is the same limitation as in Morisita's method in that for those breeding more than once cumulative mortality till the mean reproductive age must be taken.] We may call those convex curves showing the survival of more than 50% before the reproductive age type A, those with straightline reduction starting at least before reaching 90% cumulative mortality (apart from somewhat high mortality at a very early stage) type B, and those reversed J-shape curves showing very high mortality of several orders in logarithmic scale in early stages (excluding the period of severe reduction at the end of the life-span) type C (Fig. 2.3).

Fig. 2.3. Three types of survivorship curve. Commencement of reproduction is at $x=0$.

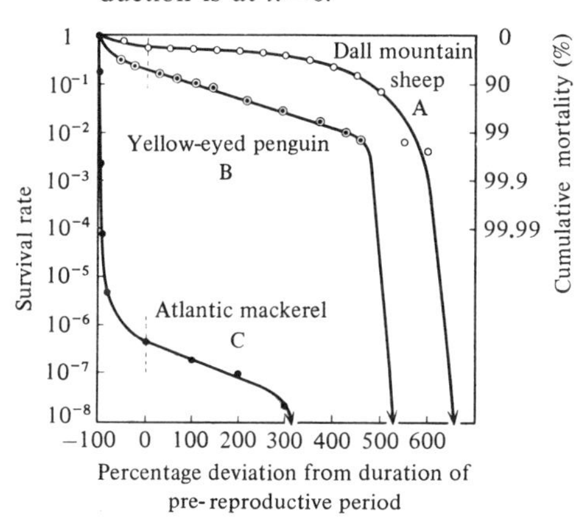

Fig. 2.4. The relationship between the number of eggs laid and the survivorship curve (from Itô, 1975–76).

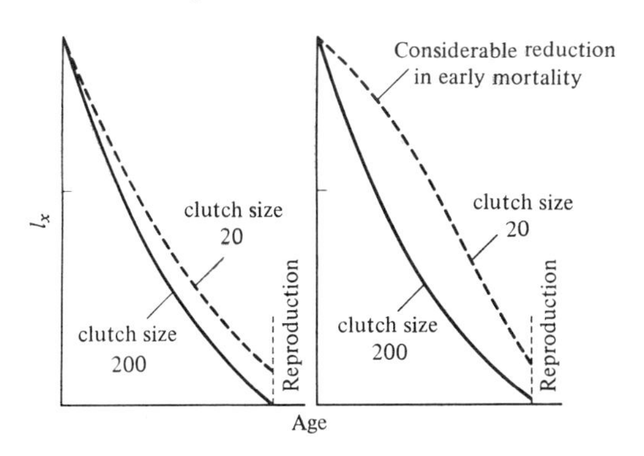

Incidentally, neither Pearl nor Deevey considered causes for the three types of survivorship curve. I pointed out in the first edition of this book (1959) that these three types reflected the evolutionary trend for parental protection in the order of C→B→A.

If the population is stable then the number of young produced is inversely correlated to the survival-rate up to the reproductive age. For example, an animal population with a sex-ratio of 0.5 and laying 200 eggs per female must lose 99% of the individuals before the reproductive age. If the mortality is lower than this the population will increase while if it is higher it will decrease. On the other hand, the mortality of animals which lay twenty eggs per female must be 90%. This change of mortality is not distributed to all age classes as shown in Fig. 2.4 (left), but normally takes a form of swelling up in the top left part of the curve as shown in Fig. 2.4 (right). This is because the reduction in egg number is tied, above all, to the evolution of parental care and as a result reduces mortality especially in the early age, instead of distributing it equally to all age classes. In other words, the three types of survivorship curve reflect the degree of parental protection of the young.

Of course the survivorship curve is a reflection of population dynamics and even within the same species its shape is different between the increasing phase and the decreasing phase of the population. It is also different depending on whether the population occupies a favourable habitat or occurs on the fringe of the species' distribution. In particular, the survivorship curve of a species population under strong climatic influences or other environmental factors may fluctuate between type B and type C (see Itô, 1958 for a review of changes in survivorship-curve and life-table parameters during the outbreaks of insects). However, the type of survivorship curve in a stable population reflects the characteristics of the life cycle of the species and similar results are obtained by different workers studying the same or closely related species. In this book consideration of the life-table data during the period of violent population fluctuation or on the fringe of distribution is avoided except where such discussion is warranted in relation to population fluctuation. Thus in most cases the survivorship curves presented in this book may be considered to represent that of a particular species being discussed under average conditions.

Survivorship curves of fishes

The marine teleosts are the most fecund group of animals. As we have seen the egg diameter of fishes that lay several hundred thousand or more eggs is very small compared with the body length of the parent, and the fry hatching out of such eggs generally have a shape different from that of adults and lead a planktonic life. We may expect such species to have high early mortality.

To estimate early mortality of fishes is an extremely difficult task; there was little information available for drawing survivorship curves of these fishes when the first edition of this book was written. In Fig. 2.5 the survivorship curves of the Atlantic mackerel *Scomber scombrus* and the Pacific sardine *Sardinops melanosticta* are shown. The former is a result of a large-scale investigation of Sette (1943); this study for the first time produced the survival-rate of young fish by sampling with plankton nets. According to this study, as many as 99.9996% of the eggs laid are lost within 70 days of laying. In this graph the proportion of eggs that survives till reproductive age (after one year) is less than one in a million. Accordingly it is necessary for a female to lay two to three million eggs, which conforms to our empirical knowledge that this species spawns twice or three times, laying half a million to a million eggs each time. The latter information is a result of large-scale cooperative research conducted by the Ministry of Agriculture Fisheries Research Institute in connection with a severe decline in sardine catches from Japanese waters. According to this study, the sardine population loses 97% of the fry

Fig. 2.5. Survivorship curves of the Atlantic mackerel *Scomber scombrus* (from Sette, 1943) and the Japanese sardine *Sardinops melanosticta* (from Nakai, 1962). In both species fish are assumed to start spawning at their first birthday (abscissa 100 = 365 days).

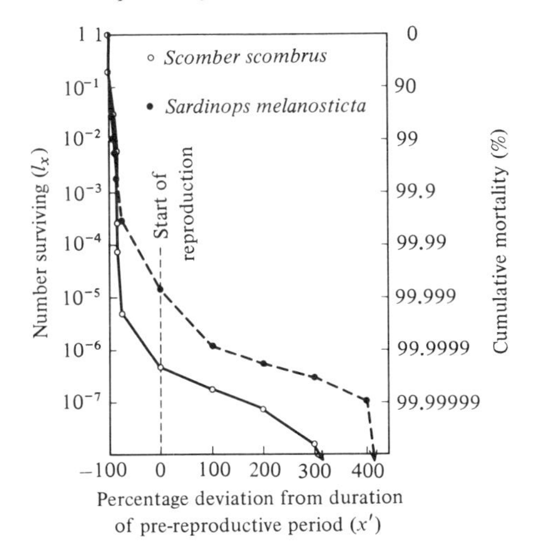

within 10 days of hatching and 99.97% within 70 days. Similarly, the anchovy *Engraulis japonica* studied at the same time lost about 99% of hatchlings within the first 10 days and 99.98% within 60 days (Nakai *et al.*, 1955). In the sardine the survival-rate up to the first-year spawning season is less than one in 50 000, which corresponds to the number of eggs laid (50 000–80 000).

Such an extremely high early mortality is, of course, derived from the planktonic stage (the species mentioned so far lay pelagic eggs and the fry have planktonic life for a period after hatching). On the other hand, the mortality of young fish after completing metamorphosis is relatively low. In the sardine even four-year-old individuals are found from time to time (Aikawa, 1949).

Fig. 2.6 shows survivorship curves for the pike *Esox lucius* of the Great Lakes and the plaice *Pleuronectes platessa*. The plaice suffers pre-breeding (the first three years of life) mortality of about 99.999% and the pike about 99.99–99.999% (the first two years of life). The mortality of adults, on the other hand, is up to about 50% for either species. Examining such detailed life-table data one finds that the survivorship curve is not simply a reversed J-shape (the pike, though a freshwater species, is a large fish living in huge eutrophic lakes comparable to seas and lays a large number of small eggs).

Fig. 2.6. Survivorship curves of the pike *Esox lucius* of the Great Lakes (from Kipling & Frost, 1970) and the plaice *Pleuronectes platessa* (from Gulland, 1971). The commencement of breeding was assumed to be at two years of age for the pike and three years of age for the plaice.

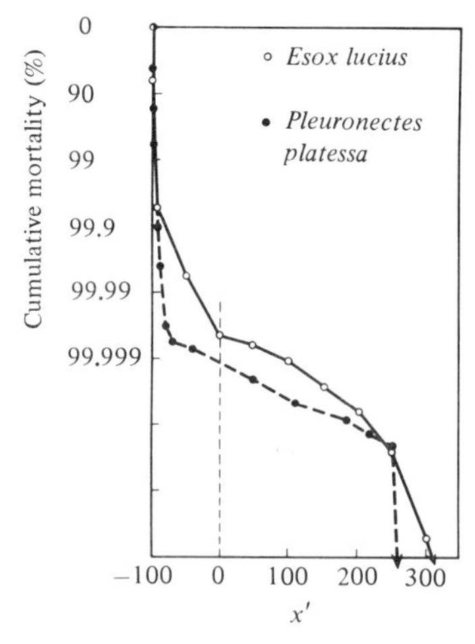

The salmonid fishes spawning in freshwater have a habit of burying their large eggs in small numbers. This simple form of parental protection is expected to lower the early mortality (Fig. 2.7). The survivorship curve of the sock-eye salmon *Oncorhynchus nerka* is based on the detailed study of Foerster and his co-workers in Canada over ten years. The salmon descend to the sea in their second year of life and return to the rivers in their fourth year. The number of individuals involved has been counted accurately:

The number of eggs laid in 1925		17 470 000	
The number of fish in			mortality
downstream migration	Total	197 494	98.87%
1926		12 500	
1927		183 272	mortality
1928		1 722	97.18%
upstream migration	Total	5 563	
1929		4 463	
1930		1 100	

In effect 99.97% of the eggs laid do not become adults but, as the mean number laid is 4500, the population will maintain its equilibrium with this mortality.

The lefthand side of Fig. 2.7 gives two examples of freshwater trout.

Fig. 2.7. Survivorship curves of salmonid fishes spawning in rivers; the rainbow trout *Salmo gairdnerii* (from K. R. Allen, 1951, in Weatherly, 1972), the brown trout *Salmo trutta* from Frost & Brown, 1967) and the sock-eye salmon *Oncorhynchus nerka* (from Foerster, 1934). The first two species live only in inland waters and spawn several times from the second year (rainbow trout) or the third year (brown trout).

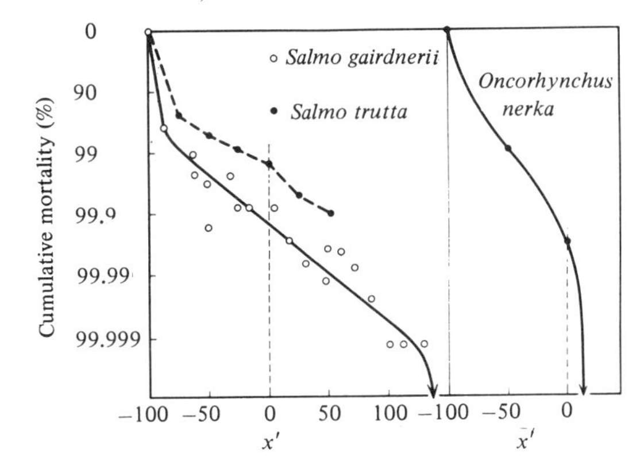

Table 2.2. *Age composition of a natural population of the black bass,* Huro salmoides

Age in years	Number of individuals
0	18 374
0 (Carnivorous period)	229
1	25
2	10
3	105
4	7
5≦	9

After Eschmeyer (1939).

Both the rainbow trout and the brown trout have a long life-span and spawn many times. The curve for the rainbow trout is what Weatherly (1972) has summarized from Allen's detailed study of a population introduced into New Zealand, while that for the brown trout is what Morisita (1975) has drawn from the book of Frost & Brown (1967). The number of eggs laid by either species is several hundreds but the early mortality is only between 99% and 99.9%.

Eschmeyer (1939) killed all individuals of the largemouth black bass *Huro salmoides* in a lake, by poisoning, and determined the number for each age group. The results are reproduced in Table 2.2. The number of individuals given against the age class 0 is not the number of eggs laid but the number of individual fish under one year of age. Thus the number laid would be about ten times this number. The reason for the large number of three-year-old fish might be either that the cohort started with a specially large number of eggs or that the mortality of eggs

Table 2.3. *Results of mark-release-recapture of the yellowtail* Seriola quinqueradiata

	No. recaptured					
No. released	Year of release	2nd year	3rd year	4th year	5th year	Total
993	206	75	20	7	1	309
Ratio to previous year	0.21	0.36	0.27	0.35	0.14	

Based on data of Tanai (1940).

Table 2.4. *Age composition of the wall-eye pollack* Theragra chalcogramma *in Korean and Japanese waters (Japan Sea)*

	Age							
	3	4	5	6	7	8	9	10
Korea	0.1	1.1	1.1	14.7	47.9	21.1	13.0	1.1
Japan	25.0	40.1	13.7	12.8	5.7	2.0	0.6	0.1

After Aikawa (1949).

or fry was specially low for some reason. In any case it is clear that this fish belongs to the type with a high early mortality.

On the whole, the survivorship curve of fishes is characterized by high early mortality and by the very low mortality, with little fluctuation from year to year, of those individuals reaching a certain age (except for fishes with a special mode of life life salmon). Tanai (1940) examined the recapture data of mark-released yellowtails, *Seriola quinqueradiata*, before the Second World War. As shown in Table 2.3 the survival-rates from year one to year two, two to three and three to four were high, ranging from 0.27 to 0.36, and were stable. According to Tanai (1940), the yearly survival-rate for the scomber was 40% from year two to three and 21% from three to four, whereas for the blue-fin tuna it was 30% up to two years old and 75% on average, remaining fairly constant, from three to about eight years old. According to Aikawa (1949) the wall-eye pollack *Theragra chalcogramma* of inshore waters of Korea and Japan (the Japan Sea side) had the age distribution shown in Table 2.4. The abundance of older fish off Korea was said to be a result of migration of the pollack from the Japanese waters. It can be seen from the table that the expectation of life is long once the pollack has reached adulthood.

Because of the above features a fish population may exhibit an age distribution with predominance of a particular age group (cohort) in the catches lasting over many years. This happens following the addition of a specially large number of individuals to the population (the new age group in the catchable population). This phenomenon is particularly conspicuous in herring (Fig. 2.8). In the adult population mortality is not only low but also fluctuates little from year to year.

Survivorship curves of aquatic invertebrates, amphibians and reptiles

There are very few life-table data of the aquatic invertebrates. For the Alaska king (Japanese crab) *Paralithodes camtschatica* there are excellent studies made by Marukawa (1933) and Wang (1937) before the

Second World War. Fig. 2.9 is constructed from Marukawa's data. The female of this species carries eggs in the abdomen and discharges them when they reach the zoea stage. The larvae suffer 97% mortality before they complete the thirteenth moult to become small crabs, and by the time they are one year old 99.2% will have died. After that the life expectancy of adults is long and Marukawa (1933) claims that some live over twenty years. However, as the number of eggs laid by a five-year-old crab is 200,000–300,000 (Aikawa, 1949) the survival rate is still too high in this survivorship curve. It is probable that the mortality in the pre-crab stages was underestimated.

Fig. 2.8. Changes of age distribution in the herring *Clupea harengus* (after Hjort, 1926).

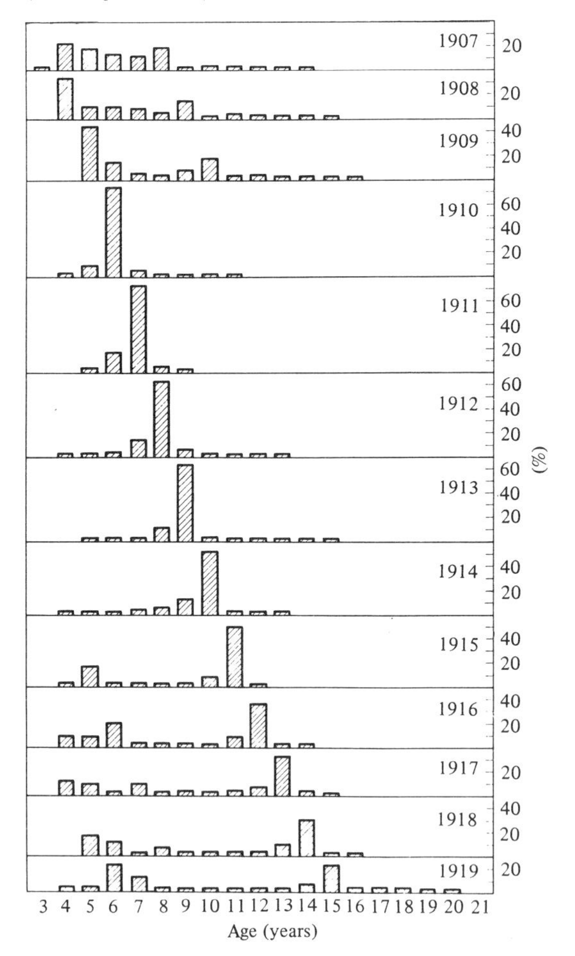

Among the crabs the population dynamics of the fiddler crab *Uca lactea* living in the tidal zone has been studied by Takao Yamaguchi (personal communication, 1976) of Kumamoto University. The mortality of this species from the egg through the zoea stage to the young crab appearing on the seashore is between 99.75% and 99.99%. Afterwards the mortality is somewhat high in the first 50 days (60%) and then becomes low. The adult life is fairly long and some are estimated to be seven years of age or older. Unlike the Alaska king, egg mortality is included in the mortality during the planktonic stage, which is estimated to be over 99.9%. The number of individuals maturing (usually at the age of two) is one in every 10 000 eggs laid and this corresponds roughly to the number of eggs laid at a time (if the sex ratio is 1 : 1, one in every 5000 eggs must reach two years of age).

Welsh (1975) studied a population of a small freshwater species, the grass shrimp *Palaemonetes pugio*. This species lays only a small number of eggs, the mean number being 434 per clutch, and spawns one to three times during one breeding season (most females die after one year). In Fig. 2.9 the number of eggs laid was estimated, assuming that the female lays 1.5 times on average. Note that this species with a large number of individuals surviving to the reproductive age has a survivorship curve with a convex shape in the early part of its life cycle.

A rare example of life-table studies on copepods is seen in the work of

Fig. 2.9. Survivorship curves of three species of crustaceans: the Alaska king *Paralithodes camtschatica* (from Marukawa, 1933), the grass shrimp *Palaemonetes pugio* (from Welsh, 1975) and the fiddler crab *Uca lactea* (from T. Yamaguchi, unpublished). Reproduction begins at the age of seven in the Alaska king, at one in the grass shrimp and at two in the fiddler crab.

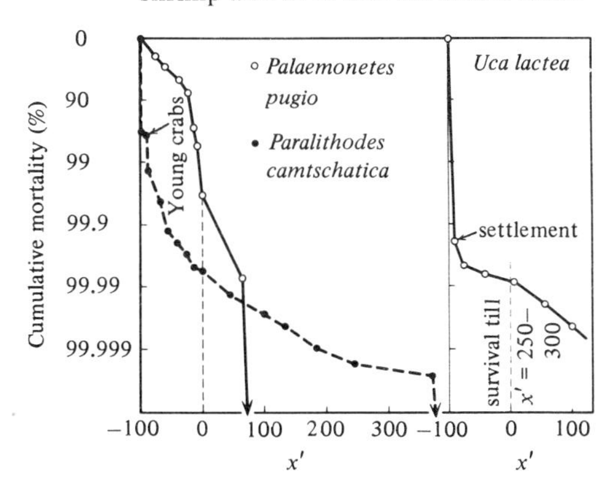

Gehrs & Robertson (1975). In the species studied, *Diaptomus clavipes*, about 99% die before attaining the adult stage (Fig. 2.10).

For an example of molluscs Hancock's (1971) study of the edible cockle *Cardium edule* (Cardiidae) gives a detailed description. As shown in Fig. 2.11 more than 99.9% die between the egg stage and the settlement of young in beach sand as bivalves (on average 15 000 eggs are laid). Thereafter mortality is relatively low and mobility is limited, permitting recapture of marked individuals for some years. In this diagram the mean values over eleven years are used to draw the curve for the first two years (below $x'=0$) and the ratios of the second year to the

Fig. 2.10. Survivorship curve of the copepod *Diaptomus clavipes* (drawn from Gehrs & Robertson, 1975).

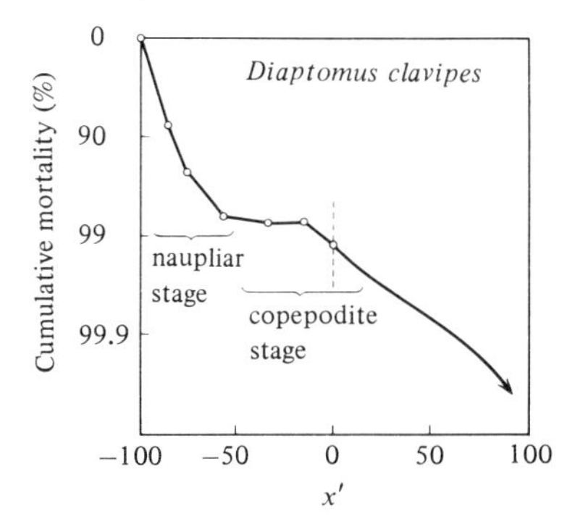

Fig. 2.11. Survivorship curve of the edible cockle *Cardium edule*, based on the calculation of Dempster (1975) from Hancock's (1971) data. For values $x' \geqq 50$ the ratios of the second year to the older groups were used for estimation.

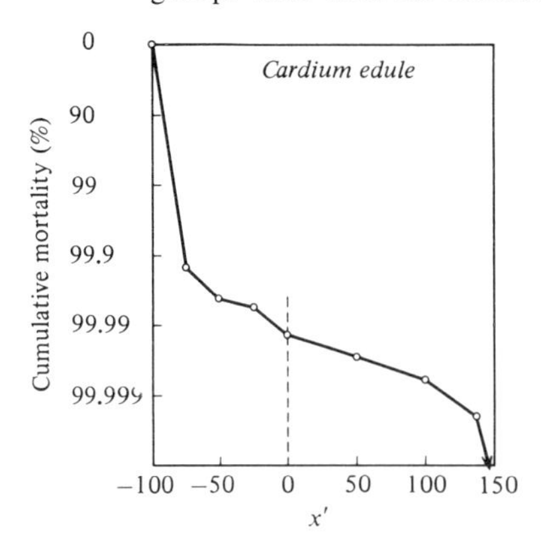

third year and older shells is used to estimate the values for $x' = 50$ (three years of age) and greater.

Among the freshwater invertebrates, populations of leeches have been studied in England. Fig. 2.12 (*A*), constructed from the study of *Glossiphonia complanata* by Mann (1957), shows high early mortality. In *Erpobdella octoculata* studied by Elliott (1973) the early mortality is not as high as in *Glossiphonia* (Fig. 2.12, *B*). According to Mann *Erpobdella* lacks a toothed sucker and swallows food such as chironomid larvae. It builds a strong cocoon in which to produce twenty-three young on average. Mann (1953) stated that both species breed twice but according to Elliott (1973) most individuals of *Erpobdella octoculata* die immediately after the first breeding. *Glossiphonia* is a blood-sucking leech taking the body fluid of invertebrates and builds a thin cocoon. The mean number of young produced is twenty-six and about half the number of adults survives to breed for a second time in the following year. Thus its breeding capacity is a little more than 1.5 times that of *Erpobdella* and is on the *r*-strategic side.

Coming to the amphibians we find a report of a very interesting series in the salamanders belonging to the genus *Desmognathus*. Organ (1961) studied the population dynamics of four species of this genus and found that *D. quadramaculatus* had the greatest fecundity, the longest larval stage and an entirely aquatic life. At the other extreme was found *D. wrighti*, which was almost entirely terrestrial. The female builds a nest underwater and lays eggs in it, but the larvae hatch at a very advanced stage and appear on land almost immediately. Between these extremes

Fig. 2.12. Survivorship curves of two species of leech. *A*, *Glossiphonia complanta* (from Mann, 1957); *B*, *Erpobdella octoculata* (from Elliott, 1973). Arrows indicate the period of breeding in *A*.

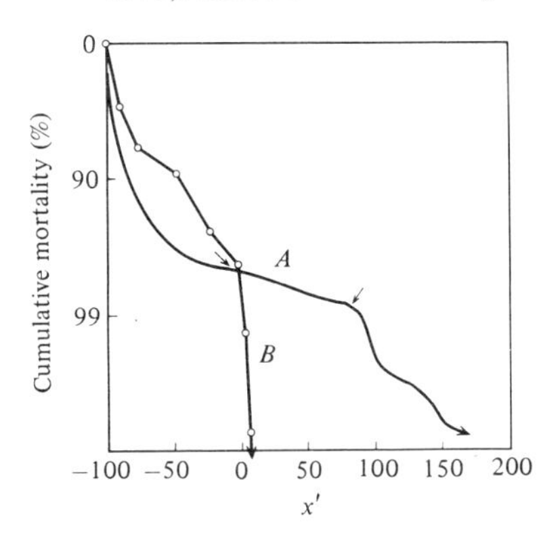

we find *D. monticola* and *D. ochrophaeus*, with the former closer to *D. quadramaculatus*. In the following the relationship between the number of eggs laid and their survival-rates (to two different developmental stages) is compared among the four species.

	D. quadra-maculatus	*D. monti-cola*	*D. ochro-phaeus*	*D. wrighti*
Mean no. of eggs laid	94	81	30	18
Survival rate to x' = about -30	23.6	22.1	33.9	49.5
Survival rate to $x'=0$	5.3	4.9	12.0	20.0

From this comparison we can see that the more terrestrial the species the less fecund it is and the lower its cumulative mortality to the reproductive age. The reduced mortality appears specially in the larval stage. The survivorship curve of *D. wrighti* approaches type A (Fig. 2.13). According to Organ, *D. quadramaculatus* is the most primitive morphologically and *D. wrighti* the most specialized. Thus the reduction in the number of eggs laid and the change of survivorship curve towards type A are considered to have occurred in parallel with the invasion of land.

Although frogs and toads are excellent materials to use for ecological studies very little work has been published to date on the life-table. Calef

Fig. 2.13. Survivorship curves of three species of salamander (from Organ, 1961). *Desmognathus quadramaculatus*; aquatic, fecund, with longest larval stage: *D. wrighti*; most terrestrial, with parental care and no larval stage in water: *D. ochrophaeus*; intermediate.

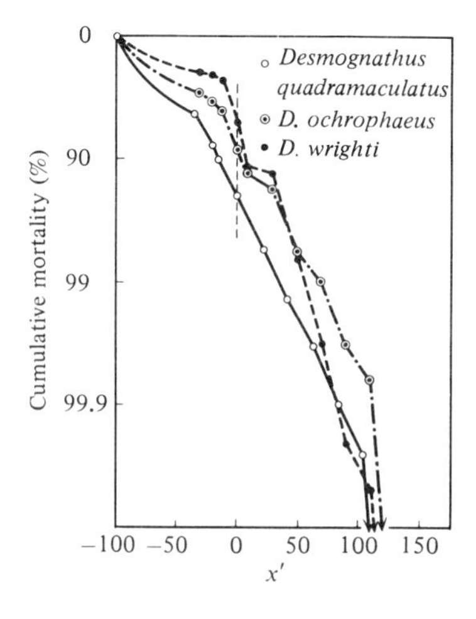

Table 2.5. *Survival rates of eggs and tadpoles in* Rana aurora

Developmental stage	No. of individuals	Survival rate within each stage
Eggs (early May)	329 000	
		9·1%
Hatchlings (early May)	300 000	
		33%
Tadpoles (early June)	100 000	
		65%
Tadpoles (late June)	65 000	
		69%
Tadpoles (early July)	45 000	
		44%
End of tadpole stage (late July)	20 000	

Data from Calef (1973).
Survival-rate from egg to the end of larval stage = 6%

(1973) investigated mortality of eggs and tadpoles in *Rana aurora* (Table 2.5). In this species an egg mass contains 531 eggs on average (probably one to two egg masses are produced in one breeding season). Though no data are available for adults it is considered that the first breeding takes place after one year and the life-span is several years. This means that for the population to be stabilized further mortality of 97–98% must occur in the remaining nine months of the first year. Most of it may be occurring during the fry stage, soon after they appear on land. In any case early mortality is considered high during the aquatic life, when predation by salamanders is said to be important.

Pearson (1955) studied the survival-rate of the spade foot toad *Scaphiopus h. holbrooki* after metamorphosis (Fig. 2.14). The result shows a relatively high survival-rate of this species after metamorphosis. This

Fig. 2.14. Survivorship curve (solid line) and mortality (broken line) of the spade foot toad *Scaphiopus h. holbrooki* excluding the period of egg and tadpole stages (from Pearson, 1955). Time intervals: F, February; M, May; A, August; N, November. The breeding season is indicated by bars.

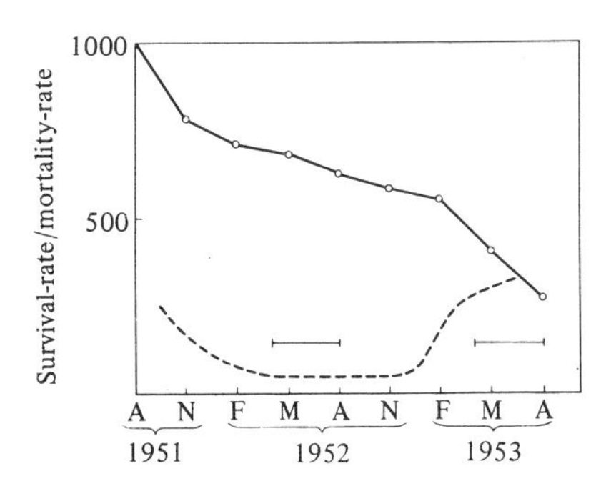

toad has a special life history, living in the forest without permanent aquatic habitats, such as creeks or ponds, and laying in synchrony in temporary puddles of water after heavy rain. The larvae metamorphose into toads within several days. Perhaps the survival of young is good if the puddle does not dry up. Mortality due to such drying up and the inhibition of laying due to drought may be controlling the population size.

Population dynamics studies of reptiles have been much advanced recently in America. Fig. 2.15 shows the survivorship curves of two species of ovoviviparous lizards, *Sceloporus undulatus* and *S. virgatus*. The former lives in woodland on river banks and has a mean clutch of 6.5 eggs. It breeds several times in one season, producing about 20 eggs per female in a year (Tinkle, 1972). The latter lives in the desert of Arizona and produces an average of 9.5 eggs once a year. Breeding starts at the age of two in the former and one in the latter. Reflecting these characteristics the survivorship curve of *S. virgatus* tends to be of type A. For the maintenance of the population the survival-rate of *S. undulatus* is on the low side while that of *S. virgatus* is on the high side. It is known that the population of *S. undulatus* was investigated during its decline, but this was not known for the population of *S. virgatus* which might have been on the increase at the time of the study.

Ballinger (1973) gives the life-table of two other ovoviviparous species belonging to the same genus. One of them, *S. poinsetti*, matures in 16–17 months and produces 4–7 young at a time. In another species, *S. jarrovi*, 60% of the first-year lizards mature and produce four young per female on average. From the second year onwards an average of ten young is

Fig. 2.15. Survivorship curves of egg-laying lizards, *Sceloporus virgatus* (from Vinegar, 1975) and *S. undulatus* (from Tinkle, 1972).

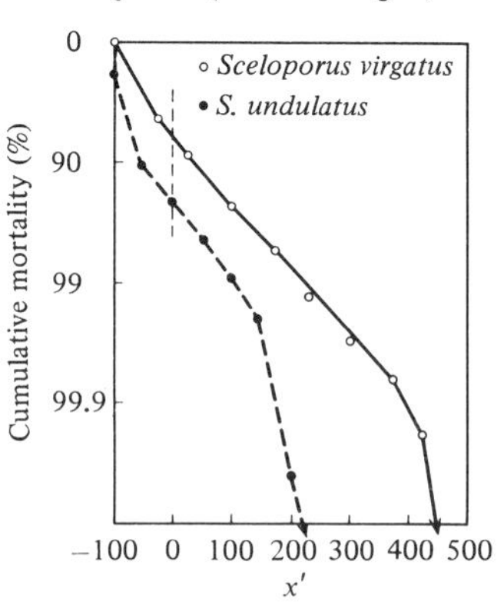

produced. Ballinger considered that *S. jarrovi* was more *r*-strategic than *S. poinsetti* because of its first-year maturation, but in the survivorship curve it is *S. jarrovi*, rather than *S. poinsetti*, that has the curve slightly closer to type A (Fig. 2.16). However, the degree of difference in litter size (clutch size) is probably smaller than the effects of population phase differences (increasing or declining) and sampling errors. Although *S. jarrovi* has a shorter life-span than *S. poinsetti*, an increase of litter size from the second year may balance its population.

Xantusia vigilis studied by Zweifel & Lowe (1966) is an ecologically interesting species. It is ovoviviparous and said to produce a litter of one

Fig. 2.16. Survivorship curves of ovoviviparous lizards, *Sceloporus poinsetti* and *S. jarrovi* (from Ballinger, 1973) and *Xantusia vigilis* (drawn from Zweifel & Lowe, 1966, in Turner *et al.*, 1970).

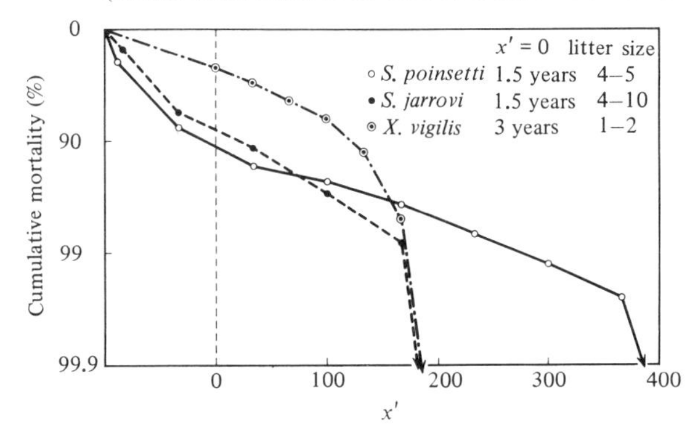

Fig. 2.17. Survivorship curves of males and females of the copperhead *Agkistrodon contortorix* (ovoviviparous; litter size, 5.3) (from Vial *et al.*, 1977).

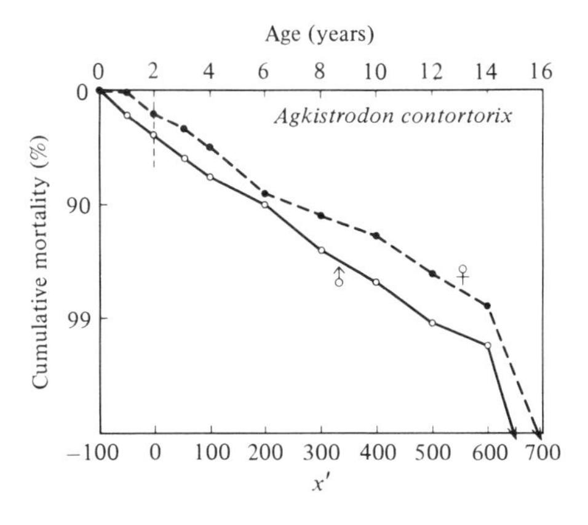

in a year (probably 1–2 young as there is a value of 1.8 in the literature). Further, it takes three years to mature and lives eight or nine years. For a small species it has a long life-span. Its survivorship curve calculated by Turner *et al.* (1970) from the data of Zweifel & Lowe is given in Fig. 2.16. It forms a curve of type A.

In lizards there are many other examples of life-table data, but *Xantusia* is exceptional in having a survivorship curve of type A; most species of lizards have type B curves.

Fitch (1960) investigated in detail the ecology of the copperhead *Agkistrodon contortorix*, a common American ovoviviparous, venomous snake, and Vial *et al.* (1977) constructed its life-table based on this work. This is shown in Fig. 2.17 (as the survival-rate is different between the sexes they are treated separately). In this figure the population would increase if the mean litter size was 5.3. Perhaps early mortality is higher than shown in the figure. Snakes may have such survivorship curves of type B in common.

Turtles are the symbol of longevity in Japan, with a saying that 'turtles live ten thousand years'. What are their survivorship curves like in reality? Recently Wilbur (1975) published a life-table of a freshwater turtle, *Chrysemys picta*, which exhibits longevity not contrary to the proverb (Fig. 2.18). However, the early mortality is very high. This is due to the high mortality of egg masses (two egg masses, each with six or seven eggs, are laid usually in one breeding season) deposited in the sand

Fig. 2.18. Survivorship curve of a freshwater turtle, *Chrysemys picta* (from Wilbur, 1975).

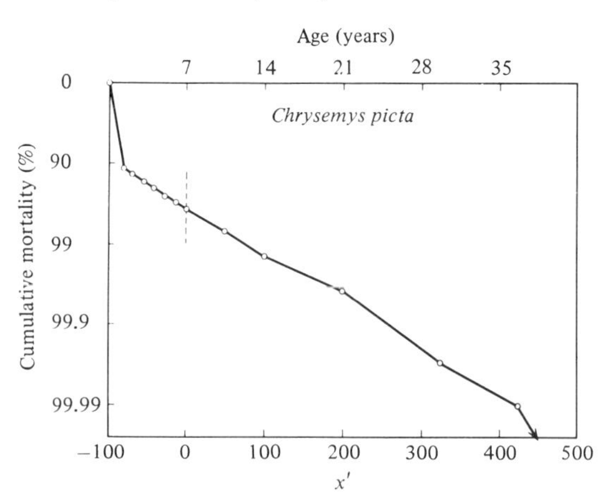

of lake shores. The cause of mortality is largely predation by racoons. The turtles may be compensating for this loss by laying many times during the long life-span. Wilbur interprets the evolution of this life history as an adaptation to a highly uncertain probability of nest success and an effective predator defence in the adult stage. [Sea-turtles are more fecund: the loggerhead *Caretta caretta gigas*, 120–150 eggs (diameter 40–45 mm); the green turtle *Chelonia mydas japonica*, 120–200 eggs (30–45 mm); the hawksbill *Eretmochelys imbricata squamata*, 150–200 eggs (35–40 mm). It is not difficult to imagine that the survivorship curves of these species would show even higher early mortality.]

An interesting fact from this point of view, though not quantitative, has been reported for the Nile crocodile *Crocodylus niloticus* (Pooley & Gans, 1976). The female of this huge crocodile lays sixteen to eighty small eggs in a year (older animals lay more eggs). Hatchlings weigh as little as one-four thousandth of the adult's body weight but feed themselves. The female crocodile protects her eggs till hatching and for a short period thereafter, but the mortality of hatchlings is considered very high. Mortality probably becomes low once they are several years old and commencement of breeding is at the age of twenty. The fact that the aquatic reptiles, such as crocodiles and turtles, lay more eggs than small terrestrial species is worth noting in relation to the theory presented in the previous chapter.

Survivorship curves of birds

Lack (1954a) has assembled information on the survival-rates of eggs and nestlings for many European and American species, part of which is presented in Tables 2.6 and 2.7. From his table Lack has concluded that the passerines with open nests in trees have survival-rates of 22–59% (average 45%) from the egg to the fledging stage while the hole-nesting passerines have a survival-rate of 67% until fledging. However, in the nidifugous birds nesting on the ground, such as gallinaceous birds, ducks and geese, the survival-rate is much lower and only about 25% of the eggs hatched become fully fledged young. Among the ground-nesters the low mortality of gulls, which are nidicolous, is noteworthy.

Lack (1946a, 1949) and others found that three-quarters or more of the early mortality in passerine birds is due to predation and the rest is caused by starvation, parasites and a small percentage of unfertilized eggs. Predation is probably high in the case of the gallinaceous birds (reduction in the number of chicks is great during the first few days of hatching). Thus the differences in mortality among the passerine open-

Table 2.6. *Survival-rates of eggs and young in passerine birds and other nidicolous species which usually nest in trees or other high places*

Species	Country	Eggs hatching (%)	Hatched young flying (%)	Young flying from eggs laid (%)	Mortality (%)
Species with open nests (natural sites)					
Horned lark[a]					
Eremophila alpestris	U.S.A.	77	58	45	[illegible]5
Song thrush					
Turdus philomelos	G.B.	71	78	55	[illegible]5
Blackbird					
Turdus merula	G.B.	64	79	51	[illegible]9
Turdus merula	N.Z.	34	88	30	[illegible]0
Turdus merula	Shetland	—	—	62	[illegible]8
Cedar waxwing					
Bombycilla cedrorum	U.S.A.	77	70	54	[illegible]6
American goldfinch					
Spinus tristis	U.S.A.	65	74	49	[illegible]1
5 passerine species	U.S.A.	48–72	61–79	33–51	[illegible]9–67
10 passerine species	Australia	73	79	57	[illegible]3
Species with nests in covered sites or holes in trees					
Two woodpeckers					
Dryocopus martius & *Dendrocopus major*	Finland	86	89	77	[illegible]3
Robin					
Erithacus rubecula	G.B.	71	76	54	[illegible]6

Nest-box studies					
Great tit					
Parus major	Netherlands	76	86	66	34
Parus major	Netherlands	—	—	56–73	44–27
Parus major	G.B.	85	86	73	27
Blue tit					
Parus caeruleus	G.B.	82	94	77	23
House wren					
Troglodytes aedon	U.S.A.	60	81	48	52
Tree sparrow					
Passer montanus	Germany	60	74	44	56

From Lack (1954b).
[a]Ground-nesting species.

Table 2.7. *Survival rates of eggs and young in non-passerine species which nest on ground (survival-rates of eggs and young from hatching to fledging)*

Species	Country	Eggs hatching (%)	Hatched young surviving (%)	Young surviving from eggs laid (%)	mortality (%)
Mallard					
Anas platyrhynchos	U.S.A.	49	52	26	74
Blue-winged teal					
Anas discors	U.S.A.	54	55	30	70
Shoveler					
Spatula clypeata	U.S.A.	70	—	—	—
Tufted duck					
Aythya fuligula	Finland	56	46	26	74
Ruffed grouse					
Bonasa umbellus	U.S.A.	59	37	22	78
Partridge					
Perdix perdix	G.B.	74	48	35	65
Pheasant					
Phasianus colchicus	U.S.A.	32	56	18	82
Phasianus colchicus	U.S.A.	18	88	16	84
Coot					
Fulica atra	G.B.	35	67	23	77
Herring gull					
Larus argentatus	Canada	71	51	36	64
Larus argentatus	G.B.	92	41	38	62
Lesser black-backed gull					
L. fuscus	G.B.	94	55	52	48

From Lack (1954b).

nesters, the passerine hole-nesters and ground-nesting nidifugous birds are caused mainly by the different degrees of protection against predators.

Lack considered that the regulation of bird populations could not be explained in terms of the early mortality alone, and further examined the total mortality of young till breeding age. According to this, the mortality of the song sparrow was 36% till fledging and then 79% till breeding, only 13% of the eggs laid becoming adults. The yearly mortality of adults was 45%. Similarly, 14% in the robin, 8–9% in the great tit and 18% in the swift survived till breeding.

In effect, about 80–93% of the eggs laid are lost before attaining adulthood in birds. The percentages are far smaller than the corresponding percentages of 99–99.99 found in fishes, but are still greater than many mammals. According to Lack (1954b), the yearly mortality of adult birds obtained from ringing or age-distribution studies is 40–60% for Passeriformes, Anseriformes and Galliformes, 30–40% for Ciconiiformes and Charadriiformes, 20% for Apodiformes, 10% for Sphenisciformes and 3% for Diomedeidae. The expectation of life for those attaining adulthood is one to two years in passerines, two to three years in herons and gulls, four to five years in swifts, ten years in penguins and thirty-six years in albatrosses.

The above is only a small part of the enormous amount of data available on the mortality of birds. But there are not many life-tables constructed for the entire life of birds. The representative survivorship

Fig. 2.19. Survivorship curves of the red grouse *Lagopus lagopus scoticus* (calculated from Jenkins *et al.*, 1963, with the ratio of young to adults 1:1 in June and commencement of breeding at the age of two) and the lapwing *Vanellus vanellus* (from Lack, 1954b, with the commencement of breeding at the age of two).

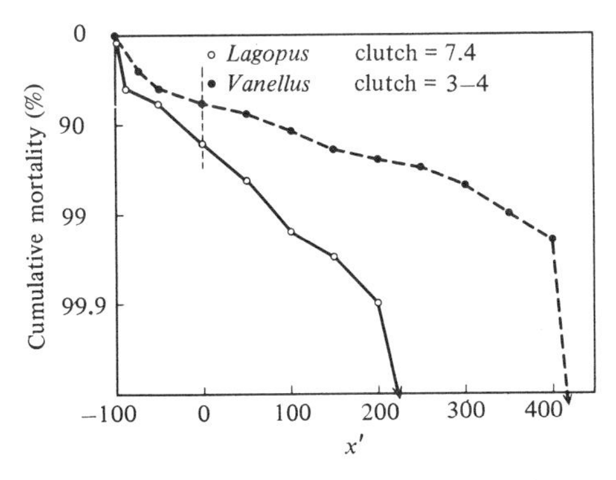

curves given below and many isolated survival-rate data (these verify the representativeness of the survivorship curves in showing each mode of life) demonstrate that the survivorship curves of nidifugous birds are of type B with relatively high early mortality, those of passerine birds are of type B with somewhat low early mortality and those of raptors, penguins and pigeons approach type A reflecting low fecundity and parental care. These depend on the differences in the reproductive rate among birds which have been examined in Chapter 1.

Fig. 2.19 compares the survivorship curves of two nidifugous species; the red grouse *Lagopus lagopus scoticus* with a large clutch size and the lapwing *Vanellus vanellus* with a small clutch size. In Fig. 2.20 that of the Canada goose *Branta canadensis* (another nidifugous bird) and that of

Fig. 2.20. Survivorship curves of the yellow-eyed penguin *Megadyptes antipodes* (from Richdale, 1957, pp. 144–59; with commencement of breeding at the age of 2.5) and the Canada goose *Branta canadensis* (drawn from Chapman *et al.*, 1969, with commencement of breeding at the age of three).

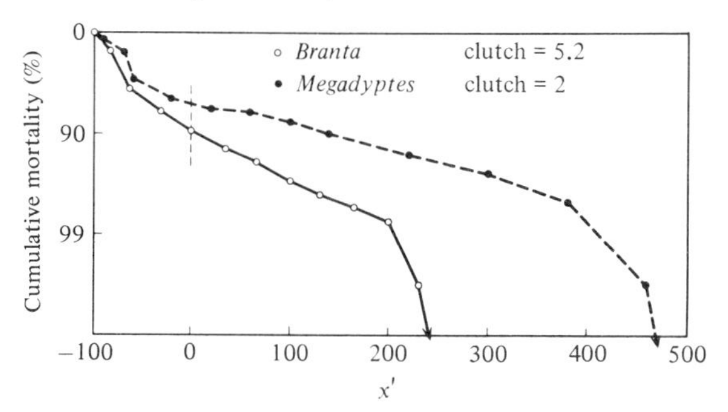

Fig. 2.21. Survivorship curves of the great tit *Parus major major* (from Kluijver, 1951, with commencement of breeding at the age of one) and the song sparrow *Melospiza melodia* (drawn from Tompa, 1964, with commencement of breeding at the age of one).

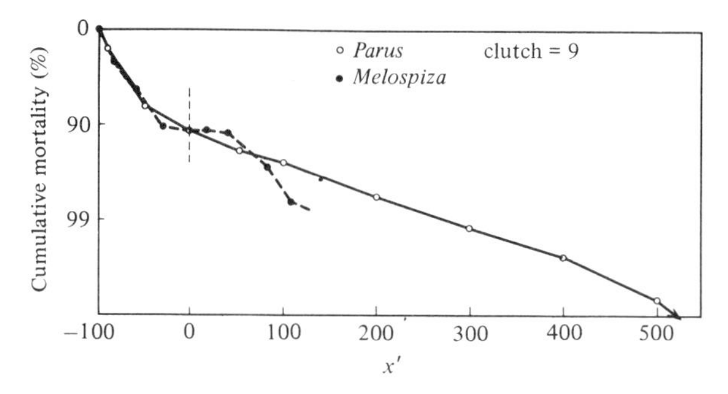

the yellow-eyed penguin *Megadyptes antipodes* (a nidicolous species with a clutch size of two, nesting on coastal cliffs) are compared. Early mortality of the red grouse is highest followed by the Canada goose, thus the number attaining reproductive age in either species is 10% or less out of the number of eggs laid. Their survival afterwards is relatively high and there is a linear decrease in l_x when plotted along a logarithmic scale (constant survival-rate from year to year). By contrast, the survivorship curve of the yellow-eyed penguin is similar to those of the tawny owl and the wood pigeon (see Fig. 2.22). In Fig. 2.21 the survivorship curves of two tree-nesting passerines are presented. Among the nidicolous birds the great tit has a large clutch size for a passerine and its cumulative mortality till the commencement of breeding reaches 90%, but note that its early mortality is lower than that of the red grouse.

As an example of predators, the tawny owl *Strix aluco* has a survivorship curve of type A (Fig. 2.22). It has a small clutch and with parental care the survivorship curve shows an evolutionary change of type C→type B→type A. I would expect the golden eagle and large owls to have type A survivorship curves. Albatrosses and petrels would also have type A curves. The wood pigeon *Columba palumbus* has a clutch of two eggs but the mortality till the reproductive age is somewhat high at 90%. As mentioned earlier the pigeons have a small clutch of two eggs as a rule but the wood pigeon nests 2.5 times a year on the average (maximum four clutches). This fact may compensate for the relatively high early mortality.

Common among the above examples is the constancy of the survival-

Fig. 2.22. Survivorship curves of the tawny owl *Strix aluco* (drawn from Southern, 1970, with commencement of breeding at the age of 1.5) and the wood pigeon *Columba palumbus* (drawn from Murton *et al.*, 1974, and Dempster, 1975, with commencement of breeding at the age of one).

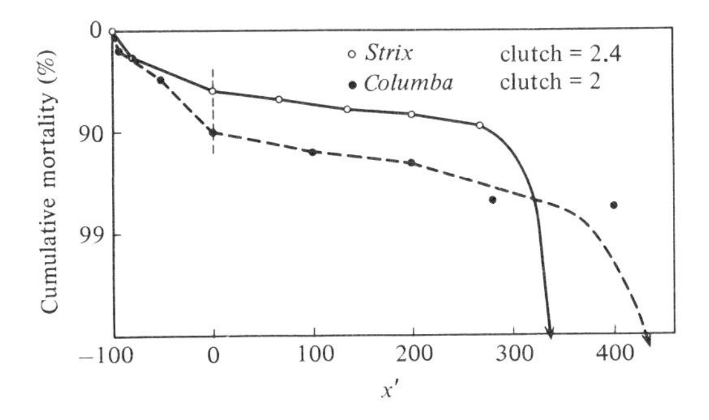

rate of adults except for the final period. In fact the q_x curve often parallels the x-axis for a period after maturation.

The survivorship curve of oceanic birds with a clutch size of one is unfortunately not available. It is conceivable that their early mortality is relatively low and their longevity in the order of tens of years assures repeated reproduction.

Snow & Lill (1974) reported that the manakins of Central and South America, famous for their lekking behaviour in trees, have a long life-span. In *Pipra erythrocephala* the minimum life-span determined by ringing varied from 6.5 to 12 years with a majority over ten years (the actual life-span is greater as birds are expected to live for a time after the last recapture). In *Manacus manacus* it ranged from seven to fourteen years. There are many other data showing long life-spans of small birds in tropical rain forest. Snow & Lill consider that higher yearly survival-rates have developed because tropical rain forest has a rich biota and in it food is available all year round and there is no need for migration. They have not touched on the survival-rate of eggs and nestlings.

Survivorship curves of mammals

The trend of survivorship curves is the same for mammals as for other groups described so far. Terrestrial species (except large ungulates)

Fig. 2.23. Survivorship curves of the field mouse *Peromyscus maniculatus* (drawn from Howard, 1949, based on spring-born individuals and commencement of breeding at the age of twenty-five weeks), the masked shrew *Sorex cinereus* (drawn from Buckner, 1966, with commencement of breeding at the age of ten months), the pocket gopher *Thomomys bottae* (drawn from Howard & Childs, 1959, with commencement of breeding at the age of eight months) and the yellow-bellied marmot *Marmota flaviventris* (drawn from Armitage & Downhower, 1974, with commencement of breeding at the age of two years). The last two species are larger than the first two.

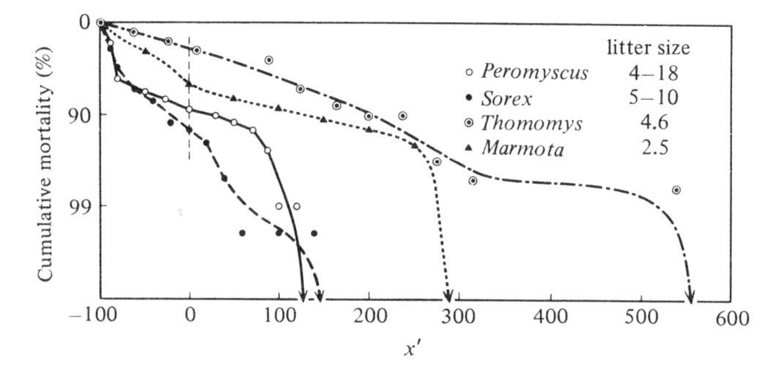

with large litters have type B and large ungulates and arboreal species with small litters show type A survivorship curves.

Fig. 2.23 shows the survivorship curve of one species of Insectivora (a shrew) and three species of Rodentia. The survivorship curve of *Peromyscus maniculatus* resembles those of birds, corresponding to a litter of 4–18. This is based on the spring-born individuals; the autumn-born individuals have a higher survival-rate (Howard, 1949). Many other

Fig. 2.24. Survivorship curves of the varying hare *Lepus americanus* (from R. G. Green & C. A. Evans, 1944, in Deevey, 1947) and the feral rabbit *Oryctolagus cuniculus* in Australia (drawn from Myers, 1971, the data from the arid zone of New South Wales and early mortality of 60% assumed).

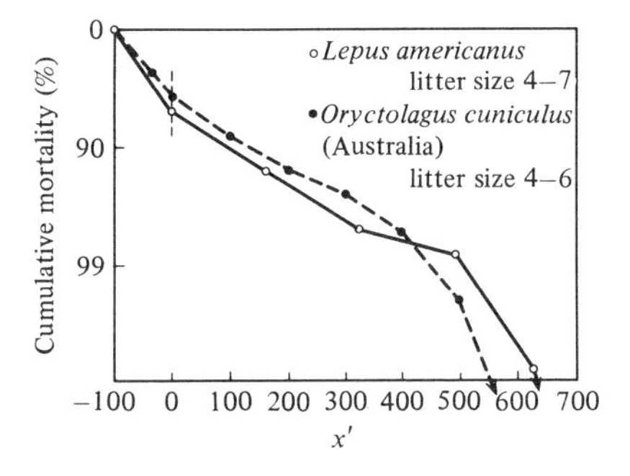

Fig. 2.25. Survivorship curves of the American red squirrel *Tamiasciurus hudsonicus* (drawn from Kemp & Keith, 1970, with commencement of breeding at the age of one), the ground squirrel *Spermophilus armatus* (drawn from Slade & Balph, 1974, with commencement of breeding at the age of ten months) and the big brown bat *Eptesicus fuscus* (drawn from Beer, 1955, with commencement of breeding at the age of two).

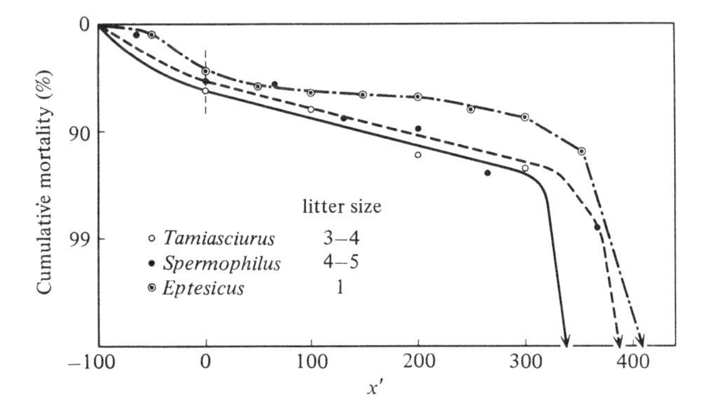

incomplete reports on rodents indicate that small rodents and small insectivores with large litters tend to have type B survivorship curves (the litter size shown in diagrams is the observed value for the generation used for the survivorship curve; if a range is given the values are taken from the literature). For example, Burt (1940) stated that most *Peromyscus leucopus* individuals die within one year of their life owing to predation and unknown causes. Tanaka (1954) also calculated that the mortality of Norway rats is 90–95% in the first year of life. On the other hand, the pocket gopher and the marmot (woodchuck), which are somewhat larger and have smaller litters than the smaller species, have intermediate survivorship curves between type B and type A. The early mortality is specially low. Is it influenced by the deep nest holes they make?

There are some data on the survival-rates of hares and rabbits available today. Reflecting the high fecundity they show basically type B survivorship curves (Fig. 2.24) as exemplified by the life-table of the varying hare *Lepus americanus* given by Deevey (1947) in his historical paper and detailed studies of feral rabbits *Oryctolagus cuniculus* in Australia reported by Myers (1971).

The squirrels have smaller litters than rats in the Rodentia and their survivorship curves approach type A. Fig. 2.25 compares the survivorship curves of the arboreal American red squirrel *Tamiasciurus hudsonicus* and the ground squirrel *Spermophilus armatus*; there is no conspicuous difference between the two.

Fig. 2.25 also presents a survivorship curve of the big brown bat *Eptesicus fuscus* which produces only one young at a time. Although it is a small animal the curve is close to the typical type A. Bats generally

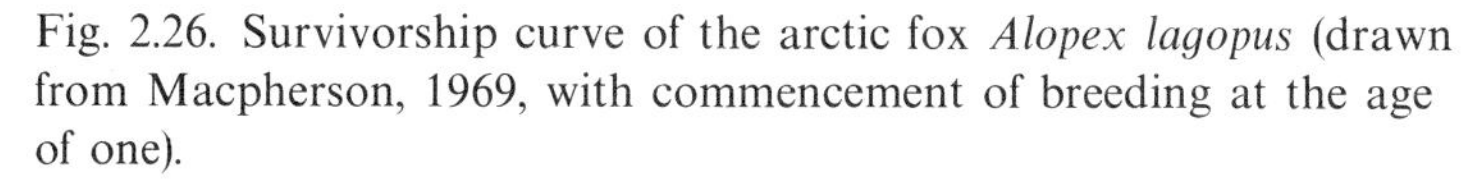

Fig. 2.26. Survivorship curve of the arctic fox *Alopex lagopus* (drawn from Macpherson, 1969, with commencement of breeding at the age of one).

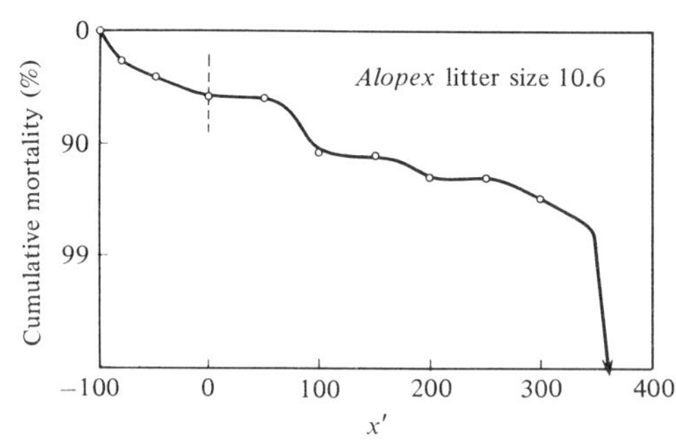

have a long life-span; in the bent-winged bat *Miniopterus schreibersii* more than half of the marked individuals lived 5.5 years and the lesser horseshoe bat *Rhinolophus hipposideros* lived seven years (Bourlière, 1947). Also, according to Cockrum (1956) the brown bat *Myotis myotis* has a mean life-span of thirteen years (Holland) or twelve years (Germany), *Myotis sodalis* ten years and *Eptesicus fuscus* nine years.

The Carnivora are predators but mostly fecund unlike raptors. Among them the survivorship curve of the arctic fox *Alopex lagopus* has been obtained by Macpherson (1969). As shown in Fig. 2.26 this is close to type A but the mortality during the nursing period is high (over 50%). In this figure, however, as many as 30% of the young born have been shown to survive till the reproductive age and the population would increase. According to Macpherson the reason for not increasing might be that the proportion of females actually giving birth was not large. For example, there were about 260 foxes in the first year (130 females if the sex ratio is 1:1), but the estimated number of litters produced was fifteen out of forty females surviving. The total numbers of litters produced by this generation including the third- and fourth-year animals was ninety-seven, with a mean litter-size of 10.6, which would amount to 1037 new individuals. Thus the population was considered to maintain its stability. [Such effects of non-breeding adults are ignored in this chapter.]

Laws (1962) constructed a survivorship curve of the finback whale *Balaenoptera physalus*, which of course belongs to type A. It is incomplete, particularly as it represents the beginning of a sharp decline in number, and is not reproduced here.

As mentioned in the previous chapter, the wild pigs among the

Fig. 2.27. Survivorship curves of the warthog *Phacochoerus aethiopicus* (drawn from Spinage, 1972, with commencement of breeding at the age of two) and the red deer *Cervus elaphus* (drawn from Lowe, 1969, with commencement of breeding at the age of two).

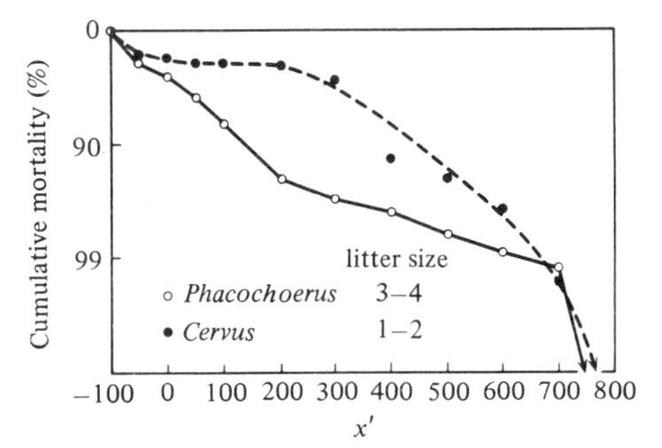

ungulates have high fecundity. The survivorship curve of the warthog *Phacochoerus aethiopicus* constructed by Spinage (1972) is of type B in contrast with that of the red deer *Cervus elaphus* which is a typical small ungulate (Fig. 2.27).

The survivorship curves of the black-tailed deer *Odocoileus hemionus columbianus* and the roe deer *Capreolus capreolus* shown in Fig. 2.28 are of type A, basically the same as for the red deer. Among other sur-

Fig. 2.28. Survivorship curves of the black-tailed deer *Odocoileus hemionus columbianus* (drawn from Taber & Dasmann, 1957, for the females in chaparral), the roe deer *Capreolus capreolus* (drawn from Andersen, 1953, in Taber & Dasmann, 1957, for females) and the bighorn sheep *Ovis canadensis* (drawn from Buechner, 1960). All assumed to commence reproduction at the age of two.

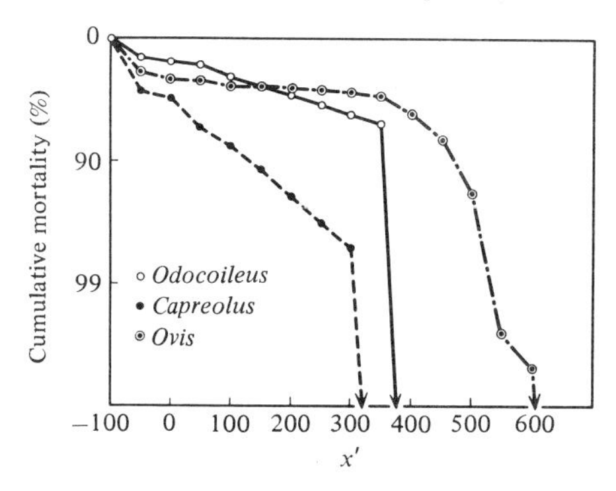

Fig. 2.29. Survivorship curves of the African elephant *Elephas (Loxodonta) africanus* (drawn from Petrides & Swank, 1966), the zebra *Equus burchelli* (drawn from Spinage, 1972, for females) and the defassa waterbuck *Kobus defassa ugandae* (drawn from Spinage, 1972). Commencement of breeding at the age of fifteen for elephants and three for zebras and waterbucks.

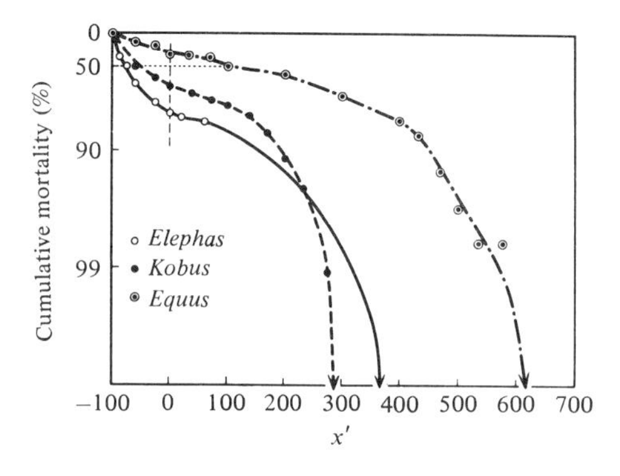

vivorship curves drawn for deer, one for the red deer by Taber & Dasmann (1957) based on the data of H. Evans (1891) and another for the caribou *Rangifer arcticus* given by Banfield (1955) show the same basic pattern. In Fig. 2.28 the survivorship curve of the big-horn sheep *Ovis canadensis* is also presented. The life-table of this species is given by Woodgerd (1964), as well as by Buechner (1960) which is used here. Like the famous survivorship curve (see Fig. 2.3, p. 51) of a related species (the Dall mountain sheep *Ovis dalli*) studied by Murie (1944), they all belong to type A.

Fig. 2.29 shows the survivorship curves of the African elephant *Elephas africanus*, the zebra *Equus burchelli* and the defassa waterbuck *Kobus defassa ugandae*. Although the elephant has a specially long pre-reproductive period, its survivorship curve is basically of the same type as the others. Among the antelopes the wildebeest *Connochaetes taurinus* and the impala *Aepyceros melampus* exhibit type A (Spinage, 1972) as do those of the Himalayan thar *Hemitragus jemlahicus* acclimatized in New Zealand (Caughley, 1966) and the African buffalo *Syncerus caffer* (Spinage, 1972).

The marsupials radiated widely in Australia have only been studied recently with respect to population dynamics. It is probable that the highly fecund marsupial mice have type C and the large macropods, equivalent to ungulates, have type A survivorship curves. As far as I am

Fig. 2.30. Survivorship curves of the ringtail possum *Pseudocheirus peregrinus* (drawn from Thomson & Owen, 1964) and the brush-tailed possum *Trichosurus vulpecula* (drawn from Tyndale-Biscoe, 1973, for females). Both assumed to commence breeding six months after leaving the pouch (at the age of one).

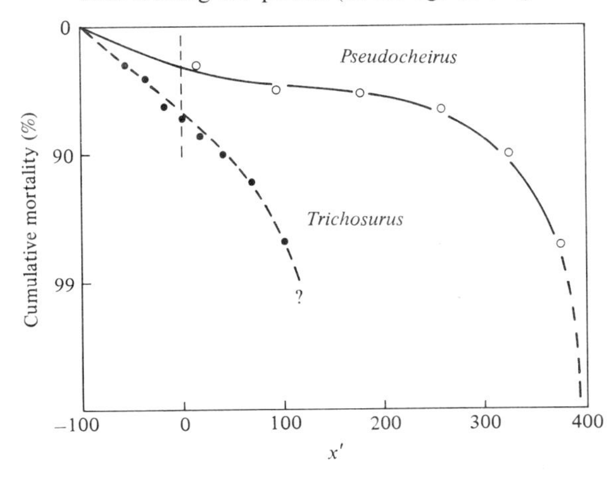

aware the survivorship curve is known only for two species of possum, which are arboreal and have relatively low fecundity. As shown in Fig. 2.30 the curves are of type B to type A. The ringtail possum *Pseudocheirus peregrinus* lives in a forest habitat and builds a nest while the brush-tailed possum *Trichosurus vulpecula* lives in the woodland of eucalypts and uses tree holes. Both have a litter of two on average (range 1–3), but why the survival-rate till the commencement of breeding is different between the two is not known. Brush-tailed possums live in a more fluctuating environment than the ringtail possum and this may adversely affect their survival, but to compensate for the greater mortality they must have a slightly larger litter than the ringtail possums. Further studies are needed to elucidate this point. The survivorship curve of the koala *Phascolarctos cinereus*, a typical arboreal species with a litter of one, if available, would be type A as expected for the sloth, which is phylogenetically different but ecologically similar.

As for the primates, in spite of the marked progress in research during the past twenty years the accumulation of data is not sufficient to permit construction of a life-table. Information available on age distribution strongly suggests that the survivorship curve is of type A.

The survivorship curve of man has changed in historical time from type B to type A (Fig. 2.31). This is because infant mortality was reduced as a result of advancement in civilization, which however did not lengthen the physiological longevity (about 100 years).

Survivorship curves of insects

In the previous sections I described the relationship between reproduction and death mainly in vertebrates. Here let us examine that of the insects, which occupy diverse niches in nature.

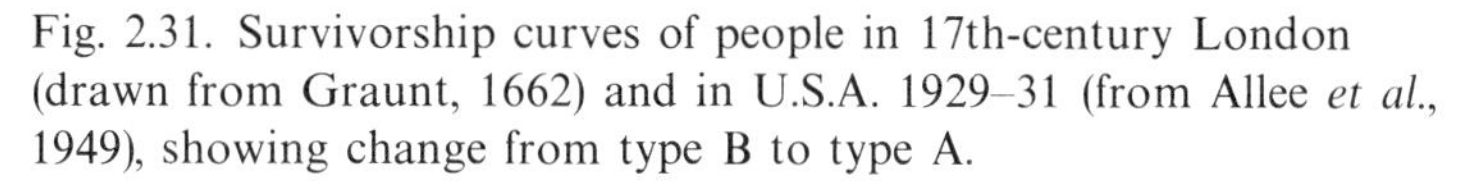

Fig. 2.31. Survivorship curves of people in 17th-century London (drawn from Graunt, 1662) and in U.S.A. 1929–31 (from Allee *et al.*, 1949), showing change from type B to type A.

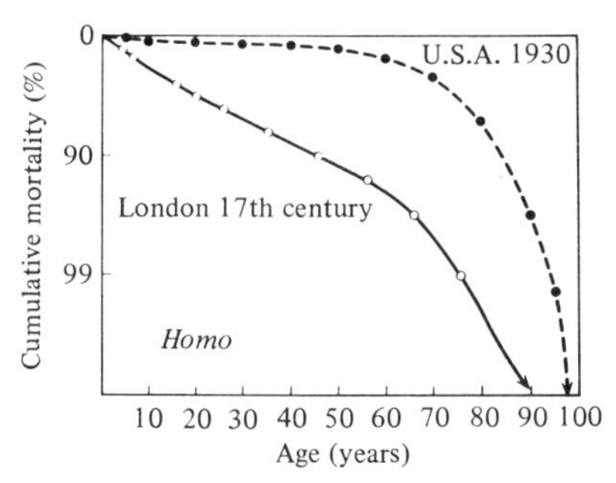

The class Insecta contains more species than all the other animals put together and has radiated into all modes of life. In size they vary from a minute parasitic wasp, *Trichogramma*, which emerges from an egg (0.5 mm in diameter) of a moth, to the giant Goliath beetle. Their life-span ranges from a few days to seventeen years and their life history is equally diverse; the tsetse fly spends the larval stage in the female's body and the young pupates as soon as it is 'born' whereas the mayfly's proboscis is degenerate and the adult never feeds. Thus it is difficult to compare the survivorship curves of insects. There are groups, such as the Hymenoptera, which demonstrate evolutionary sequences of development in parental care but even among closely related species there are cases where the mode of life is quite different and hence the survivorship curve differs. In particular, unlike the free-living animals we have dealt with so far, the fact that many insects spend part of their life cycle in plant tissues complicates the problem.

A pioneer work on the life-table of insects was conducted by E. Ballard, A. N. Mistikawi and M. S. Zoheiry who under the guidance of Bodenheimer studied the desert locust *Schistocerca gregaria*. They kept locusts in an enclosure of wire-netting in the field and investigated their survival every 5 days. They expressed the result with a survivorship curve which was close to type C in contrast to the type A curve obtained for the population kept in the laboratory. Their work is significant in that it was aimed at producing a life-table, though it was still artificial in that predation was prevented as a result of enclosure. Bodenheimer devoted one chapter of his *Problems of Animal Ecology* published in 1938 to life-tables. This book stimulated Hidetsugu Ishikura's study on the life-table of the larvae of the rice-stem borer in postwar Japan.

Fig. 2.32. Survivorship curve of the lycosid spider *Geolycosa godeffroyi* (drawn from Humphreys, 1976). Average number of eggs laid, 600; breeding twice a year.

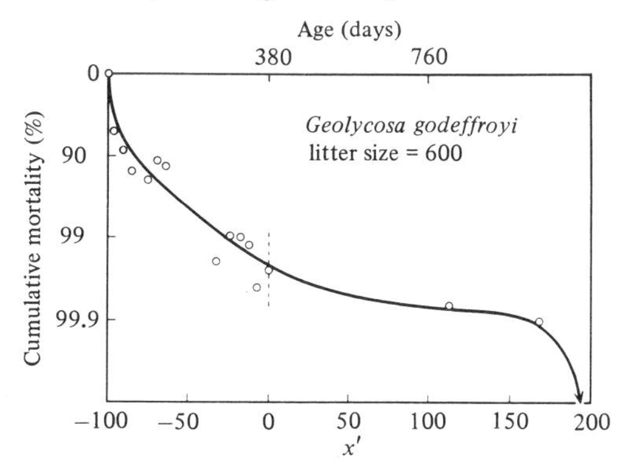

Before treating insects let me describe a survivorship curve of a spider. [The spiders do not pupate and both young and adults are predatory. Ecologically they are close to the typical Hemimetabola discussed below.] Fig. 2.32 shows the survivorship curve of the lycosid spider *Geolycosa godeffroyi* which does not make a web but has a habit of wandering. This species starts laying at the age of about twelve months and breeds twice in one season, each time laying 600 eggs on average. A certain degree of parental care exists as the egg mass is carried on the back of the parent. However, young falling off the back of the parent, the parent's death and death during dispersal cause high mortality during the juvenile stage, and the survivorship curve approaches type C.

Insects may be divided into the Hemimetabola (having incomplete metamorphosis) and the Holometabola (having complete metamorphosis) according to the life cycle. The former lacks the pupal stage. In reality the Hemimetabola includes those that keep moulting after the commencement of breeding and those that have a pseudo-pupal stage such as scale insects. More important than these subdivisions of developmental characteristics is the distinction between those that show differences of habitats occupied by juveniles and adults (e.g. mayflies, dragonflies, cicadas) and those that do not (e.g. cockroaches, grasshoppers, pentatomid bugs). The former should perhaps be called the 'ecological Holometabola' regardless of their morphological characteristics in development.

Fig. 2.33. Survivorship curves of the Moroccan locust *Dociostaurus maroccanus* (from Dempster, 1957, for the 1955 generation) and the grasshopper *Mecostethus magister* (drawn from Takai *et al.*, 1963, survival rate from eggs to the first instar is an average of the 1961 and 1962 generations, thereafter the 1961 generation). Adult mortality is based on the mark-recapture data in both species. Nymphal instars are numbered and adult eclosion is indicated by A on the curves.

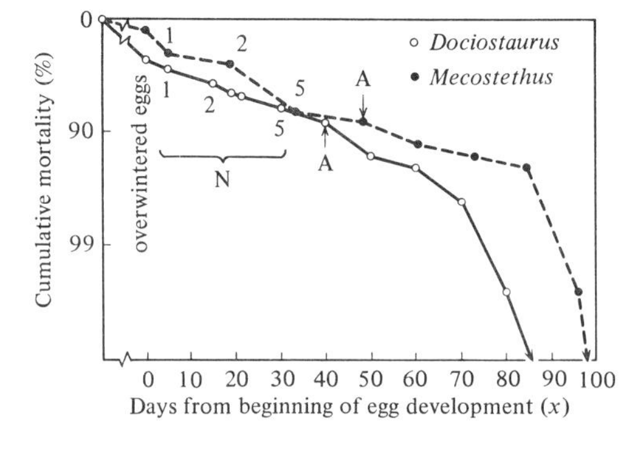

There are about ten species of grasshopper (metamorphosis incomplete both ecologically and morphologically) for which the life-table has been constructed. Of these the survivorship curves of the Moroccan locust *Dociostaurus maroccanus* studied by Dempster (1957) and *Mecostethus magister* studied by Takai *et al.* (1963) are presented in Fig. 2.33. These examples are taken from irrupted populations but the survivorship curves of non-irrupting populations of grasshoppers, *Chorthippus* (two species) studied by Richards & Waloff (1954) and *Parapleurus alliaceus* (Nakamura *et al.*, 1971) show basically the same pattern. That is, the number surviving decreases at a constant rate after hatching and the trend continues to the adult stage. The surviving females lay once a week for several weeks. The number of eggs in an egg mass ranges mostly from ten to fifty depending on the species and rarely exceeds 100. Thus the survivorship curves of the grasshoppers are of type B with slightly higher pre-breeding mortality. In other words, they are intermediate between type C and type B. Without the pupal stages the grasshoppers do not have the stepwise curve as seen in the Holometabola.

Since most insects die as soon as they breed it does not help to plot x' on the abscissa. Also they have inactive periods, such as egg, pupa and diapause, and mortality during these periods cannot be compared meaningfully with that during the active periods. Thus on the abscissa

Fig. 2.34. Survivorship curves of the spittlebug *Neophilaenus lineatus* (drawn from Whittaker, 1967) and the delphacid planthopper *Javesella puellucida* (drawn from Raatikanen, 1967). The life-span of the spittlebug was estimated from a related species, *Philaenus spumarius*, to be a maximum of two month after emergence (Whittaker, 1973). Mortality of *Javesella* from the time of emergence to sexual maturity was due to adult parasites. Times of hatching (H), nymphal instars (numbered) and adult eclosion (A) are indicated on the curves.

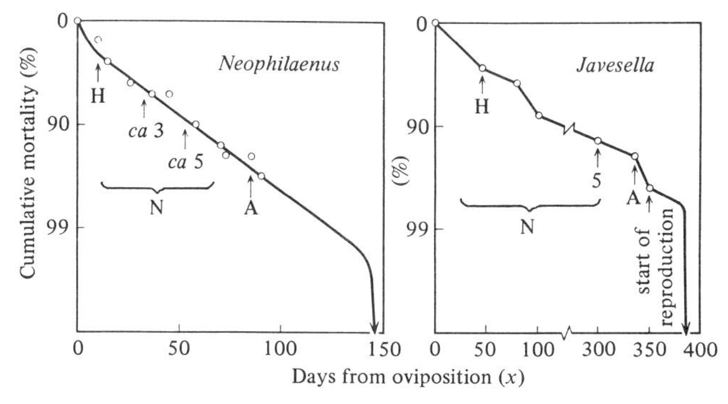

the time scale is given in days or in developmental stages and the time of emergence is indicated on the curve. The number of days for the inactive periods are reduced in the graph.

Fig. 2.34 shows the survivorship curves of the spittlebug (a froghopper) *Neophilaenus lineatus* and the delphacid planthopper *Javesella pellucida*. For the former the data were supplemented by the life-span of a related species *Philaenus spumarius*. They are both type B curves with mortality slightly on the higher side.

Among the bugs the survivorship curves were obtained by Kiritani and his collaborators (Kiritani & Hokyo, 1962; Kiritani *et al.*, 1963, 1967; Kiritani & Nakasuji, 1967; Kiritani, 1971) from the southern green stinkbug *Nezara viridula* which became a pest insect on rice plants in Wakayama Prefecture after the war. Fig. 2.35 shows the second generation only from their detailed life-tables.

The three species of Hemiptera given above share the same habitat and food between nymphs and adults (the nymphs of spittlebugs, though living on the same plants as adults, are found in the spittle). All have type B survivorship curves with somewhat high mortality. Among other plant-sucking insects the survivorship curve is available for a psyllid, *Arytaina spartiophila* (Watmough, 1968) and the viburnum whitefly *Aleurotrachelus jelinekii* (Southwood & Reader, 1976); these also exhibit type B–C curves.

On the other hand, the sugar cane cicada *Mogannia minuta* (= *M. iwasakii*) in the same homopteran group has quite a different sur-

Fig. 2.35. Survivorship curves of the southern green stinkbug *Nezara viridula* (drawn from Kiritani & Nakasuji, 1967; for the second generation of 1962, supplemented with the adult life-span data obtained in a field cage, from Kiritani *et al.*, 1967) and the sugar cane cicada *Mogannia minuta* (*iwasakii*) (drawn from Nagamine *et al.*, 1975). Stages of eggs (E), nymphal instars (numbered), adult eclosion (A) and oviposition (O) are indicated.

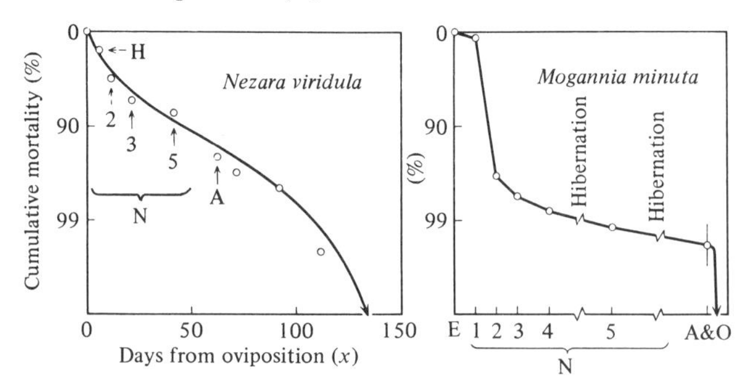

vivorship curve. In this species more than 95% of the eggs laid die during the first instar between the time of hatching and the time of penetration into the ground. Mortality during the two years of subterranean life is very low. When they emerge adults mate, lay eggs and die within a few days. The number of eggs laid per female is estimated at 420 on average (Nagamine *et al.*, 1975). The survivorship curve of this species (Fig. 2.35) is of type C. Other cicadas probably share the same trend of mortality. The unique mortality pattern of cicadas among the plant-sucking insects is probably due to the fact that the adults and nymphs live in entirely different habitats. There is high mortality when the nymphs move into a new habitat. Corresponding to such high mortality the cicadas lay a large number of eggs compared with other members of the Homoptera.

Like cicadas, dragonflies shift their habitats completely in their life cycle. Females lay on the surface of water or, in some species, on water weeds. The nymphs are aquatic predators while the adults, emerging after a year in water, are aerial predators. They usually lay about 1000 eggs. According to Corbet (1962) dragonflies lay a large number of eggs; those scattering eggs on the water may lay as many as 5000 and many of those laying on plants produce about 1000. However, some species even laying on the water are reported to lay only about 500. Fig. 2.36 shows the survivorship curves of two species of dragonfly *Ladona deplanata* and *Celithemis fasciata*. The survival-rate during the egg stage is not known but it is assumed to be 0.5. In this graph the number of eggs needed to maintain the population would be about 1000 for *Ladona* and 400 for *Celithemis*. Such low fecundity may be unrealistic, possibly a result of the

Fig. 2.36. Survivorship curves of two species of dragonfly *Ladona deplanata* and *Celithemis fasciata* (drawn from Benke & Benke, 1975; the number of eggs laid was assumed to be twice the estimated number of young hatched). E, eggs; N, nymph; LN, mature nymph; A, adult eclosion.

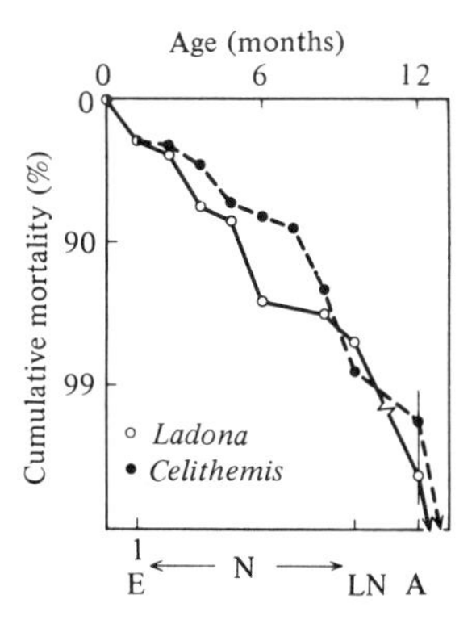

underestimation of egg mortality (for example, dragonflies without suitable territories may lay on temporary water and their young may all die). In any case the number emerging is one in several hundreds of eggs laid and this is different from the survival pattern of other Hemimetabola.

Among the Holometabola the larvae and adults of midges occupy entirely different habitats, rather like those of dragonflies. Fig. 2.37 is the survivorship curve of *Chironomus plumosus* studied by Borutzky (1939). There is comparatively high mortality in the egg stage and the curve is a typical type C. The mortality between the egg stage and the reproductive stage reaches an unusually high rate (>99.9%) among the insects, but the number of eggs laid also reaches several thousands. Similar results to this pioneer study have been obtained recently by Yamagishi & Fukuhara (1971).

For terrestrial species the survivorship curves of the Holometabola are taken from the Lepidoptera which show stepwise decrease in survival if the logarithmic scale is not used for the ordinate. That is, rapid decreases occur generally from the eggs to the first instar larvae, at the end of the last larval instar (usually the fifth, occasionally sixth or seventh) and during the mid- and late-pupal stages (Fig. 2.38). These derive from the high mortality of newly hatched larvae establishing on host plants, and deaths caused by the larval parasites emerging at the end of the larval life

Fig. 2.37. Survivorship curve of the midge *Chironomus plumosus* (drawn from Borutzky, 1939, for the second generation 1935). Stages of development are marked as E (eggs), H (hatching), Mi (mid-instar) and Li (late-instar), L (larvae), P (pupae), F (floating), Em (emergence) and O (oviposition).

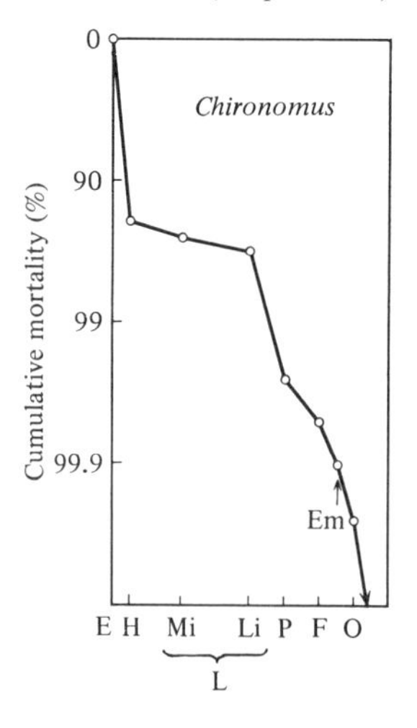

and the pupal parasites in mid- to late-pupal stages. [Parasites parasitizing the mid-instar larvae and emerging from the pupae may be regarded as a mortality factor of larvae or pupae. Here the host mortality is given to the developmental stage in which the parasites emerge from the host.] In many lepidopterans the mortality from the second to the fourth instars (early larval life) is very low. In other words, larval and pupal parasites (strictly they are both parasitoids) as the cause of mortality are far more important to the lepidopterans than to other insects. The stepwise change of survivorship curves has been pointed out by Itô (1960a) but as this is common some workers consider it as a fourth type of survivorship curve or a subtype of B (Odum, 1971).

Apart from the stepwise decrease of survival there is another question worth noting in the survivorship curve of lepidopterans. First of all the general rule holds for the Lepidoptera that those with high fecundity have type C survivorship curves and those with low fecundity have type A curves. Representative species with a type C survivorship curve include the cabbage armyworm as shown in Fig. 2.38 (see also Fig. 2.39), the

Fig. 2.38. Survival-rate (solid line) and mortality-rate (broken line) of the autumn generation of the cabbage armyworm *Mamestra brassicae* (Itô, unpublished). Mortality was calculated per thousand over four days. Note that the ordinate is not a logarithmic scale. Stages of development: E (eggs), L (larval instars 1–6), P (pupae) and A (adult eclosion).

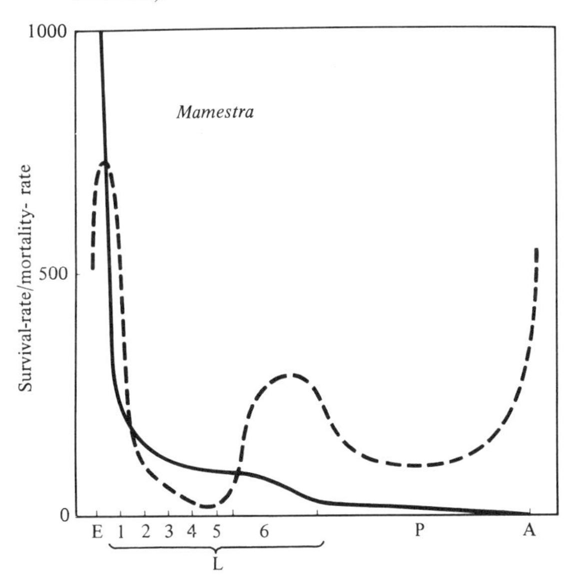

winter moth *Operophtera brumata* and the pine moth *Dendrolimus spectabilis* (Fig. 2.39). They all lay large numbers of eggs and the larvae are surface feeders with low degrees of aggregation. On the other hand, the leaf rollers, which deposit single eggs on new buds, and the casebearers,

Fig. 2.39. Survivorship curves or the cabbage armyworm *Mamestra brassicae* (drawn from Oku & Kobayashi, 1973), the gypsy moth *Porthetria dispar* (drawn from Bess, 1961, for the 1941 generation), the fall webworm *Hyphantria cunea* (drawn from Itô & Miyashita, 1968, the first generation), the winter moth *Operophtera brumata* (drawn from Embree, 1965) and the pine moth *Dendrolimus spectabilis* (drawn from Anon., 1965, for the 1964/65 generation in Korea). Stages of development: E (eggs), H (hatching), L (larvae) with instars 1–6, S (establishment) and ML (middle larvae), PP (prepupae), P (pupae) and A & O (adult eclosion and oviposition).

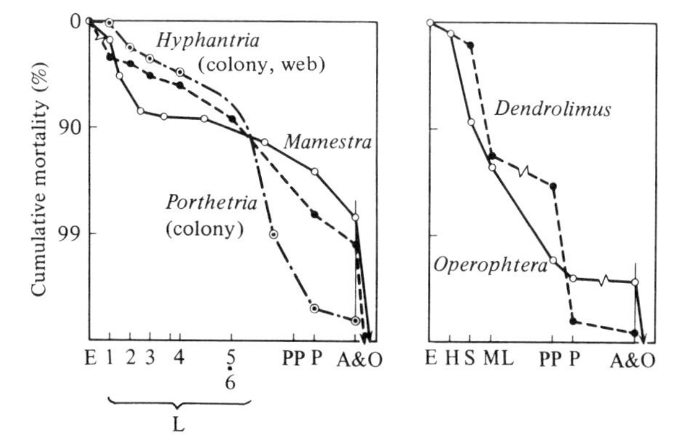

Fig. 2.40. Survivorship curves of the leaf roller *Leucoptera spartifoliella* (drawn from Agwu, 1974; for the 1964/65 generation, egg number 20–40) and the pistol case bearer *Coleophora serratella* (drawn from LeRoux *et al.*, 1963; egg number about 20). Stages of development: E (eggs), L (larvae) with instars 1–5 and M (mature larvae), P (pupae) and A (adult eclosion).

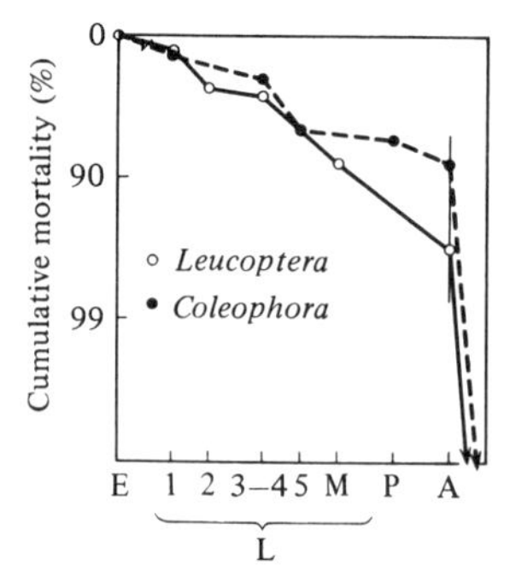

which leave eggs in cases, lay small numbers of eggs and their survivorship curves are of type A. In Fig. 2.40 the leaf roller *Leucoptera spartifoliella* and the pistol casebearer *Coleophora serratella* are used as examples of this type. [The pistol casebearer is not a psychid moth; here the case-bearing habit is a result of convergence. Unfortunately there is no life-table available in the true case moth.] The leaf roller was studied in Britain by Agwu (1974) who presents data for the 1964 and 1965 generations. As the laying of only 20–40 eggs per female in the 1965 generation would cause the population to decline rapidly, the value for the 1964 generation was used here, which would keep the population fairly stable.

Many survivorship curves of species with intermediate fecundity (though the number of eggs in the ovaries is great the female laying one egg at a time is considered to produce not very many eggs in a life time) fall between the two types described above. The swallowtail *Papilio xuthus* is a typical example (Fig. 2.41). This butterfly produces four generations a year in central Japan, but the survivorship curve is much the same for all generations (Watanabe, 1976). Other lepidopterans with intermediate survivorship curves include the apple leaf roller *Archips argyrospilus* (Paradis & LeRoux, 1965; number of eggs per female,

Fig. 2.41. Left: survivorship curves of two species of tent caterpillars; *Malacosoma neustra* (drawn from M. Shiga, unpublished) which constructs a nest web and *M. disstria* (drawn from Witter *et al.*, 1972) which does not. The egg number is about 200 for both species. Right: survivorship curves of the cinnabar moth *Tyria jacobaea* (drawn from Dempster, 1971, for the 1967 generation) which has very low larval mortality and the swallowtail *Papilio xuthus* (drawn from Watanabe, 1976, the fourth generation) which is an intermediate type. Stages of development: E (eggs), H (hibernation), L (larvae) with instars 1–5, P (pupae), A (adult eclosion) and O (oviposition).

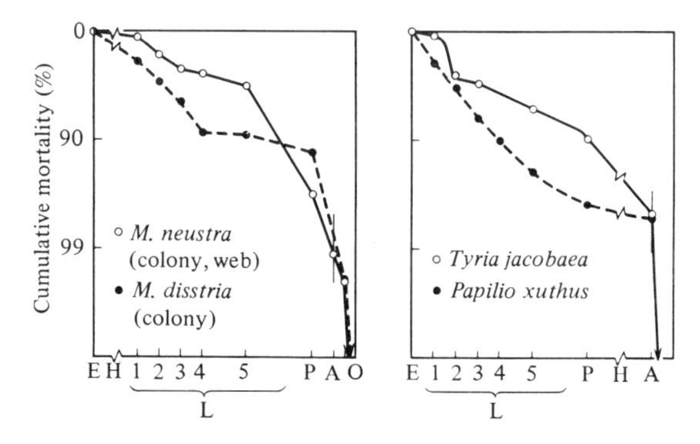

50–60) and the spruce budworm *Choristoneura fumiferana* (Morris & Miller, 1954; number of eggs per female, 70).

Now, there is another group of lepidopterans whose survivorship curve is entirely different from the three types described above. That is the group with high fecundity yet accompanied by very low early larval mortality. The fall webworm *Hyphantria cunea* in Fig. 2.39 and the Japanese tent caterpillar *Malacosoma neustra* in Fig. 2.41 are representatives of this group and in both the larvae live in aggregations surrounded by the nestweb which they make out of silk. During the first three instars they survive well, being protected by this nest. However, later on the fall webworm suffers severe predation by *Polistes* wasps and birds between the fifth instar and the seventh instar, and the mortality during this period rises to 98–99% of those reaching the fourth instar (this species has seven instars in the larval stage but does not construct the nestweb during the last two instars). In the Japanese tentworm the fifth instar larvae stop feeding and disperse in search of a place to pupate. Predation by birds starts then and continues till they make cocoons and become pre-pupae. The mortality during this period is about 70%.

Fortunately life-table data are available for the Japanese tent caterpillar and a closely related species, the forest tent caterpillar *Malacosoma disstria* (Witter *et al.*, 1972), which in spite of its name does not construct a nestweb. The latter has much greater early mortality than the former (see Fig. 2.41 left). The difference may be due to the protective effect of the web in *M. neustra.* Observations made by many people of other lepidopterans living in aggregation within the nestweb, e.g. *Spilarctia imparilis, Aporia hippia japonica* and *Nymphalis xanthomelas,* suggest that they have type A survivorship curves.

However, this is not the only problem. Gregarious insects are conspicuous and we might think that nestwebs are constructed to prevent predation. But there are many species, particularly among the saw-flies (Hymenoptera), which in spite of tremendously large aggregations and without the nestweb do not suffer any significant predation. Even in the tentworm the fourth and fifth instar larvae often appear on the surface of the web and 'sun-bathe' without attracting predators. Are they aposematic? The tussock moths, for example, could well be aposematic though they are not entirely immune to predation (if not eaten at all there would have been no need to evolve more venomous hair on caterpillars). The tent moths, however, are attacked by birds as soon as they disband the aggregation in search of a place to pupate. It is difficult to imagine that their palatability changes in such a short time. Might the large aggregation of caterpillars be seen by birds as a huge object instead of as an aggregation of individual caterpillars?

Another example of lepidopterans with low early mortality is the cinnabar moth, *Tyria jacobaea*, belonging to the Arctiidae. This moth lays about 150 eggs per female but larval mortality is very low. The population is regulated by starvation as a result of depleting its food supply (Dempster, 1971). The larvae have warning coloration of black and orange stripes and are probably unpalatable. Dempster in his eight years of study has never seen one taken by a bird.

The Coleoptera (beetles) may be divided into two groups: the chrysomelid beetles that creep on the surface of plants like lepidopteran larvae and the others that live in protected places such as under bark, inside wood and underground. The former tend to aggregate. In this group the survivorship curve is known for the Colorado beetle *Liptinotarsa decemlineata* (Harcourt, 1971), a herbivorous coccinellid, *Epilachna vigintioctomaculata*, and its relatives *Epilachna* spp., as well as the walnut leaf-beetle *Gastrolina depressa* and the broom beetle *Phytodecta olivacea* used in Fig. 2.42 (left). The walnut leaf-beetle represents this group in having high early mortality like many lepidopterans while the broom beetle has lower early mortality and a low proportion of breeding adults. Among

Fig. 2.42. Survivorship curves of the walnut leaf-beetle *Gastrolina depressa* (drawn from Matsura, 1976; population assumed to be stable for calculating the number of ovipositing adults), the broom beetle *Phytodecta olivacea* (drawn from Richards & Waloff, 1961; for the 1955/56 generation, the actual number of eggs was taken as 27% of the expected number), the bark beetle *Scolytus scolytus* (drawn from Beaver, 1966, for the 1962 data) and the buprestid beetle *Agrilus viridis* (drawn from Heering, 1956, for the sample from the seedlings). Stages of development as in previous figures.

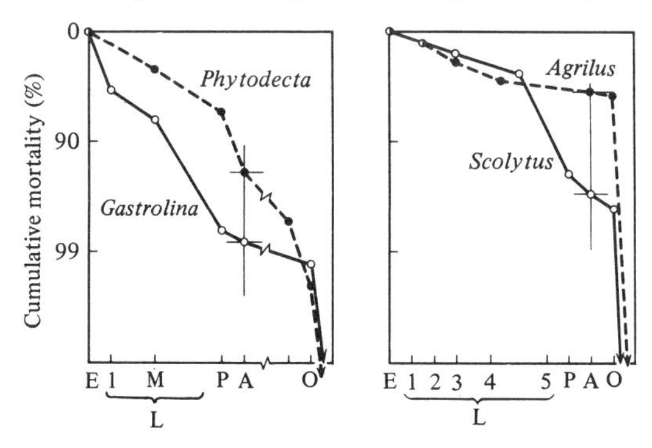

this group there are others with relatively low early mortality such as *Epilachna vigintioctomaculata*. The gregarious and slow-moving larvae of chrysomelid and coccinellid beetles are not eaten much by spiders, and chrysomelid larvae are too small (mostly less than 5 mm long when mature) to be significant prey for birds. The high larval mortality of the walnut leaf-beetle is due to the specialized coccinellid predator, *Ithono mirabilis*. Most leaf-beetles and coccinellid beetles overwinter as adults and the low larval mortality may compensate for the mortality of adults in winter. However, why their larvae are not preyed upon is not known.

Among the beetles those that protect young are found in the group living inside wood or under bark. Survivorship curves of this group are taken from the buprestid beetle *Agrilus viridis* and the bark beetle *Scolytus scolytus* (Fig. 2.42, right). These clearly show type A curves. Apart from Beaver's (1966) study used in the figure, type A survivorship curves of bark beetles have been reported by Thalenhorst (1958) for *Ips typographus* and DeMaris *et al.* (1970) for *Dendroctonus brevicomis*. The proportion of eggs reaching the stage of adult emergence in bark beetles is similar to that in leaf beetles but the bark beetles have a greater

Fig. 2.43. Survivorship curves of wasps and the honey-bee in various stages of social evolution: the pine sawfly *Diprion pini* (drawn from Thalenhorst, 1953), the sphecid wasp *Ectemnius rubicola* (drawn from Ohgushi, 1953), the sphecid wasp *Sceliphron assimile* (drawn from Freeman, 1973; egg numbers, mean maximum 32.7, mean 6.7), the potter wasp *Eumenes colona* (drawn from Taffe & Ittyeipe, 1976; the bank nesting colony, the number of adults laying is calculated from the egg number 30 and assuming stability of population) and the honey-bee *Apis mellifera* (drawn from Sakagami & Fukuda, 1968, for the workers of July). Stages of development: E (eggs), H (hatching); S (small), M (medium) and L (large) L (larvae); C (cocooning), P (pupae), A (adult eclosion) and O (oviposition).

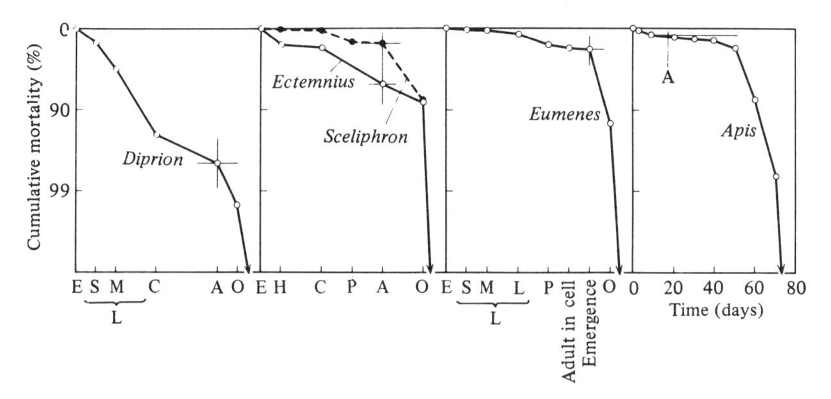

proportion of adults surviving to breed than the leaf beetles. In both species adults drill a hole in the bark and deposit eggs in it. Specially in the bark beetle the female penetrates the bark and digs out a maternal chamber (mother gallery) under the bark to lay eggs at equal distances in it (to avoid competition among the larvae). Thus early mortality is expected to be low. The bark beetles lay about 50 eggs per female whereas the mean number of eggs laid by a female of *Agrilus* is only 6.4.

The evolution of parental care has reached its extreme in the hymenopterans among insects. Fortunately the life-table is known for several species with different stages of social evolution. Comparison of these species show that the stages from high fecundity/no protection to low fecundity/protection are paralleled by the changes of the survivorship curve in the order of types C→B→A, which confirms the conclusion reached from the examination of other groups. In Fig. 2.43 the survivorship curve of the pine sawfly *Diprion pini* is of type B. This species feeds and literally lives on the leaf surface and lays more than 100 eggs per female. The only parental protection provided is that the eggs are singly spaced in the plant tissue. By contrast the sphecid wasps *Ectemnius rubicola* and *Sceliphron assimile*, which construct nests and provide a bulk of food (mass provisioning) for the larvae, have very low early mortality. Further advanced in parental care is the potter wasp *Eumenes colona*, which feeds larvae during their development (progressive provisioning) and its early mortality is almost nil. Population explosions do not occur in these species because parents often meet death during nest building, hunting and carrying prey for the larvae. Disappearance of nests (desertion by the parent, predation of whole nest) also occurs but this factor has not been taken into account in the above life-table study. Reflecting the cost necessary for the parental care the number of eggs laid is only about thirty in these species.

The apex of this evolution is occupied by the honey-bee *Apis mellifera*, which has the most delicate stabilizing mechanism of a colony outside of man. Its survivorship curve is identical to man's! Unfortunately the survivorship curve shown in Fig. 2.43 is that of worker bees but that of the queens would be much the same. In the honey-bee the number of eggs is increased again in this evolutionary sequence and is in the order of tens of thousands. However, the queen of the honey-bee is not really an independent individual but may be considered a reproductive cell of an 'individual' known as the honey-bee colony. If we accept this point of view we can conclude that the theories of Lack (1954a) and Itô (1959) hold for the survivorship curves of insects.

Finally let me touch on an apparently peculiar survivorship curve obtained for a species of Hymenoptera, the chestnut gall-wasp *Dryocosmus kuriphilus* (Fig. 2.44). This wasp reproduces parthenogenetically and lays an egg mass inside the bud of a chestnut tree. The hatched larvae apparently secrete a substance which stimulates the tissue of the chestnut tree and the bud in the following sprouting season transforms into a spherical gall. The larvae eat the inside of this gall, and pupate and emerge inside it. The number of eggs in the ovaries is over 300, yet 10% of them become adults along a type A survivorship curve. This appears contradictory. The result of an investigation suggested, however, that the females of this wasp laid only 1.2–2.9% of the eggs found in the ovarioles (Miyashita *et al.*, 1965). Also, oviposition was a slow process in this species and a maximum of only two or three score could be laid in a day. Predation by spiders was high during the laying period and adults disappeared quickly. Thus the chestnut gall-wasp, though fecund, in reality has the mode of life that produces a small number of protected eggs.

Discussion on the evolution of reproductive rates and survivorship curves

In Chapter 1 and in this chapter I have reviewed the evolution of reproductive rates and mortality on a large taxonomic scale. Here we shall examine possible mechanisms that cause such evolutionary changes.

The fecundity of animals differs greatly according to species. Among closely related species and even within the same species it differs between

Fig. 2.44. Survivorship curve of the chestnut gall-wasp *Dryocosmus kuriphilus* (drawn from Miyashita *et al.*, 1965, for the 1963 generation). Mortality of ovipositing adults is very high but that of the larvae inside the gall and pupae is low. Stages of development as in previous figures.

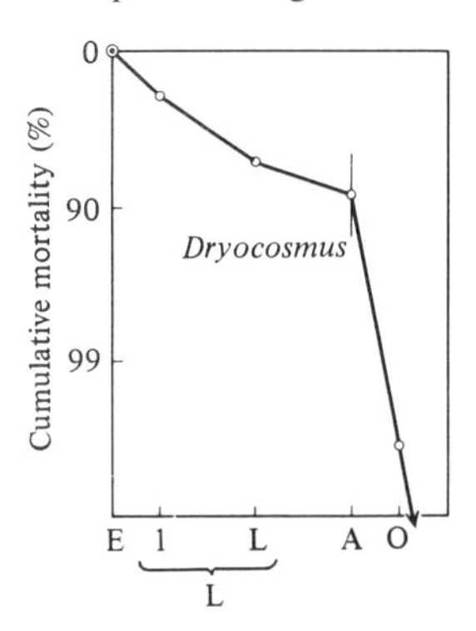

populations. For example, the clutch size of the great tit *Parus major* varies from three to twelve while the number of eggs produced by the copepod *Arctodiaptomus bacillifer* ranges from six to sixty.

In the past such variations have been explained teleologically in terms of mortality. Examples of such explanation are that migratory birds have a large clutch size in order to compensate for the great loss during migration and that the elephant has a litter of one because it has a long life-span. However, as stated earlier, according to this theory all species should have evolved towards increasing the number of eggs. The fact is that there are rather more examples contrary to this.

Lack (1954a) accepted this evolutionary tendency as a 'fact' but considered the 'explanation' wrong; he claimed that 'the reproductive rate of animals, like other characters, is a product of natural selection, hence that each species lays that number of eggs which results in the maximum number of surviving offspring ... With broods above normal size, fewer, not more, young are raised per brood'. The way to maximize the number of surviving offspring is, as stated many times already, not restricted to increasing the number of eggs laid. Some animals took the high-fecundity road while others took the low-fecundity road.

In Chapter 1 (p. 7) I said that the fishes living in the eutrophic sea laid large numbers of small eggs whereas fishes living in the oligotrophic environment of upper rivers laid small numbers of large eggs and provided some sort of parental care. Mizuno's (1961) report is interesting in this respect. He noticed that what has been known as the common freshwater goby *Rhinogobius brunneus*, a river bottom species, in fact consists of two forms; the river form that completes the life cycle within the upper waters and the migratory form that lives in the lower waters and spends a larval stage in the sea. The migratory form that lives in the eutrophic environment of lower waters lays several thousands of small eggs (diameter 0.6–0.7 mm) and the fry at hatching are in a planktonic stage of development. The fry are carried to the estuary by the current and after a period of growth in the sea they return to the river. On the other hand, the river type that lives in the oligotrophic environment lays only a few eggs (about 100), much larger in diameter (1.5–2 mm) than in the migratory form. The male defends the eggs, and the fry hatch at a much advanced stage of development and immediately start bottom living. Mizuno on these grounds classified the river form as *Rhinogobius flumineus* [he described it as a different genus, *Tokugobius*, but later (Miyadi *et al.*, 1976) put it in the same genus as the migratory form]. It is thought that among the freshwater gobies some became adapted to the oligotrophic conditions and differentiated into *R. flumineus*. Further, Mizuno & Niwa (1961) found a similar phenomenon in another stream-dwelling bottom fish, the Japanese sculpin *Cottus hilgendorfi*. In this species there is also a migratory form that lays a large number of small eggs. These eggs hatch at an early stage of development. The river form attaches a small number of large eggs to the undersurface of a rock. After spawning the male stays there to defend the eggs. The fry do not go down to the sea but spend the early juvenile stage under the rock. After about ten days they begin their bottom life. Mizuno distinguishes three forms of the sculpin, i.e. the migratory form, the lake form and the river form. The lake form, which is found only in Lake Biwa, is even more

Fig. 2.45. A graph showing the relationship of egg size and habitat to the egg number among some Japanese species of gobiid (circles) and cottid (triangles) fishes. Those filled in black indicate species living only in rivers, the black dot inside a circle indicates an estuarine or sea-living species and the rest are river species whose young spend some time in the sea.

Gobiidae: 1, *Eleotris oxycephala*; 2, *Tridentiger obscurus*; 3, *Rhinogobius giurinus*; 4, *Rh. brunneus*; 5, *Rh. flumineus*; 6, *Chaenogobius annularis*; 7, *Ch. isaza*; 8, *Leucopsarion petersi*; 9, *Luciogobius guttatus*; 10, *Acanthogobius flavimanus*; 11, *A. lactipes*; 12, *Sicyopterus japonicus* (spawns in the river but its young spend a long time as pelagic fish in the sea); 13, *Odontobutis obscura*.

Cottidae: 14, *Cottus fasciatus*; 15. *C. pollux* (migratory form); 16, *C. pollux* (land-locked form); 17, *C. hilgendorfi* (lake form); 18, *C. hilgendorfi* (land-locked form).

Data taken from Miyadi *et al.* (1976) and N. Mizuno (personal communication, 1977).

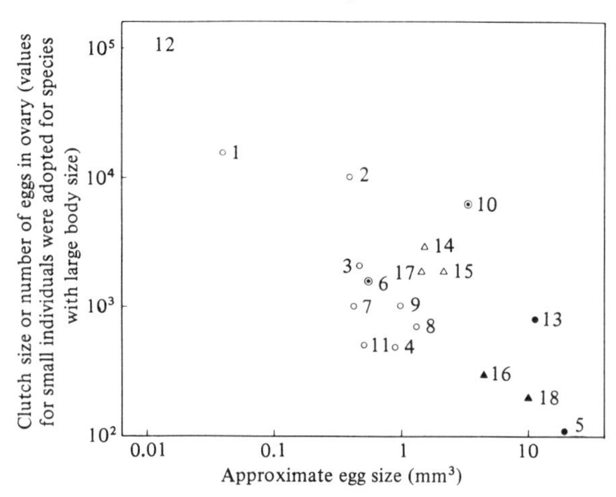

fecund than the migratory form and it has the smallest egg diameter. Evolution of such parental care as seen in the river form may be thought of as an adaptation to the oligotrophic environment, in which selection has operated within the population that has expanded into this environment. Similar phenomena have been reported for the wrinklehead sculpin *Cottus pollux* of northern Japan.

Fig. 2.45 shows in graph form the habitat, egg size and the number of eggs (values taken from similar-sized individuals) of Japanese fishes belonging to Cottidae and Gobiidae. As the two families are phylogenetically not close their adaptation to bottom living is considered to be a result of convergence (N. Mizuno, personal communication, 1977). In either family those that live in the middle and lower waters of the river and enter the sea and those that live entirely in the sea produce large numbers of small eggs, whereas those that live in the upper waters and do not go down to the sea lay small numbers of large eggs. Besides, there is a negative correlation between the logarithm of the egg volume and the logarithm of the egg number. In the case of Gobiidae freshwater species are considered to have been derived from marine species so that this may be an example of the occupation of an oligotrophic environment having been made possible by the low-fecundity/protection strategy. The history of the differentiation within the Cottidae is not understood, but some may have taken the road to high fecundity (see examples of North American darters, p. 213).

Ueda (1970) presented data on the egg size (length and diameter) and

Fig. 2.46. A graph showing the relation between the egg length and the number of eggs carried by some Japanese freshwater species of atyid (circles) and palaemonid (dots) shrimps, Drawn from Ueda (1970) and Shokita. [Shokita (1973a) for X, Shokita (1973b) for Y and Shokita (1976) for Z.] X, *Caridina brevirostris*, Y, *Macrobrachium shokitai; Z, Caridina denticulata ishigakiensis*. These three species hatch at extremely advanced stages of development.

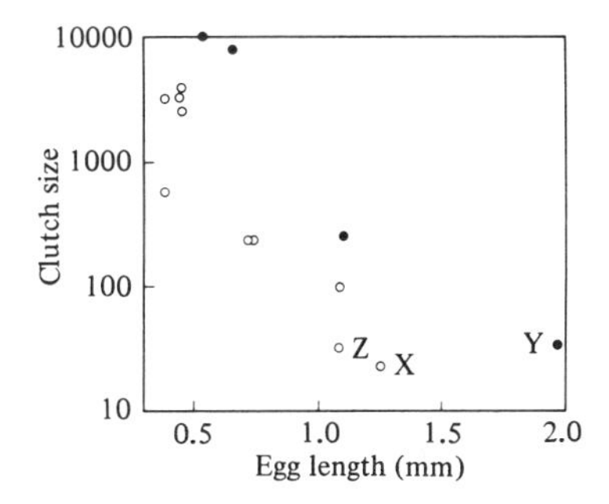

the number of eggs carried by freshwater shrimps in Japan. Fig. 2.46 is drawn from his data. In both Atyidae and Palaemonidae the larger the egg the smaller the number of eggs carried by the shrimp. Also, in general the large eggs hatch into advanced larvae (shortening the zoea stage). In the Atyidae the species with a small egg diameter mainly live in lower waters and the species with a large egg diameter live in upper waters or in small inland ponds. Most members of Atyidae are amphidromous and they enter the sea in the zoea stage and return to the river as shrimps, but some species remain in freshwater. Among them is a land-locked species, *Caridina brevirostris*, which is found on Iriomote Island, Okinawa. They lay 10–30 large eggs and the larvae hatch at a well advanced state (Shokita, 1973b). Another freshwater species on Ishigaki Island, *Caridina denticulata ishigakiensis*, also lays 20–40 large eggs and the larvae become bottom-dwellers as soon as they hatch (Shokita, 1976).

Many species of palaemonid shrimps also enter the sea. Among these *Macrobrachium nipponensis* living in Kasumigaura Lake and other eutrophic lower waters lays the largest number of eggs and *Palaemon paucidens*, which lives in small bodies of water such as ponds and moats, carries 262 eggs on average and its egg diameter is larger than 1 mm. *Macrobrachium shokitai*, which occurs only in inland waters of Iriomote Island, has an egg diameter close to 2 mm and its larvae are shaped like an adult when they hatch. The number of eggs in a clutch is smallest (20–50) among the Palaemonidae (Shokita, 1973b). The species most closely related morphologically is *M. formosense* found in Formosa. This species is amphidromous and lays a few thousand eggs. In Formosa there is another species, *M. asperulum*, which lays a small number of large eggs. It lives in the mountain streams, and Shokita (1975) predicts that this is a land-locked species without the zoea stage in juveniles. The life history of *M. hendersonyanum*, a completely land-locked species in India, shows an even more shortened larval period (Jalihal & Sankolli, 1975).

Marshall (1953) compared the characteristics of fish eggs found in the Arctic Sea, Antarctic Sea and deep seas with those of related species found in other regions. Some of these are reproduced in Table 2.8. From this table one can see that the arctic species lay larger but fewer eggs and their larvae hatch at more advanced stages than the related species ranging south. The same is probably true for the antarctic species. In Marshall's table the species endemic to the Antarctic Sea are shown to have a body length at hatching of about 1 cm; 5–10% of the body length of the parent. The deep-sea species lay large eggs, which hatch at advanced stages of development, compared with related species found in

Table 2.8. *Characteristics of eggs of fishes living in the Arctic Sea, boreal waters, and Antarctic and Falkland waters*

Family	Region	Egg diam. (mm)	Size of newly hatched larvae (mm)	No. of eggs produced per female
Liparidae	Arctic and chiefly arctic	2.13–4.59	7.2–16.0	300–434
	Arctic boreal	1.35–1.67	*ca* 5.4	—
	Boreal	1.03–1.19	3.3–3.8	793
Cottidae	Arctic	1.8–4.0	9.0–11.5	57–680
	Arctic boreal	1.95–2.51	7.4–8.6	2742
	Boreal and north boreal	1.51–2.04	4.4–5.5	1586–28 735
Gadidae	Arctic	1.5–1.9	4.3–5.3	2×10^4–6.7×10^4
	North boreal	1.13–1.67	3.7–4.0	17×10^4–934×10^4
	South boreal and endemic boreal	0.97–1.32	3.0–3.8	6×10^4–826×10^4
Nototheniidae	Antarctic	1.5–3.0[a]		
	Falkland	1.2–1.7[a]		

From Marshall (1953).
[a]Calculated from Marshall's Table 2.

shallow waters. Marshall attributed these differences to the scattered distribution of plankton aggregations in the Arctic and the Antarctic Seas, which is considered to result in natural selection for large larvae with strong swimming abilities. In the deep seas, the ability to swim under strong water pressure would be required. The large-egg/low-fecundity strategy seems to have permitted these fish to expand into the severe environment of polar regions and the deep sea.

According to Jeannel (1956) the females of the blind silphid beetles *Speonomus longicornis* and *S. pyrenaeus*, true cavernicoles, lay only one egg each, of almost adult size, in crevices of limestone inside caves. Their larvae grow and pupate inside the egg shell without taking any food from outside. They emerge from the egg shells as adults! This extraordinary life cycle has developed under practically no predation pressure. Very few die in the pre-reproductive period and adults probably lay less than five eggs during their life of several months (or a few years?). The species adapted to caves or underground water (e.g. some shrimps) are known to have low fecundity. I view this as a result of adaptive extension into an environment in which food is difficult to obtain for the young and there are only few natural enemies. This low fecundity of the cavernicoles contradicts the theory of MacArthur & Wilson (1967) and

Cody (1966), which states that the severer the competition the lower the fecundity.

Fig 2.47 plots mortality to emergence (escape from the nursery cells) against egg size (some taken from related species) of four species of wasp and bee, whose survivorship curves are described in Fig. 2.43. The ordinate on the left shows a scale of k developed by Varley & Gradwell (1960) for evaluating mortality – the value obtained by subtracting in logarithm the number emerged from the number of eggs. From this diagram it can be seen that pre-reproductive mortality and the egg size have a strong inverse correlation. As stated in Chapter 1 the egg size of wasps and bees depends on the degree of parental protection which in turn is reflected in mortality.

Up till now we have seen examples of low-fecundity/protection strategies as being selected for in the animals that colonized the environment in which food is not easily procurable. What about the opposite case? One of the examples is found in turtles though quantitative data are wanting. The turtles, particularly sea-turtles, belong to the most fecund group among the reptiles. They have a long life-span, but their early mortality is high and their survivorship curve is broken into two lines (see Fig. 2.18). Although the survivorship curve is available only for a freshwater species today, the sea-turtles are expected to have even greater early mortality.

The arctic fox *Alopex lagopus* is thought to be derived from an ancestral temperate species which gradually adapted itself to the tundra

Fig. 2.47. Egg size and pre-emergence (up to leaving the nest chamber) mortality of four species of wasp and bee. Mortality is taken from Fig. 2.43. 1, *Diprion pini*; 2, *Ectemnius rubicola*; 3, *Eumenes colona*; 4, *Sceliphron assimile*. The egg length for *Eumenes colona* was substituted by the mean of three Japanese species of the same genus and that for *Sceliphron assimile* was substituted by the mean of two Japanese species of the same genus, taken from Iwata (1955a, 1958a).

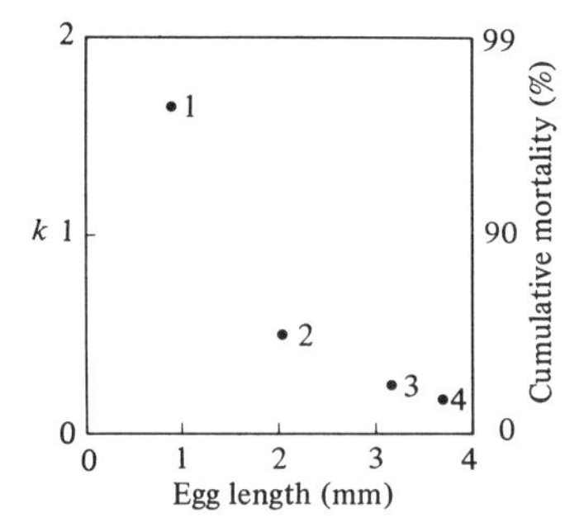

and snow fields of the north. Macpherson (1969) presented a graph showing relations between the litter size of foxes and the latitudes in North America where the species are found (Fig. 2.48). According to this the grey fox *Urocyon cinereoargenteus* found in temperate deciduous forest, has a litter of four, the red fox *Vulpes fulva*, found in boreal coniferous forest, has five to six and the arctic fox has ten on the average. Further, the litter size of the arctic fox varies greatly depending on food supply. During the decreasing phase of the lemming population on which they prey, they produce only four or five young per litter, but in the increasing phase the litter size reaches twelve. In the simple ecosystem of tundra near the north pole, vegetation is uniform everywhere and young foxes can hunt prey easily when lemmings are abundant. At the same time the tundra is an unstable, severe environment and the lemming population fluctuates periodically. In such an environment the high-fecundity strategy would have been advantageous. Although there appears to be an opposite trend within the same species, Macpherson concluded that the arctic fox was able to colonize the polar region by having high fecundity.

Finally, by way of summary I present a graph (Fig. 2.49) showing the relation between clutch size (egg number) and a special mortality-ratio for the animals in which both the egg number and the life-table are known. Mammals are excluded as they have little variation in litter size; sampling errors and variations due to other causes are greater than any trend which might exist. The ordinate plots a ratio of mortality in percentages obtained by dividing the mortality k_1 for the first 20%

Fig. 2.48. Relation between the litter size of three species of fox and the latitudes in North America in which they occur; the grey fox *Urocyon cinereoargenteus*, the red fox *Vulpes fulva* and the arctic fox *Alopex lagopus* (from Macpherson, 1969).

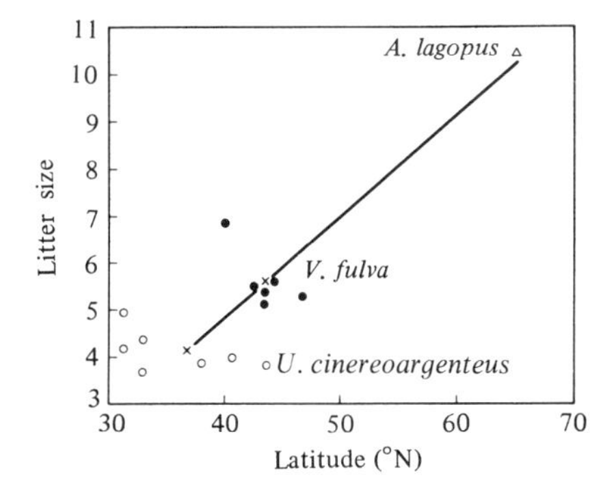

($x' = -80$) of the period from birth ($x' = -100$) to the commencement of breeding ($x' = 0$) by the mortality (k_α) for the whole period up to the commencement of breeding ($x' = 0$). This value indicates the weight of early mortality within the total mortality up to the commencement of breeding, which must naturally correlate with the egg number. As can be expected from the types of survivorship curve, the greater the egg number (litter size) the greater the ratio k_1/k_α, which is higher than 80% for the marine animals laying more than 100 000 eggs. However, it looks as though one group composed of fishes and aquatic invertebrates and another group composed of birds and reptiles fit two different lines. This is because in the latter the number of clutches is large due to longevity as well as breeding more than once in a season in many species. Thus their egg number is not directly comparable to the egg number of fishes. One aberrant species here is the freshwater turtle *Chrysemys picta*, which has a particularly high ratio of early mortality among the reptiles. As turtles have a very long life-span its clutch size may not be meaningfully compared with that of other reptiles or birds, but the mean number of clutches of this species will not exceed a few times that of other reptiles. Also, increase in the number of clutches has been shown to be of little significance in population dynamics (Cole, 1954). We would require the life-table of more fecund sea-turtles to solve this problem, but in the case of *Chrysemys picta* the adaptive strategy of longevity seems to be reflected in the relation of high early mortality to clutch size.

Fig. 2.49. Relation between the mortality-ratio expressed in percentages and the number of eggs or young produced per female: open circles, aquatic invertebrates; solid circles, fishes; solid triangles, reptiles; open triangeles, birds. The mortality ratio $(k_1/k_a) = (k$ value to $x' = -80)/(k$ value to $x' = 0$, commencement of reproduction).

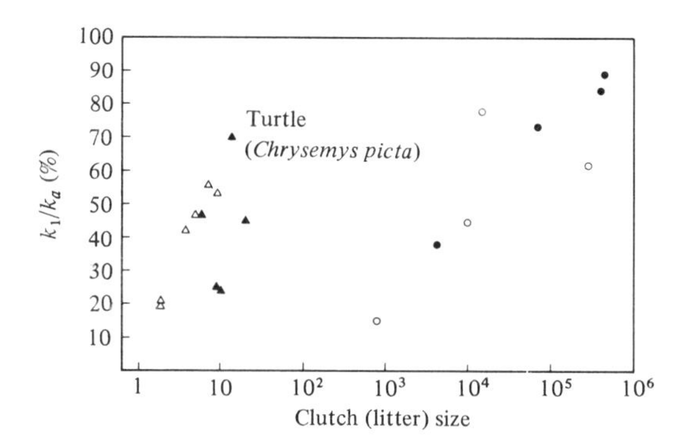

3

Population fluctuations

A perspective of theories concerning the fluctuations and stability of animal populations

(1) There are two aspects to animal numbers: the first is that animal numbers usually fluctuate from year to year or from generation to generation, the second is that in spite of such fluctuations most populations do not increase to the extent that their food supply is exhausted, nor do they become extinct; instead, a characteristic population level is maintained for each species.

These two aspects, though naturally inter-related, may be examined independently of each other. In the history of controversy in population dynamics these two aspects have often been confused.

(2) Probably the most striking example of such fluctuations is the rapid increase of numbers seen in pest insects during an outbreak (Fig. 3.1). The prediction of these fluctuations has been a matter of importance.

When we consider the question of factors leading to the irruption of pest insects, the first thing that comes to mind is the fluctuation of climatic factors. In the autumn of 1976 there occurred such an outbreak of millipedes in Yamanashi Prefecture that a train was stopped by them. The newspaper reporting on this event speculated that its cause might be the 'cold wave'. An unusual change in the number of animals seems to make people think of climatic influences. At the turn of the century the electric oven equipped with a thermostat (a bimetallic strip) was developed and this became a tool of many biological investigations. At the same time the control of humidity became possible with use of over-saturated solutions of various salts. Thus entomologists began experimenting on the development and mortality of insects in relation to temperature and humidity. This kind of work was advanced by Bremer (1929) in Germany and by Shelford (1927, 1929, 1932) in North America. In these studies they emphasized the importance of climatic factors in the outbreaks of insects. Following this Bodenheimer (1930) and Uvarov (1931) asserted that annual fluctuations of pest numbers and the geographical distribution of pests were determined mainly by the climate. Both workers demonstrated, with many examples, that specific climatic factors at specific stages of development modify the mortality of insects or the speed of development to cause asynchrony with the conditions of the host plants, thus influencing the number of insects emerging. They also rejected the concept of equilibrium (see below) in nature.

Bodenheimer published *Problems of Animal Ecology* in 1938 and in it he devoted a whole chapter to the discussion of population fluctuations. He said that there were three theories in population regulation, namely (1) the climatic theory, (2) the biotic equilibrium theory and (3) the mathematical analysis theory, and claimed that none of these was correct by itself. However, the contents of his theory were close to the climatic theory. It is true that he valued the distinction made by the biotic school between density-dependent and density-independent factors. He also recognized the possibility that density-dependent predation plays an important role in the control of prey at the end of an outbreak. But he

Fig. 3.1. Group flight of the desert locust *Schistocerca gregaria* in Africa. Copyright: Centre for Overseas Pest Research. Photographer: D. L. Gunn.

claimed that it is normally the climate that controls animal numbers, and in fact most insects are killed by climatic factors during the very early stage of development when the density effect is not operating.

Bodenheimer's book made a greater impact on Japanese entomologists than on workers in other countries. For example, the programme of pest outbreak prediction, founded in 1941 as a result of an outbreak of planthoppers reducing the rice crop, dictated that its major task should be the calculation of correlation coefficients between various climatic factors and the number of pest insects.

The defects of the climatic theory are: (i) on the basis of the knowledge gained in physiological studies it is difficult to explain abundance of insects in terms of climate, and the predictability of models based on trial-and-error type correlations with climatic statistics is not high; (ii) in connection with (i) the theory lacks stability as may be seen in cases where the outbreaks or low densities that have been correlated with specific climatic factors until certain periods repeat themselves thereafter under entirely different climatic conditions; for example, outbreaks of planthoppers in Japan were thought to be caused by abnormally high temperatures and low precipitation from late spring to early summer (wet season), with which they were apparently associated, but the rainy season of 1935 (an outbreak year) was normal. Besides, the outbreaks were not always preceded by low rainfall during the wet season (Ishikura, 1950). The outbreaks of the rice-stem borer *Chilo suppressalis* before the war occurred in years when the June and July temperatures were low and a physiological explanation was offered for this correlation, but it was discovered later that a low-temperature year following an outbreak did not necessarily cause another outbreak, and this could be interpreted as a density-dependent effect (Miyashita, 1955); (iii) although it may be established that the outbreaks occur under certain climatic conditions as in the above example, *the reason for not having outbreaks* under the same climatic conditions found in other years cannot often be explained. The above are all concerned with aspects of 'changes' in animal numbers, but the greatest shortcoming of the climatic theory of population dynamics is that (iv) it fails to explain the reason why the population fluctuation of some species is mild.

(3) Apart from the climatic theory there was an entirely different idea being developed concerning the process of determination of abundance. Like the climatic theory, this stream of thought emerged out of the use of natural enemies in pest control; so-called biological control.

The idea of using natural enemies for pest control is old. For example, ants' nests were sold in the third-century A.D. Canton for the control of pest insects of citrus trees (Konishi & Itô, 1973). However, the major development in pest control techniques using natural enemies took place in the United States. Many crop plants have been introduced into the New World from Europe accompanied by unwanted pest insects. These insects attained much higher densities than in the countries of origin and caused severe damage to crops. Soon it was found that this was due to the absence of their natural enemies, and the release of natural enemies was tried. Since the great success in the control of the citrus pest, the scale insect *Icerya purchasi*, by the release of the predatory lady beetle *Rodolia cardinalis*, there have been many trials of natural-enemy introduction.

From the beginning people who founded applied entomology in America, such as Forbes, Howard and Fiske, were much interested in the interaction of organisms in nature. Forbes while studying freshwater communities published a small article called 'The lake as a microcosm' (1887) and emphasized the holistic aspect of biotic communities whereas Howard & Fiske (1911) classified the mortality factors of insects into two categories: those called 'catastrophic factors' that would reduce population size irrespective of its density and those called 'facultative factors' that would vary in effect with the density. It was these two categories that H. S. Smith (1935) later called the *density-independent factors* operating independently of the density of the pest in question and the *density-dependent factors* varying in their effect in relation to the density of the pest. Smith further considered that the density-dependent factors were mainly biotic factors and only these could determine the equilibrium of the population.

Nicholson, an entomologist in another new continent, Australia, also developed the biological theory. He (1933) claimed that there was a fixed quantitative relation between the insect and its parasite (strictly, its parasitoid). Two years later he and Bailey, a mathematician, published a mathematical model to express this theory, which is known among ecologists today as the Nicholson–Bailey model. The important point of Nicholson's idea is not the relation between the pest insects and their natural enemies. Nicholson (1933) thought that all populations were ultimately controlled by intraspecific competition. It was the existence of regulation itself that was the main point of Nicholson's model. Thus the idea of Forbes–Smith–Nicholson might be more accurately named as the regulation theory or the saturation theory rather than the biotic theory.

In Germany a similar idea was developing from a different angle. It began with Möbius (1877, cited by Allee *et al.*, 1949) who considered the importance of interactions among species in a community and established the concept of Biocönose (biocoenosis). The holistic idea grew among the forest researchers, represented by Escherich (1914), Friedrichs (1930, 1937) and Schwerdtfeger (1941), that the organisms are members of a community in which the outbreaks of species populations are checked automatically by the adjustment of the community.

As we have seen, the climatic theory and the regulation theory, though they appear to address the same problem, do not in fact treat the same theme. What the climatic theory tries to explain is the change in the number of organisms and what it claims is that the change is attributable to climate. On the other hand, the regulation theory pays little attention to the change in number; the change is viewed merely as an oscillation around the average density. The first thing to be noted in this theory is that populations have self-regulatory functions.

(4) Experience taught us that the climatic theory, whilst attempting to explain the fluctuation of numbers, does not really explain the fluctuation at all. In the field proof was being obtained that reproduction was limited under a high density – in other words, a 'density effect' or 'crowding effect' was seen. But the idea that all animal populations are in equilibrium also appears to simplify reality too much. In reality many species are maintained normally at densities too low to cause intraspecific competition, yet extinction occurs seldom even without regulation. It is possible that species on the way to cxtinction may be considered mistakenly to be in equilibrium if the process is slow.

Thus there appeared people who, while opposing the simplicity of weather-determinism, were not satisfied with the regulation theory either. Such attitudes led to 'comprehensive theories', to use Solomon's (1949) terminology.

Among the proponents of comprehensive theories there are naturally people whose views are (relatively) close to the climatic theory and others with views close to the regulation theory. Thompson (1928, 1939, 1956) is an early representative and Milne (1957a, b) is a recent representative of the former.

Thompson, with his mathematical inclination, was involved in biological control and never rejected the role of biotic factors nor their density-dependence. He was, however, opposed to the idea of saturated populations. According to him the reason why a population does not become extinct or increase to the extent of eating out its food supply is that the natural environment is complex and has a patchy structure. Under such conditions no special regulatory mechanism is needed to explain the stability of populations. That is, species have their own requirements and in an environment in which these are met an increase in number is always possible. This occurs in very few places and there the surplus individuals over the carrying capacity move out to unfavourable places in the neighbourhood where they are subject to strong limitations. In unfavourable places local extinction occurs frequently but recolonization from favourable places is also common. However, the suitability of the areas may be reversed in some years and on the whole the population is maintained at a not very high density.

It is clear that Thompson recognized density-dependent processes (including emigration) operating to prevent the population from attaining too high a density. He considered, however, that at a medium or low density such processes do not operate, yet extinction is prevented in effect by chance fluctuations of climatic factors and by immigration. This idea is akin to the scheme presented later by Milne (1957a, b).

Schwerdtfeger expanded his holistic view when he published a monumental paper in 1941 on the fluctuation of pest insects in the pine forests of Germany. His views may be summarized as follows: (i) The parasite theory (stating that only the specialized parasitic insects can be related to population fluctuations) and the biocoenosis theory are too simple to be true. In reality many parasitic insects have several alternative hosts. (ii) Bodenheimer's view that weather kills more than 90% of the insects is not correct. In the pine beauty (a noctuid moth) such mortality is less than 25%. (iii) The theory of Eidmann (1937), that the insect population is always tending to irrupt but the resultant shortage of food reduces its size, causing the number of forest pests to fluctuate, does not explain why the population levels at the peak of outbreaks vary and how the lower peaks are checked. (iv) There are no cases where a single factor is invariably predominant. Many factors are operating at the same time and which one becomes a dominant factor depends on the characteristics of the species concerned and also on the year. Neither the biological theory nor the climatic theory is correct on its own. The only thing common to all fluctuations is that the factors that cause population change are tremendously varied.

Varley (1949), who brought Schwerdtfeger's views to the notice of British ecologists, said, 'Thus, after a useful critique of a number of theories, Schwerdtfeger seems to relapse into obscurantism.' However, I think that the above mentioned opinions themselves are absolutely correct and it seems necessary for us to start from such a standpoint at one stage.

Voûte (1943) in Holland reclassified the regulatory factors of population density and in it he criticized the idea of equating biotic factors with density-dependent factors and abiotic factors with density-independent factors. Climate may operate density-dependently through the reduction of shelters or by weakening disease resistance whereas the action of generalist predators or parasitoids may not depend on the density of a specific prey or host species. In 1946 Voûte discussed the problem of pure stands of plantations being more susceptible to the outbreak of pest insects than the natural forest. [The same point was emphasized by Yano (1922) and Kamiya (1938) in Japan and Graham (1939) in America.] Voûte proposed the 'escape theory' of outbreaks. According to his theory the polyphagous predators, such as birds, spiders, ground beetles and frogs, are biotic factors and yet operate density-independently as they are not influenced by the density of any particular prey species. The predators keep the density of pest populations low in natural forests, but if this equilibrium is lost as a result of a hurricane or other severe conditions pest populations start increasing. If this increase is below a threshold level the population will be brought under control by predation, etc., but if it goes beyond the threshold level the population 'escapes' from the predation pressure and breaks out. The outbreak is finally terminated by parasitoids, diseases and intraspecific competition that respond or increase density-dependently.

Solomon (1949, 1953, 1957) in England synthesized these theories and made the position of the comprehensive theory steadfast. His paper 'The natural control of animal populations' published in 1949 is outstanding in the completeness of the literature reviewed and the broadness of discussion attempted, and marks a turning point in the dispute of this problem. In this paper he stated that the comprehensive theory was born in the dispute between the 'biotic theory' including the intraspecific competition theory that developed from it and the 'physical theory' including the various theories on cyclic population fluctuations. He emphasized that animal populations in nature are maintained far below the level at which the Nicholsonian process appears and the population irruption refers to the cases where such a normal population density is broken through. If climatic and other conditions are uniform, biological factors play a major role in population regulation, 'but in the great majority of cases, and particularly with insects, it is clear enough that numerical variations primarily reflect changes in the weather' (Solomon 1949, p. 8). He advanced Voûte's view and suggested that 'it is sounder to consider density-dependent or density-independent *processes* or *actions*, than to classify factors on this basis' (Solomon 1949, p. 11; italics by Itô). He further stated that 'density dependence of biotic factors may be so slight as to leave them practically density-independent' (Solomon 1949, p. 12) and claimed that the typical density-dependent factors do not usually operate on the population at its normal density.

With respect to the theory of Nicholson, Solomon's comments are:

> Natural populations, on the whole, fluctuate much more violently and erratically, responding not only to changes in associated populations (natural enemies and competitors), but to the irregular variations of the physical environment ... Many populations fluctuate chiefly within a low range of density, with occasional outbreaks ... Movements towards the mean (density) are often reversed before it is reached ... There are forces tending to reduce the population even when the density is low, and forces tending to increase the numbers even when they are high. (Solomon 1949, p. 28; words in parentheses are by Itô.)

However, Solomon accepts Nicholson's reasoning (presented in 1933 and clearly expressed in the Cold Spring Harbor Symposium paper in 1957) that in a uniform environment all species ultimately must limit their numbers by intraspecific competition. This standpoint of Solomon's is more lucid in his later writings (1957, 1969). Upon the acceptance of such a theoretical possibility he argued that many animal populations are regulated to different extents, in other words some populations are determined largely by the weather while others are under strict regulatory influences. That is to say, the extent to which the ultimately necessary regulation is realized depends on the way the density-independent factors operate.

(5) These comprehensive theories have been influenced strongly by the empirical studies in the field, which developed rapidly after the war and centred in Canada during the 1950s.

In eastern Canada outbreaks of the spruce budworm *Choristoneura fumiferana* have occurred since the 1940s and defoliated the boreal coniferous forest in an area as large as the whole of Hokkaido. [The main species defoliated was not the spruce, after which the pest insect was named, but the fir.] Canadian forest entomologists attempted a thorough investigation of the number of insects at various developmental stages as part of a plan to control these outbreaks. In a large expanse of coniferous forest dozens of permanent stations have been established and the life-table data have been accumulated according to strict sampling methods. On the basis of more than eighty life-tables analysed, Morris *et al.* (1958) clarified the causes of the outbreaks of this insect. The causes were found to be complex interactions of the stand conditions (ages of forests), climatic conditions and biotic conditions (parasitism and starvation). A voluminous report edited by Morris was published in 1963. Thus Morris *et al.* supported Solomon's comprehensive theory. Their work, though limited to the spruce budworm, presented a clear message that what was needed in the development of population dynamics was not a dispute but an accumulation of such detailed population data.

Therefore, the research of Morris *et al.*, called the Green River Project, gave strong impetus to population analysis by means of life-tables, which were independently developed in Britain (Richards & Waloff, 1954 and others; Varley & Gradwell, 1958) and in Japan (Itô & Miyashita, 1955; Miyashita *et al.*, 1956). Since then a large number of life-table studies, particularly of insects, have been published. Among these, discussions of population dynamics based on many years of life-table studies were made by Varley & Gradwell (1963, 1968) and Embree (1965) on the winter moth *Operophtera brumata* (sixteen generations and ten generations respectively), Klomp (1965) on the pine looper *Bupalus piniarius* (fourteen generations), Itô & Miyashita (1968), Itô *et al.* (1969) and Itô (1977) on the fall webworm *Hyphantria cunea* (eight generations at two localities) and, apart from insects, by Southern (1970) on the tawny owl *Strix aluco* (eleven generations) and by Hancock (1971) on the edible cockle *Cardium edule* (eleven generations). Dempster's (1975) book contains annotated data from many reports.

There are three major findings through these and other life-table studies. First, density-dependent processes often play an important role even in the field in damping population fluctuations; secondly, in spite of these processes there is no population which has been shown to be in an equilibrium state in the strict sense of the word; and thirdly there are many species in which the intraspecific interference is rather more important as a density-dependent process than the predator–prey interactions. [Varley & Gradwell (1963) and Varley *et al.* (1973) claimed that the host–parasitoid relations determined the average density in the winter moth and other moths, but it is hard to say if this is supported by sufficient evidence (Itô & Murai, 1977, chapt. 9).]

It is generally accepted today that the view of Andrewartha & Birch

(1954), who almost entirely denied the role of density-dependent processes, was wrong. Now, so far as a level is maintained by the population the existence of regulation is theoretically necessary. Yet animal populations in reality fluctuate under the influence of climatic and many other factors and are not controlled by density-dependent factors at all times. Even if the density-dependent factors are weak, species populations do not easily become extinct or eat out their food supply because of chance fluctuations of climatic factors. Thus through many generations the environment changes and the species itself also changes.

The conclusion reached by many current ecologists is that 'the truth lies in-between'. In some species the number may be fluctuating mostly at the mercy of changes in the inorganic environment, the eating-out of the food supply may be evaded by the pressure of the environment, and extinction may be avoided by migration and recolonization. In other species, however, the density effect may be operating on most generations and the effect of weather on the fluctuation of numbers may be very little. The question is the relative importance of the density-dependent and density-independent processes, which should be clarified from carefully conducted field studies.

If we follow this conclusion alone, however, we cannot even have a working hypothesis to determine the main parameters of the life-table until a detailed life-table study becomes available for each species (Viktorov, 1955, and other Russian ecologists take such a view). It has been said since the early days that for some populations the role of density-dependent processes is dominant in the favourable area near the centre of the range while the role of density-independent processes, such as the fluctuation of climate, is dominant on the fringe of distribution, and this has been demonstrated (Thompson, 1939; Richards & Southwood, 1968). This can be used as a working hypothesis in the investigation of a species population. But can we not predict the process that plays an important role in population fluctuations by knowing the mode of life of that species? As we have seen in the previous chapters various organisms have faced the choice between the two strategies of high fecundity and low fecundity/parental protection. Such a conspicuous difference in the way of life must influence the pattern of their population fluctuations. I shall examine this question in the following sections.

Population fluctuations of microbes and invertebrate animals other than insects

There is hardly any reliable information regarding the population fluctuation of microbes and small aquatic invertebrates.

Most papers dealing with temporal changes in the number of microbes are concerned with ecological succession and do not deal with fluctuations of species populations under relatively stable conditions, which we wish to examine. An example of such studies might be an ecological succession of soil microbes during one month following the application of organic fertilizer in a cultivated field. Besides, many such studies of soil or aquatic micro-organisms are of the type in which investigation is made once a month for over a year or every day for a week or so. Because bacteria and protozoans fluctuate in number greatly from day to day such fluctuations cannot be traced in once-a-month investigations; only some seasonal trends may be discovered (Waksman, 1927). In the investigation for a very short period, on the other hand, the environmental conditions related to the fluctuation cannot be clarified.

However, from various experiences we may be able to say that most fluctuations of microbial numbers are, first, extremely *irregular* and, secondly, strongly influenced by *external factors* (density-independent processes). Here density-independence does not necessarily mean 'inorganic' but includes changes in nutritional conditions, which fluctuate not as a result of predator–prey relations but are brought about by rainfall or artificial drainage.

Waksman (1927) found that some bacteria appear only at the end of a long dry period and increase rapidly after the first rain of the season. However, this is considered to occur as a result of an increase of inorganic and organic nutrients in available form rather than the direct influence of rain. For bacteria the action of water connected with nutrients is said to be generally more important than the temperature. In *Actinomyces* pH is also important, while increase in soil moisture always activates protozoans within a few hours. They encyst or die when conditions become unfavourable. However, this does not mean that the fluctuation of microbial populations is limited only by the influence of density-independent factors such as rainfall and influx of nutrients. Increase in number of these organisms begins with such external influences, but once the increase begins they outbreak in no time because of the speed of their cell divisions. In many cases rapid decline follows their eating out of the food supply. Red tide is one such example. This is a general term applied to outbreaks of blue-green algae, diatoms and flagellates which change the colour of sea-water. The ratio of densities during the outbreak to before the outbreak reaches several million or greater. In recent years the discharge of certain waste materials from factories and sewage has been causing an increase in the frequency of these outbreaks.

Long-term investigations of micro-organisms have been made under relatively stable environmental conditions in sewage beds. For example, Fig. 3.2 shows the fluctuation in numbers of some micro-organisms at a sewage farm. Here the inflow is purified along the following route: screen→detritus tank→sedimentation tank→gravel/bacteria bed→humus settlement tank. According to Barker (1942) the greatest fluctuation of numbers occurs in the sedimentation tank (not shown in the figure). Hence the fluctuation of protozoan numbers is sporadic and is limited by accidental changes in the quality of the inflow such as polluted water with a high organic content, change in metal ions and influx of rain water. The next highest in the fluctuation of the environmental factors is the gravel/bacteria bed and the most stable one is the humus settlement tank, where the fluctuation of protozoans is correspondingly slow (note in the figure that the unit for the two areas is different by a factor of ten). In the figure there is no species-by-species description of numbers but it can be seen that the number of individuals of Ciliophora or Mastigophora as a group fluctuates at least several hundred-fold. Where the quality of inflow is relatively stable, however, the number of ciliates becomes high and its further increase is said to be checked by predation by the larvae of Diptera such as psycodid flies (Barker, 1946).

Reynoldson (1955) investigated the following interacting species in English ponds: (1) *Urceolaria mitra* (epizoic protozoan), (2) *Polycelis tenuis* (planaria), and (3) bacteria.

Urceolaria (1) attaches itself to the planaria (2) and feeds on the bacteria (3) like a hydra. In Fig. 3.3 the numbers of *Urceolaria* and bacteria are plotted. According to this figure it appears that *Urceolaria* is limited by the number of bacteria which in turn is limited by the amount of rainfall. The upper two graphs are based on the data from two different ponds, but the trend of rainfall is similar and the trend of bacteria fluctuations is also similar (a positive correlation with the amount of rainfall). The fluctuation of bacteria is shown in logarithmic scale (base 10) yet the amplitude exceeds twenty-fold. The amplitude of fluctuations in *Urceolaria* was about three-fold and the two ponds showed the same trend of fluctuations except for some periods when, as investigation has revealed, one pond received an influx of other food organisms (flagellates). The number of individuals of co-habiting planaria did not influence *Urceolaria*. *Urceolaria* has no predator and its number is limited by the amount of food bacteria. Conversely, the number of bacteria is not considered to be influenced by *Urceolaria*. From this, Reynoldson has concluded that it is not appropriate in this particular

Fig. 3.2. Fluctuations of protozoan numbers in the gravel/bacteria bed (G.B.B.) and the humus settlement tank (H.S.T.) at a sewage farm (from Barker, 1942).

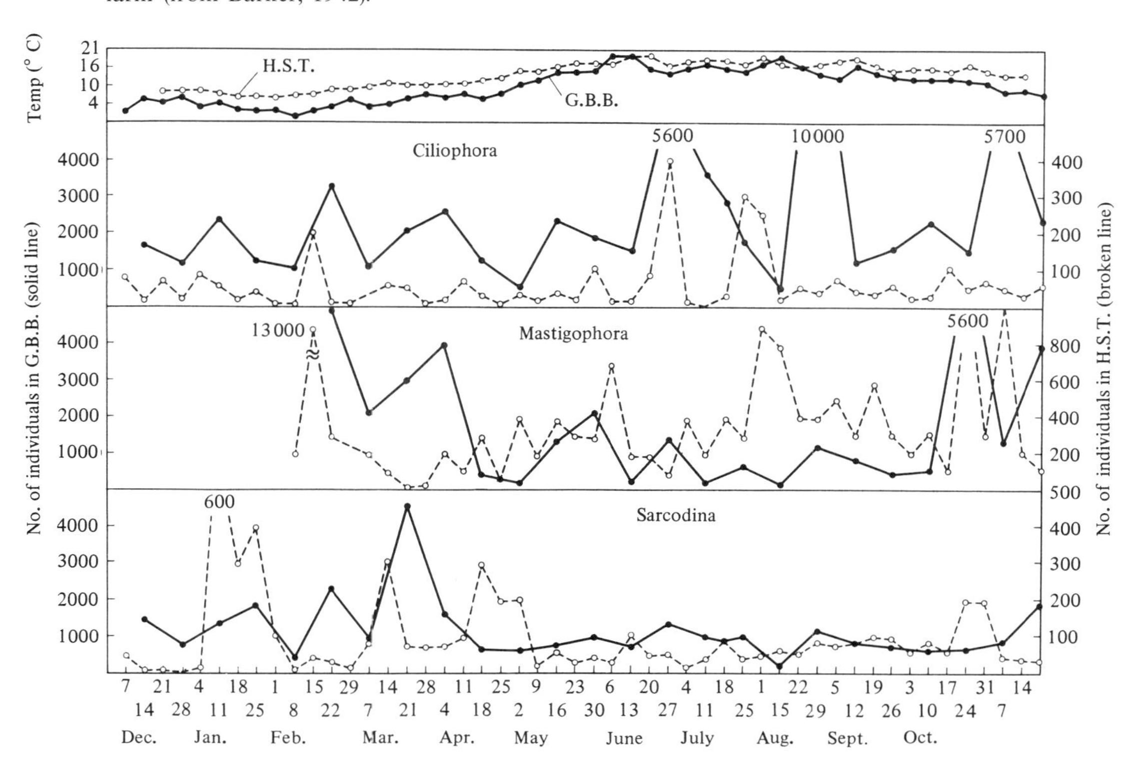

Fig. 3.3. Fluctuations of *Urceolaria*, its prey bacteria, precipitation and temperature in a pond (after Reynoldson, 1955).

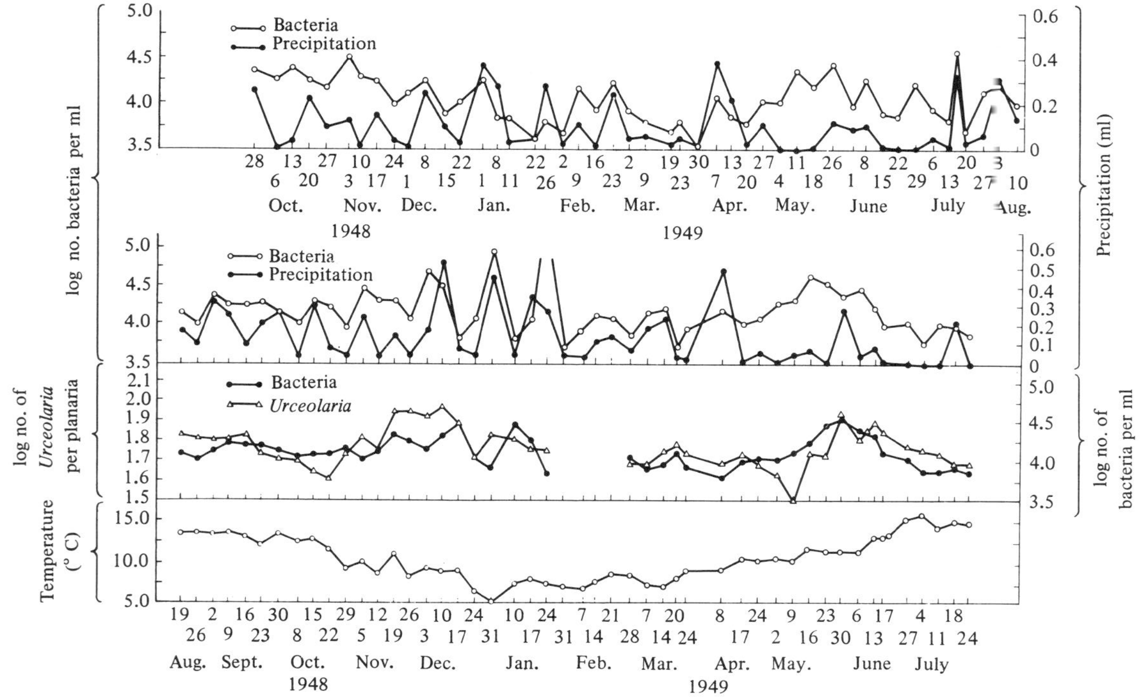

case to consider that density-dependent factors alone determine the fluctuation of numbers.

Let us now look at the density regulation mechanisms of planarians, which have a very special mode of life among the aquatic invertebrates. Reynoldson (1964) investigated three species of Planaria in England. Fig. 3.4 shows the change of numbers for two species (*Polycelis nigra* and *Dugesia lugubris*) collected at regular intervals. These are not the actual numbers of animals present but generally reflect population fluctuations. This example lacks long-term information but at least shows that the number of individuals of *D. lugubris* is very stable. In fact the impression gained by Reynoldson and co-workers over a long period of investigation was that the amplitude of fluctuations was rather small in all three species investigated.

What causes this stability? There were no significant predators. As the shortage of food was considered important Reynoldson reduced the population of *P. nigra* down to about one-thirtieth in a pool where the number was usually about 600. He found that following the removal a great many young were born during the summer (about 5.5 times the normal number) and the number reached 1200. By autumn, however, shortage of food became evident and shrinkage of body size and mortality occurred. The number declined to 760 and it was thought that the number would return to the normal level. For *Polycelis tenuis* he

Fig. 3.4. Graphs showing changes in the number of planarians collected in ponds (drawn from Reynoldson, 1964). The dots connected with broken lines show the result of a removal experiment in a small pool (removal of most planarians at a normal density of 580).

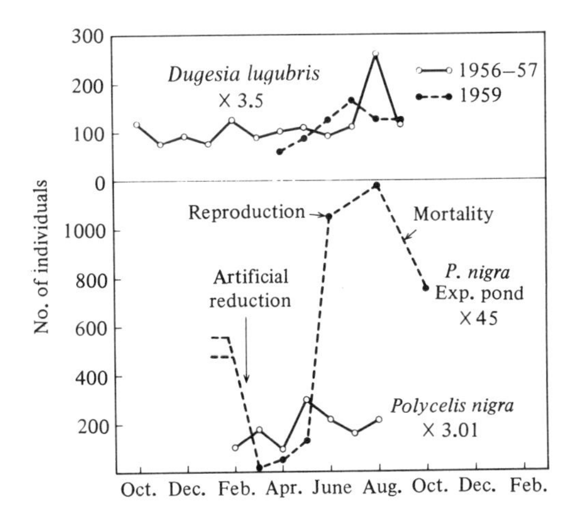

supplemented the diet. The result of this experiment is presented in Fig. 3.5. Active reproduction was evident in the pond where food was supplemented.

These planarians have a very special life cycle. The adult of *P. nigra* is about 8 mm long but the juvenile at hatching is 2.3 mm long. The adult body length of *P. tenuis* is 9 mm and that of the juvenile 2.2 mm. Of course they produce very few young. Normally the number of juveniles at hatching is about 50% of the adults present – that is one juvenile per adult female – but when extra food was given this jumped to 200–500% (4–10 young per female). Thus the number of eggs laid during one breeding season may be up to ten. As they secrete venomous liquid from their body they have practically no natural enemies, and as they live in ponds with stable water temperatures it is reasonable to consider that the number represents a saturated state. These animals can shrink their body during periods of hunger and survive several months without food (shrinkage).

After pointing out that these planarians maintain their numbers near the limit of the carrying capacity of the environment, Reynoldson (1964) stated that 'considered from a slightly different viewpoint, these three triclad species support Itô's (1961) generalization following Lack (1954b), that animals with low mortality in early life commonly produce few young and show low amplitude in population fluctuations. But they are exceptions to his suggestion that these two features are usually associated with parental care and high order of social structure'. [Itô (1961)

Fig. 3.5. Fluctuations in the composition of body length groups of a population of the triclad *Polycelis tenuis* (after Reynoldson, 1964). Upper figure, fluctuation under normal conditions; lower figure, fluctuation observed when additional food was supplied. Numerals indicate the total number of individuals: the blackened area, very small young; the stippled area, adults.

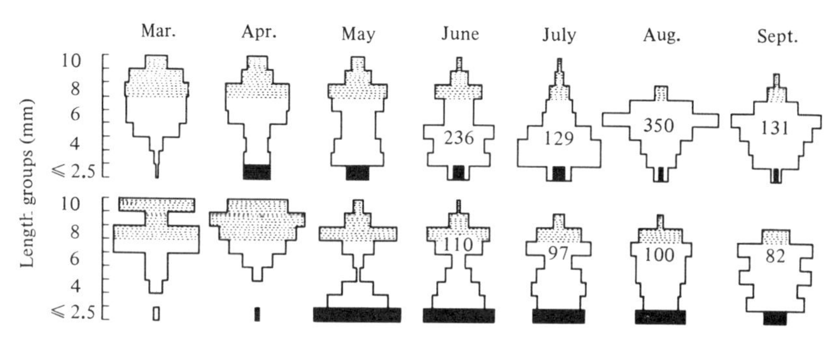

is an English summary of Chapter 2 of the first edition, *Comparative Ecology*, 1959.]

It is true that there is no parental care of young in planarians, but the fact that they produce young as heavy as 20–30% of the adult weight is a form of parental protection and these three species can be regarded as the low-fecundity/parental-protection strategists (*K*-strategists). It is interesting to note that even in the case of aquatic invertebrates species with this strategy have stable populations.

Moving to the marine invertebrates, we have already seen that many of them lay several thousand to a hundred million eggs, almost all of which are lost during the planktonic stage. On this point these lower animals differ from micro-organisms which multiply by cell divisions. This fact suggests that the number of marine invertebrates, in many cases, is limited by accidental changes of the inorganic environment or the fluctuation of food organisms unrelated to the density and that its fluctuations are irregular and very large in amplitude.

The irregularity of populations of marine animals is certainly related to the enormous number of eggs or young they produce. Thorson (1950), cited in Chapter 1, examined data collected on the fluctuations in the number of marine invertebrates and found that those species whose eggs are small but large in number and whose larvae spend a long period of planktonic life show conspicuous fluctuations in numbers whereas those species whose eggs are large but small in number and whose larvae have

Fig. 3.6. Yearly fluctuations of marine invertebrate biomass (adult biomass per square metre of bottom) showing different patterns of fluctuation according to the mode of life of the larvae (after Thorson, 1950). Study areas: A, Sallingsund; B, north coast of Fyn (Denmark). Species: IA, *Cultellus pellucidus* (long planktonic life); IB, *Nucula nitida* (short planktonic life); IIA, *Abra alba* (long planktonic life); IIB *Macoma calcarea* (no planktonic life); IIIA, *Corbula gibba* (long planktonic life); IIIB, *Nucula nitida*.

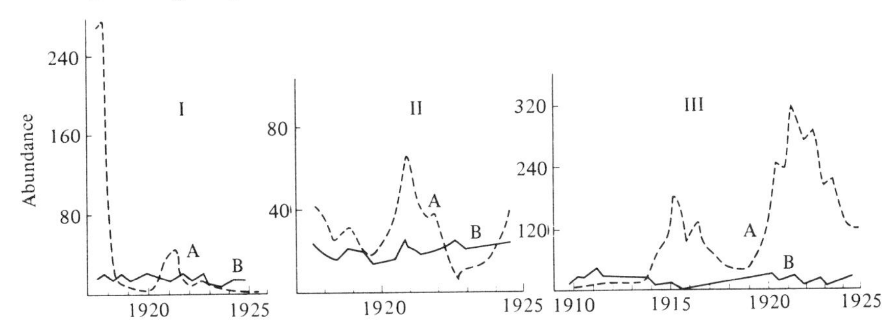

a very short planktonic life because of ovoviviparity maintain relatively stable populations (Fig. 3.6). This indicates that the main cause of fluctuations operates on the planktonic stage and this is probably the change of water temperature or quality (including nutrient organic matter). According to Kubo (1957) the octopus *Octopus vulgaris* of the (Japanese) Inland Sea hatches from the end of July to the beginning of October and becomes a young octopus from the middle of October to the middle of November. There is a negative correlation between the catch of octopus in one year and the amount of rainfall in the previous year. They are most susceptible to environmental influences during a short period after hatching.

Williamson (1961) recorded changes of plankton numbers in the North Sea (off the coast of Scotland) for eleven years from 1949 to 1959. Some of the results are given in Fig. 3.7. The upper graph shows the fluctuations of numbers in the copepods *Centropages hamatus* and *C. typicus*. The amplitude of fluctuation in the former is not very great, the maximum number being eleven times the minimum number, but in the latter the maximum is as much as 205 times the minimum. The trend of fluctuations, however, is similar; in both species the number was high at

Fig. 3.7. Fluctuation of plankton numbers in the North Sea (drawn from Williamson, 1961). Top graph; copepods, *Centropages hamatus* and *C. typicus*: middle graph; a copepod, *Temora longicornis*: bottom graph; planktonic molluscs *Clione limacina* and *Spiratella retroversa*. (Maximum–minimum ratios are given in parentheses.)

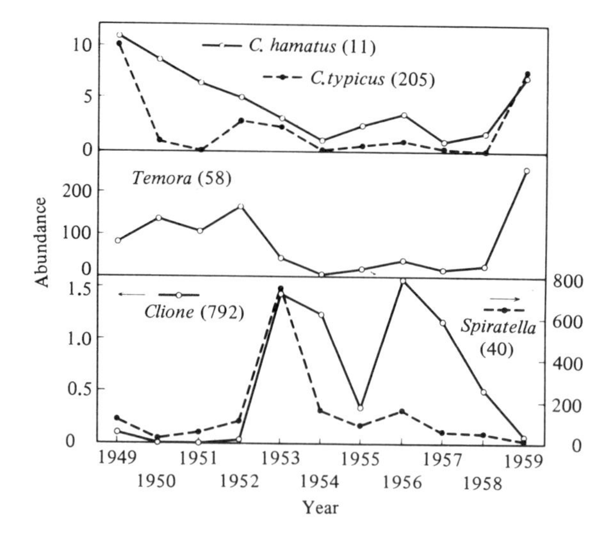

the beginning and the end of the study period and was low in 1954 and 1957. The fluctuation of another copepod population, *Temora longicornis*, shown in the middle graph is fifty-eight-fold in amplitude and reveals a somewhat similar trend to the first two species. Of the two species shown in the lower graph, the number of adults of the pteropod *Clione limacina* (a swimming mollusc that loses its shell as an adult) varied 792-fold (the number of larvae, not shown in the graph, changed 1267-fold excluding the years in which they were absent) while the other species *Spiratella retroversa* varied forty-fold. These figures demonstrate the violent nature of fluctuations in plankton populations. At the same time, the fact that within copepods or within pteropods the pattern of fluctuations is similar suggests the existence of some common external factors, such as perhaps salinity, winds, freshwater effluent and other inorganic factors, causing these fluctuations.

Some of the marine zooplankton, such as those copepods and pteropods discussed above, spend their entire life as plankton, but many others are planktonic only in the larval stages and spend their adult life as part of the benthic fauna (e.g. many bivalves, gastropods and some crustaceans). What happens to the population in the process of change from planktonic life to sedentary life has not been properly described

Fig. 3.8. Fluctuation in number of the edible cockle *Cardium edule* (from Hancock, 1971). A, the number settled in November; B, the number of first-year cockles in May; C, the total number in May. (Maximum – minimum ratios are given in parentheses.)

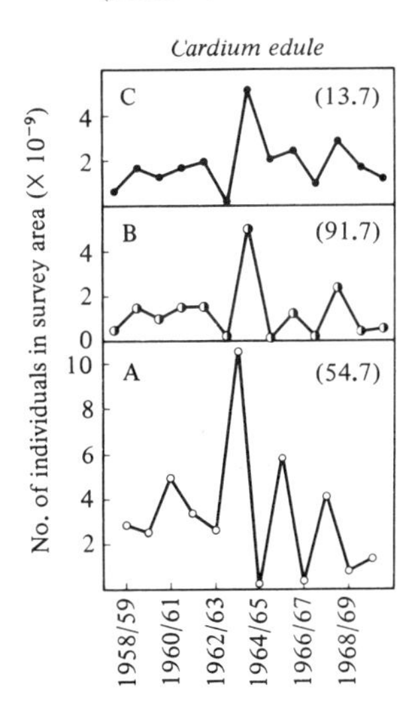

until recently. Hancock (1971) was the first to investigate populations of the edible cockle *Cardium edule* through this period; he studied the population dynamics of this species for eleven years on an extensive mudflat of Wales. Some of his results are given in Fig. 3.8.

The sexes are separate in this species and spawning occurs in early summer. The egg number is 10 000–20 000 per female (mean 15 000). The eggs hatch soon after they are laid and the larvae are planktonic until they form shells and settle on beaches in June. They live up to ten years. Hancock also described the yearly fluctuation of numbers of eggs discharged, estimated from the number of females spawning and the relation between shell length and egg number. Though the reliability of the estimates is not high the ratio of the maximum to minimum numbers estimated for eggs laid over a period of eleven years is 38.2. Over 99% of the larvae die before settlement but this mortality varies from year to year. Thus the number of cockles found in November under the age of one (bottom graph) changed as much as 54.7-fold over ten years. These cockles could be counted in the following May as the first-year cockles. The fluctuation in the number of these cockles increased in amplitude to 91.7-fold in the ratio (middle graph). However, the amplitude of fluctuations for the total number of shells was smaller (ratio 13.7) as shown in the top graph. One reason for the difference is that the life-span is long and the number of adults raises the lower values. Another reason is the density effect among adult cockles.

What is the cause of these fluctuations? The cause of fluctuation in numbers settling is not known but it is probably due to some oceanographic factors. However, there was a negative rather than positive correlation between the estimated number of eggs produced and the number of settled individuals. At the same time mortality of those under the age of one occurred when the winter temperatures were low. In the autumn of 1962, for example, more than 2600 million settled but this was reduced to about 100 million by May 1963 (survival of less than 4%). This was much lower than the average survival-rate of 65% in other years and was attributed to the cold winter. On the other hand, in 1963 as many as 10 000 million new cockles settled, in spite of the relatively small number of mother shells spawning that year, and 47% of them survived the winter, boosting the total population. The reason for the small number of first-year cockles in 1965 was that only a small number settled in 1964.

The natural enemy of cockles, apart from man, is the oystercatcher *Haematopus ostralegus*, but there was no obvious relation between the density of oystercatchers and the density of cockles. According to Hancock's (1973) re-examination of recruitment data, there was a negative correlation between the adult density and the survival-rate of settled young. Particularly in the years of high second-year cockle density, the survival of settled young was low. The first-year cockles were considered unable to obtain space if the density of the second-year cockles was high. In short, the number settled fluctuates according to the marine conditions and water temperature but after settlement the density is limited by competition among cockles.

Southward & Crisp (1956) investigated the distribution of two species of barnacles *Chthamalus stellatus* and *Balanus balanoides* in the tidal zone of England for several years. The first is a southern and the second is a northern species, overlapping in distribution around the English Channel. In the 1940s *Balanus* was dominant there but it decreased in the early 1950s and *Chthamalus* became dominant. Recently the relation

Fig. 3.9. Fluctuations in the maximum number of settled barnacles (spats) per square centimetre of rock surface and of air and sea temperatures (after Southward & Crisp, 1956). Upper graph, *Balanus balanoides* at Schoalstone Beach; lower graph, *Chthamalus stellatus* at Kallow Point, in England.

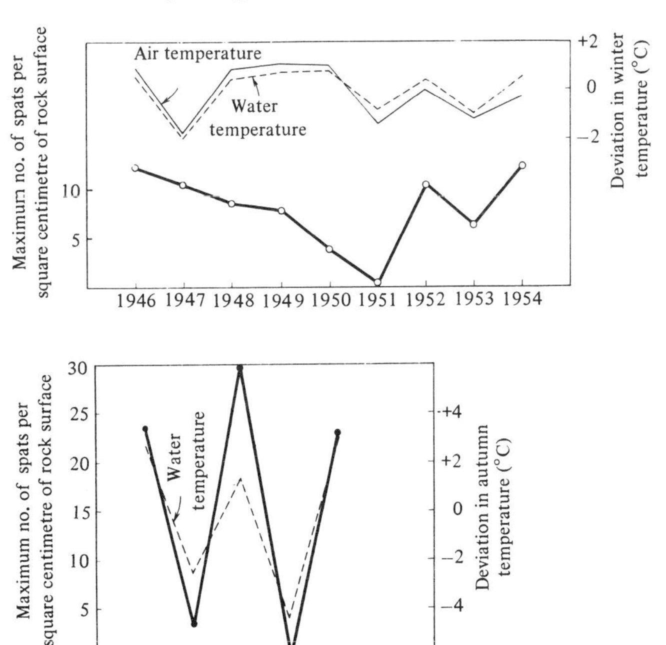

is again reversed (Fig. 3.9). When the number of individuals was counted in March–April after the barnacles had settled on rocks and the result was examined in relation to the oceanographic records, a beautiful correlation was found between the number of *Chthamalus* spats and the sea temperatures during the planktonic period of young from the discharge of nauplii to their settlement (September–December) as shown in the lower graph. Thus the population fluctuation of this species at its northern limit of distribution (as in England) was caused by the change of sea temperatures during the larval stage. On the other hand, *Balanus* did not show any obvious correlation with temperatures though, when the population was declining during 1948–50, the winter temperatures were generally higher than the mean (upper graph). The authors considered that the reason for the decrease in the number of spats in 1951 in spite of the low temperature was that the number of nauplii produced had become small as a result of continued decline till the preceding year.

In effect, the number of barnacles is limited mainly by the climate and food during the planktonic life and their food supply fluctuates not in relation to the density of barnacles but probably in relation to climatic conditions. Because the initial number of larvae is very large, if the survival-rate was high at all, dense settlement would cause density-dependent competition thus limiting the number of adults.

Population fluctuations of fishes

For the study of population fluctuations of fishes there is a long tradition since the work of Hjort and D'Ancona (who supplied data to

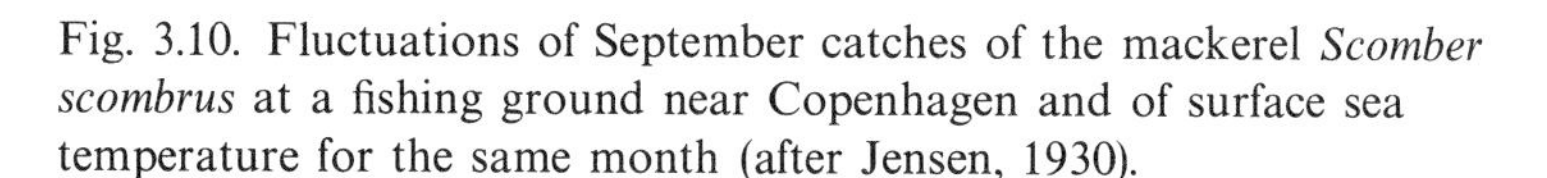

Fig. 3.10. Fluctuations of September catches of the mackerel *Scomber scombrus* at a fishing ground near Copenhagen and of surface sea temperature for the same month (after Jensen, 1930).

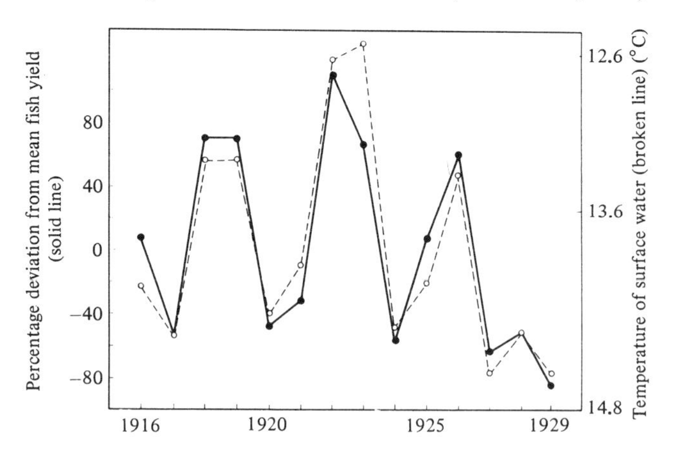

the mathematician Volterra) and there exists an accumulation of excellent material in the form of fisheries statistics over many years. However, the fact is that not enough is known about the fluctuations. This is because there are the following difficulties in the statistics of fish populations. First, the fluctuations seen in the catches do not necessarily reflect the fluctuations of species populations. In many cases the catch depends rather on whether schools have been encountered or not in the fishing grounds. Secondly, the catches are influenced by socio-economic factors. Thirdly, most fisheries statistics deal only with adult fish of the species concerned. For example, Fig. 3.10 illustrates fluctuations of mackerel catches near Copenhagen. In this graph it looks as though the population was fluctuating with a relatively short cycle but what actually happened was that schools of mackerel appeared in the fishing grounds when the cold water mass approached. Also, if fishermen refrained from fishing in order to prevent dumping when there were too many fish in the area or if they fished earnestly when the fish was scarce and the price was high, then the amplitude of catch fluctuations would be smaller than that of the actual fluctuations of fish numbers in the sea. As a method to cancel such artificial factors we use 'the catch per fishing effort'; that is, the catch per average fishing boat or per unit length of line in line fishing and so on. In the following the fluctuation of catches is used only when it is generally accepted that they reflect the density in the field. In fisheries science the study of biological aspects, such as mortality factors, appears to be lagging behind the analysis of catch statistics which has been advanced with the use of the computer for the practical requirements of catch prediction.

Fig. 3.11. Fluctuations of catches per unit fishing effort for the cod *Gadus morhua* and the haddock *Melanogrammus aeglefinus* (maximum–minimum ratios in parentheses). After Gulland (1971).

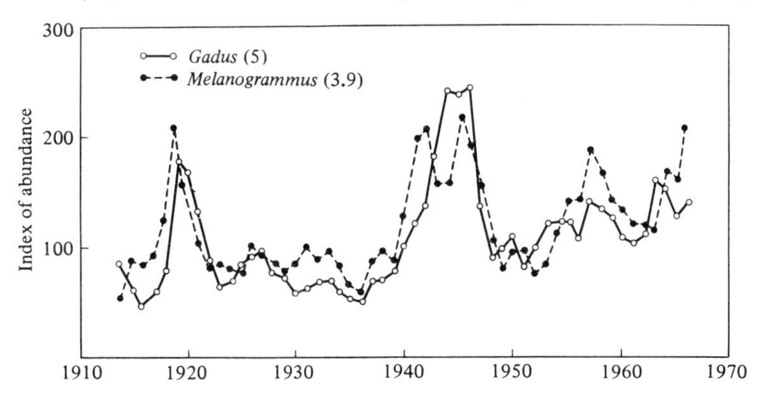

As mentioned in Chapter 1 many marine fishes lay huge numbers of eggs and their larvae suffer extremely high mortality. The oceanographic and climatic conditions are probably responsible for such mortality. Because of this I thought that the number of fish would fluctuate violently. In reality, however, the populations of many marketed fishes are stable. For example, Fig. 3.11 shows fluctuations of catches per unit fishing effort for the cod *Gadus morhua* and the haddock *Melanogrammus aeglefinus* caught by English and Scottish trawlers in the North Sea. For the unit of fishing effort, one hour of trawling by an average-sized fishing boat was adopted. This graph may give the impression that the number fluctuates greatly but the fact is that, because the low level catches are not very low, the ratio of the maximum to the minimum number caught over a period of nearly sixty years is only 5 for the cod and 3.9 for the haddock. The increases during the late 1910s and early 1940s reflect real increases in density, following the reduced absolute catches during the wars. Such small-scale fluctuations over a long period are rare in other animals.

Another point worth noting in Fig. 3.11 is the fact that the fluctuation is not a rapid alteration of high and low levels producing saw-shaped lines, nor a sporadic type, but instead it forms gentle waves over many years. This tendency is more clearly seen in the case of the sardine *Sardinella* shown in Fig. 3.12. This population fluctuated to a greater extent than the populations of the two species just mentioned but its fluctuation formed a gentle wave over many years.

As mentioned many times, marine fishes lay a colossal number of eggs; naturally their eggs are small and the young hatched from them are small and fragile, drifting in the sea as plankton. Their survival-rate is strongly affected by the changes in currents and other conditions of the sea. Why

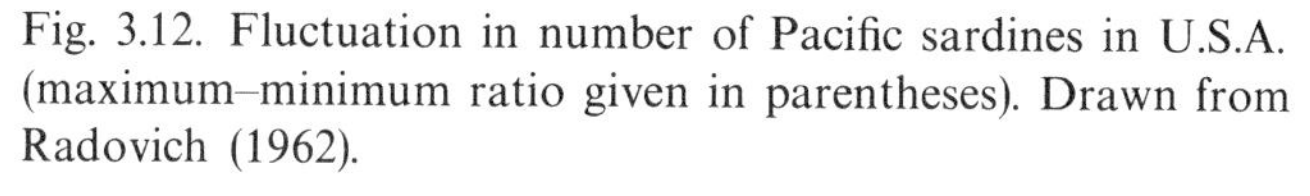

Fig. 3.12. Fluctuation in number of Pacific sardines in U.S.A. (maximum–minimum ratio given in parentheses). Drawn from Radovich (1962).

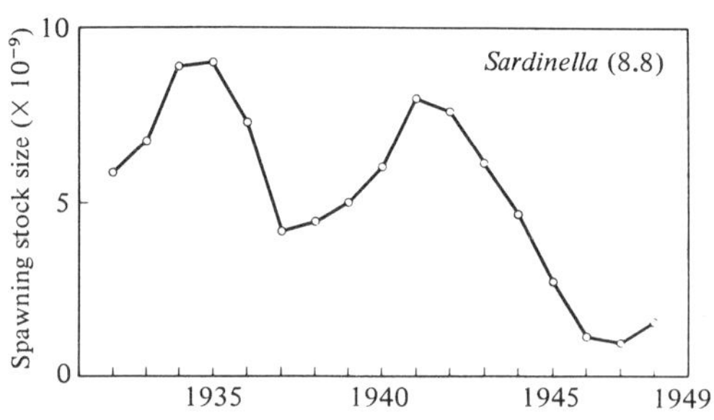

is it then that the fluctuation of the adult numbers is so small in amplitude?

One of the reasons for this is that, as mentioned in Chapter 2, though early mortality is high once they start breeding the mortality becomes low and they live many years to repeat spawning. The catch reflects the total number of adults of all age classes but, if the survival of young is high in one year and the prominent cohort appears in the population, it will continue to occupy a large proportion of the catches in several subsequent years. The influence of such a cohort would cancel fluctuation of new additions to the catchable population (see Fig. 2.8, p. 58; the three-year-old cohort of 1907 became abundant in 1908 and remained prominent until 1919 when the cohort was fifteen years of age).

As an example let us examine the Pacific herring *Clupea pallasii*, which is famous for the effect of lasting prominent cohorts. Fig. 3.13 shows yearly fluctuation of the spring catches of herring in Hokkaido and the number 'emerging' as adults in each cohort. Here the number emerging refers to the number caught of a particular cohort over the period of records. For example, the cohort born in 1907 starts appearing in the catch in 1910 and all individuals of this cohort caught in this and subsequent years are added to obtain the number emerging (age-determination techniques are well advanced for fishes, using year rings on scales, etc.). This would give us a *minimum* number of individuals born

Fig. 3.13. Fluctuations in spring catches of the Pacific herring *Clupea pallasii* in Hokkaido and of recruitment (total catch of each cohort over the period of records) (maximum minimum ratios in ten year periods are given in parentheses). Drawn from Ishida (1952).

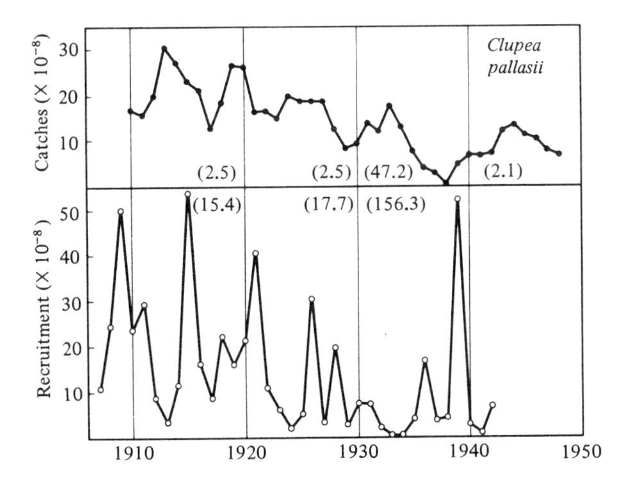

in 1907, which attained the catchable size (of course in reality the number would be much higher because of the natural mortality from year to year not taken into consideration). Then, compared with the fluctuation of total catches, the amplitude of fluctuation of the number emerging is far greater. In the figure the ratio of the maximum to the minimum is given for each ten-year period. Except for the 1930–39 period in which a very small catch was recorded, the amplitude in terms

Fig. 3.14. Fluctuations in the catches of the Japanese sardine *Sardinops melanosticta* along the Pacific coast between Fukushima and Chiba Prefectures (upper graph) and of the number of age II fish (about two years of age) in the catch (lower graph) (maximum–minimum ratios are given in parentheses). Drawn from Nakai (1962).

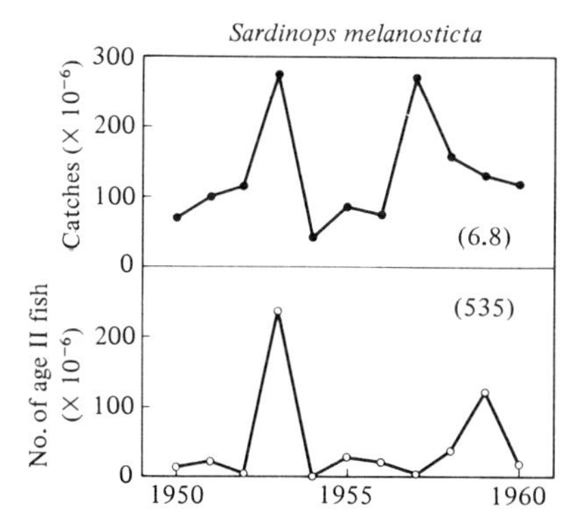

Fig. 3.15. Fluctuations of adults (upper graph) and young caught three years later (lower graph) of the haddock *Melanogrammus aeglefinus* at Georges Bank (drawn from Ricker, 1954). The catch is expressed in weight of fish caught per day by an average-sized boat. (Maximum–minimum ratios are given in parentheses).

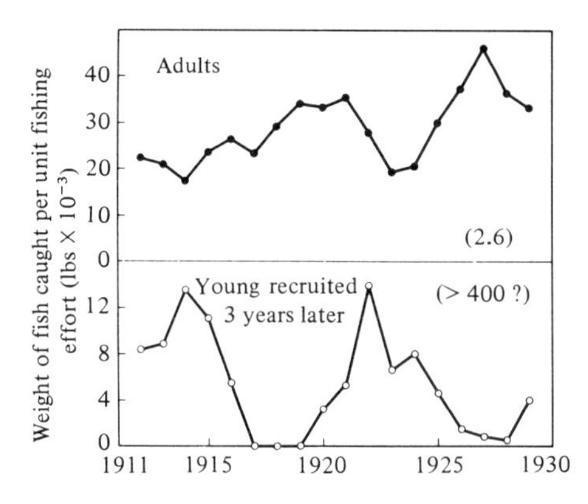

of this ratio was 2–2.5 for the total catches and 15–150 for the numbers emerging (the number reaching catchable size).

The amplitude of fluctuation in the number emerging is probably much smaller than that of young fish. The number of young emerging might be fluctuating at amplitudes hundreds or thousands of times as great as that of adults emerging; unfortunately there is no information on such values.

Similar occurrences are found in other fishes. Fig. 3.14 gives the fluctuation of catches of the Japanese sardine *Sardinops melanosticta* along the Pacific coast between Fukushima and Chiba Prefectures and the numbers of young fish about two years old in the catch. In this case, the number of second-year fish fluctuated with an amplitude of 535 in the above ratio while the total catch varied only 6.8-fold in the same period. In the case of the haddock at Georges Bank (Fig. 3.15) the number of adults caught per unit fishing effort fluctuated within an amplitude of 2.6 in ratio while the number of three-year-old young represented in the catches fluctuated at an amplitude greater than 400 in ratio (read from a graph in Ricker, 1954).

There are very few long-term records of the number of eggs laid by marine fishes. Among them are valuable materials obtained by Watanabe (1970, 1972) on the mackerel in Japanese waters (Fig. 3.16, cited by Tanaka, 1973). As can be seen in the figure the number of eggs increased gradually from about 1950 to 1964 and then decreased. The record of catches corresponding to this period is available for eleven years (1957–67), and the maximum–minimum ratio is only 2.7. In the

Fig. 3.16. Fluctuations in the catches of the Japanese mackerel *Scomber japonicus* per day per fishing boat (upper graph) and of the number of eggs laid (lower graph) (maximum–minimum ratios are given in parentheses). From Tanaka (1973).

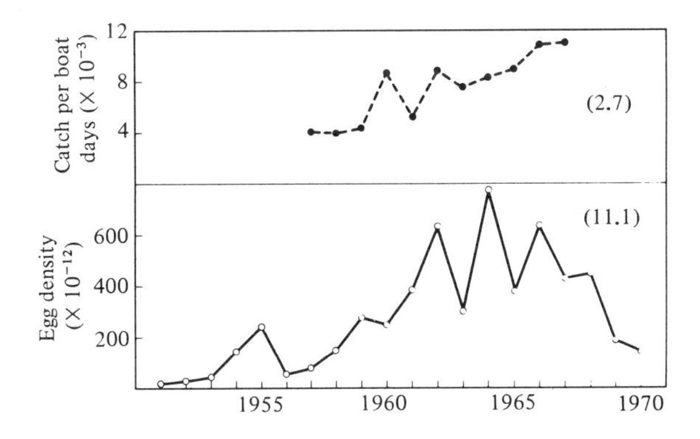

same period the ratio for the number of eggs laid is 11.1. The fluctuation is unexpectedly small in amplitude but since the adult population appears to be stable it may be reasonable that the number of eggs laid fluctuates relatively little and the greatest fluctuation occurs in the number of hatched young.

On the other hand, the salmonid fishes that migrate back to the river to spawn after a fixed period of two, three, four or five years in the sea show population cycles. Fig. 3.17 gives the fluctuation of the number of sock-eye salmon *Oncorhynchus nerka* entering the famous Skeena River in British Columbia. It shows greater fluctuation than the other marine fishes so far discussed. The frequency distribution of the intervals between peak years is inserted in the top right corner of the figure and the broken line over it is a theoretical distribution based on random fluctuations. This shows that the frequency of two-year intervals between peaks is much lower than the expected value whereas the frequencies of four- and five-year intervals are higher than the theoretical values. That is, the fluctuation of the sock-eye salmon population is not random but the probability of having a peak in the fourth or the fifth year is high. This is because most individuals return to the river in their fourth of fifth year. Including such fluctuations deriving from the migratory cycle the ratio of the maximum–minimum numbers reaches sixteen. However, the use of correlogram, a statistical method to extract cycles, did not produce a statistically significant cycle. This is probably because the frequencies

Fig. 3.17. Population fluctuation of the sock-eye salmon *Oncorhynchus nerka* returning to the Skeena River, British Columbia (after Larkin & McDonald, 1968). The inserted histogram shows frequency distribution of the interval of peaks (for explanation see text). (Maximum–minimum ratio given in parentheses.)

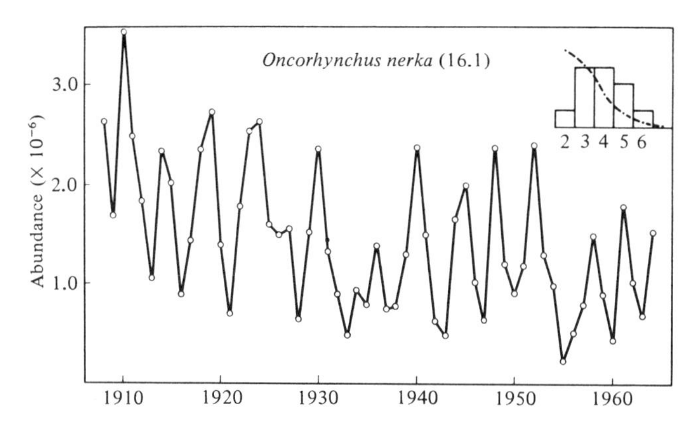

of three-, four- and five-year intervals between peaks were similar. In the pink salmon *Oncorhynchus gorbuska* which has a life cycle of two years, the fluctuation of catches in Japan is close to a two-year cycle (the amplitude of fluctuation is about fifteen in ratio).

What, then, are the factors related to the fluctuation of fish populations? For the number of young emerging there are many instances of correspondence with physical factors such as water temperature, influx of freshwater due to rainfall, winds, waves and currents carrying the young into unfavourable waters or causing mechanical damages, and other oceanographic factors. For example, during the period of low water temperatures off the coast of California, the catch of sardine along the Pacific coast of the United States decreases while that of the anchovy *Engraulis*, which is adapted to cold water, increases. The same can be said for the Japanese sardine. The Japanese sardine had almost disappeared soon after the war and its catch recovered only recently. The cause of the decline is said to be the cold water mass which appeared off the southeast coast of Honshu, destroying the spawning grounds of the sardine.

Changes in the oceanographic conditions can also influence plankton indirectly. According to Dementjeva (1957) the quantity of young herring in the Baltic Sea is influenced by the mortality of juveniles which in turn depends on the amount of plankton. The abundance of plankton is said to be affected by the qualitative changes of sea water due to the influx of

Fig. 3.18. Relation between the stock size and recruitment (young entering the population) for the sock-eye salmon *Oncorhynchus nerka* and the halibut *Hippoglossus hippoglossus* (after Gulland, 1971).

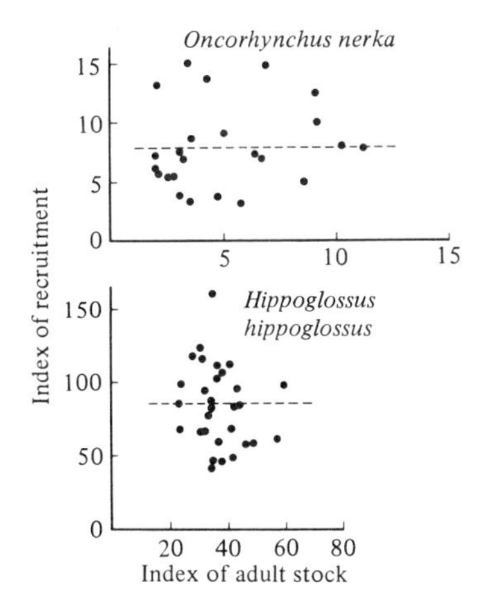

freshwater from the rivers. Similarly, Russell (1936) observed that if oceanic water rich in nutrient salts enters the English Channel it brings about a good catch of the whiting *Merlangius merlangius* through increase in phytoplankton.

Thus when we plot the density of parents and that of those young derived from the parents and entering the catchable population a few years later, no correlation between the two appears (Fig. 3.18). In the examples of the sock-eye salmon and the halibut, the number of young fluctuated greatly around the mean; this was probably caused by oceanic conditions.

However, we have to be careful about the following points in examining this diagram. If the adults always laid the same number of eggs or if the mortality of fry was constant regardless of the density of eggs, then the higher the density-index of adults the greater should be the increase in the density-index of juveniles. The reason that this was not the case in reality might be that the density effect acted on the egg number or the survival-rate or both. In fact, if we plot the number of young per adult fish (an index of survival-rate) on the ordinate we can obtain a line running downward to the right. Such a graph is often cited to show that the fish population is regulated. It is certainly possible that mortality increases as a result of food shortage in fishes that lay such a large number of eggs and that when the adult density is high the growth-rate may be retarded and the egg number may decrease (the sardine along the Pacific coast of Japan is said to have been larger during the period of the poor catch than fish of the same age before or after this period). But it is also possible that the mortality of fry is so high that the relationship between the adult and young densities has disappeared. When the 'survival-rate' is plotted along the ordinate the vertical oscillation of values in the high adult density zone (the right-hand side of such a graph) may become so small in the ordinary graph that it looks as though there were no influence of oceanographic factors on mortality.

At present, the only established case for the lowered survival-rate of young as a result of food shortage in fishes with a large egg number is that of the sock-eye salmon living in an oligotrophic lake (Johnson, 1965). What this fact shows, however, is that there is a case of a fish population with the upper limit being determined at least in part by intraspecific competition for food.

Population fluctuations of insects

(1) The fluctuation in the number of insects has been well studied for applied reasons but here it is the opposite of fisheries science

in that, in contrast to the advancement of analytical studies of mortality factors, there are very few long-term statistics of population fluctuations. The longest record of population fluctuations has been obtained by German forest scientists working on lepidopteran pests of pine forests. The larvae eat pine needles in summer and overwinter in the forest litter as larvae or puape. These have been sampled and the results are collated by Schwerdtfeger (1935a, b, c) (Figs. 3.19, 3.20).

In these figures no regularity of fluctuations can be recognized except for the pine looper *Bupalus piniarius*. In *Bupalus* the outbreaks occurred about every six years between 1881 and 1906 and then eleven years apart

Fig. 3.19. Fluctuations in the number of the pine moth *Panolis flammea* (scale on the top right), the pine hawk moth *Hyloicus pinastri* (scale on the right centre) and the pine looper *Bupalus piniarius* (scale on the left) in a pine forest in central Germany (drawn from Schwerdtfeger, 1935a).

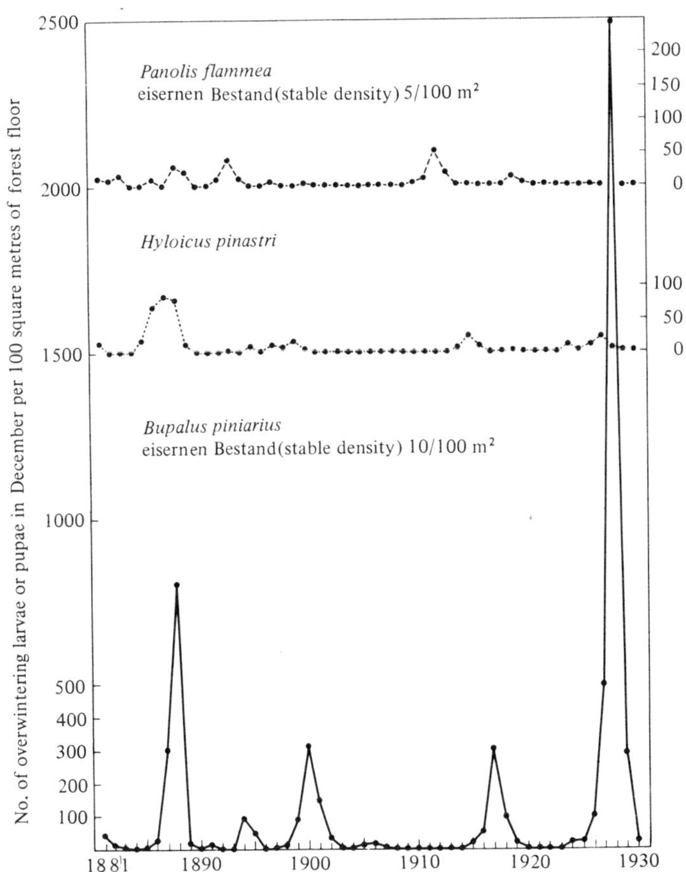

between 1906 and 1928, giving an impression of somewhat regular fluctuations. However, according to a later report (Schwerdtfeger, 1941) the next outbreak occurred in 1936 after eight years.

Another fact one notices in these diagrams is that the amplitude of fluctuations in outbreaks (the maximum–minimum ratio) is extremely large and it usually takes a few years to reach a peak and also a few years to decline. Between the outbreak periods there are small-scale fluctuations which can be seen in a log scale graph given in Fig. 3.21.

Schwerdtfeger called the low density which cannot be plotted in graphs like Figs. 3.19 and 3.20 the stable density (eisernen Bestand). Also the outbreak which appears after a few years of increase over the range of the stable density is called the gradation (Gradation). [For Japanese translations see Miyashita, 1955.]

Table 3.1 shows the characteristics of gradation in four lepidopteran species. The ratio of the maximum to the minimum density is 450 for *Panolis*, 320 for *Hyloicus*, 20 000 for *Dendrolimus* and as much as 32 000 for *Bupalus*.

Fig. 3.20. Fluctuations in the number of the large pine moth *Dendrolimus pini* in a pine forest in central Germany (drawn from Schwerdtfeger, 1935a, and Varley, 1949).

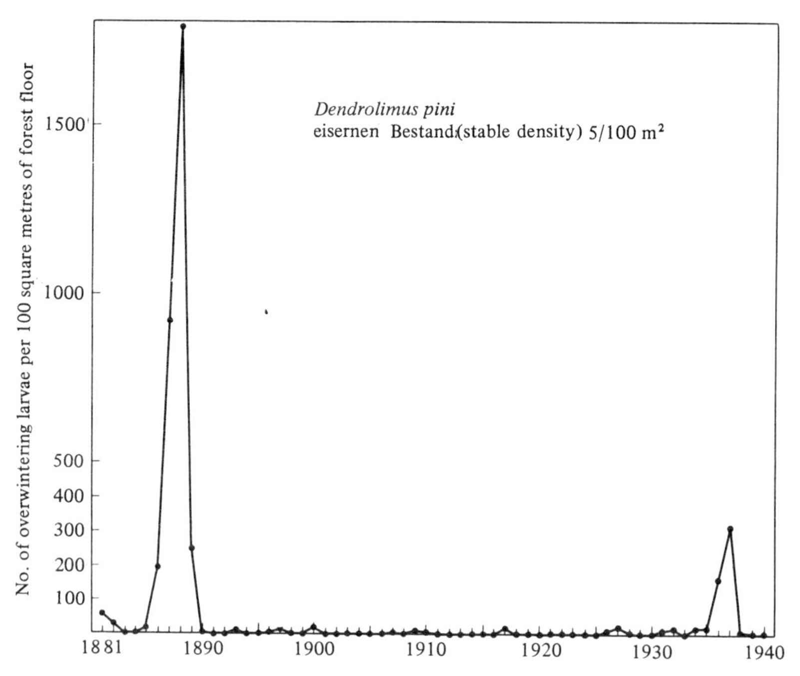

Table 3.1. *Characteristics of the gradations of four species of pine moth; the density is read from the graphs of Schwerdtfeger (1935a) and expressed as the number of overwintering individuals per 100 m^2 of forest floor*

Species	Mean no. of successive years of a gradation	No. of successive years of stable density	Maximum density *A*	Minimum density *B*	Ratio *A/B*
Panolis	6	1–14	45	0.10	450
Hyloicus	7	0–10	38	0.12	320
Dendrolimus	7	0–33	1874	0.09	20 000
Bupalus	7	1–10	2549	0.08	32 000

Such violent fluctuations of *Bupalus* do not occur everywhere. In Holland *Bupalus* is said to fluctuate less than in Germany. Klomp's (1965) result over fourteen years of study showed an amplitude of forty-five in the maximum–minimum ratio in the fluctuation of pupal density. Since random fluctuations can increase the maximum–minimum ratio as

Fig. 3.21. Log scale expression of fluctuations shown in Figs. 3.18 and 3.19 (redrawn from Varley, 1949, based on Schwerdtfeger, 1941). Horizontal lines indicate the upper limits of Schwerdtfeger's 'stable densities'.

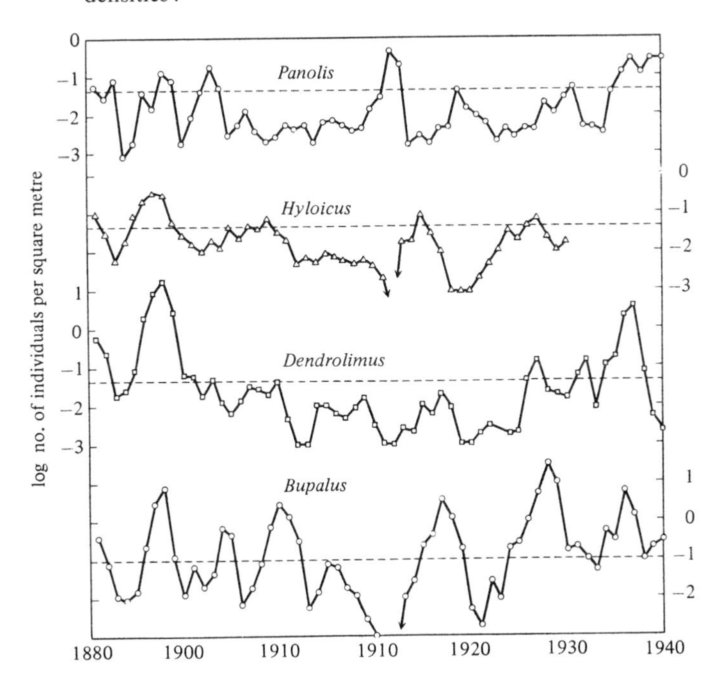

the period of observation increases, Klomp's result may not be comparable to Schwerdtfeger's. However, the *Bupalus* population studied by Schwerdtfeger fluctuated over 1000-fold in any fourteen-generation period. Besides, the reason for the small fluctuation in Holland is not due entirely to low peak densities (the maximum density in Holland was similar to the peak densities of 1888, 1900, 1917 in Germany), but appears to be due to the relatively high minimum densities.

What are the causes of these fluctuations? Schwerdtfeger (1932, 1935a, b, c) made a detailed examination of climatic conditions but failed to

Fig. 3.22. Fluctuations in the number of *Bupalus piniarius* and *Panolis flammea* in different forests in Germany, showing synchronization of gradation (drawn from Schwerdtfeger, 1935b). A–I, different localities.

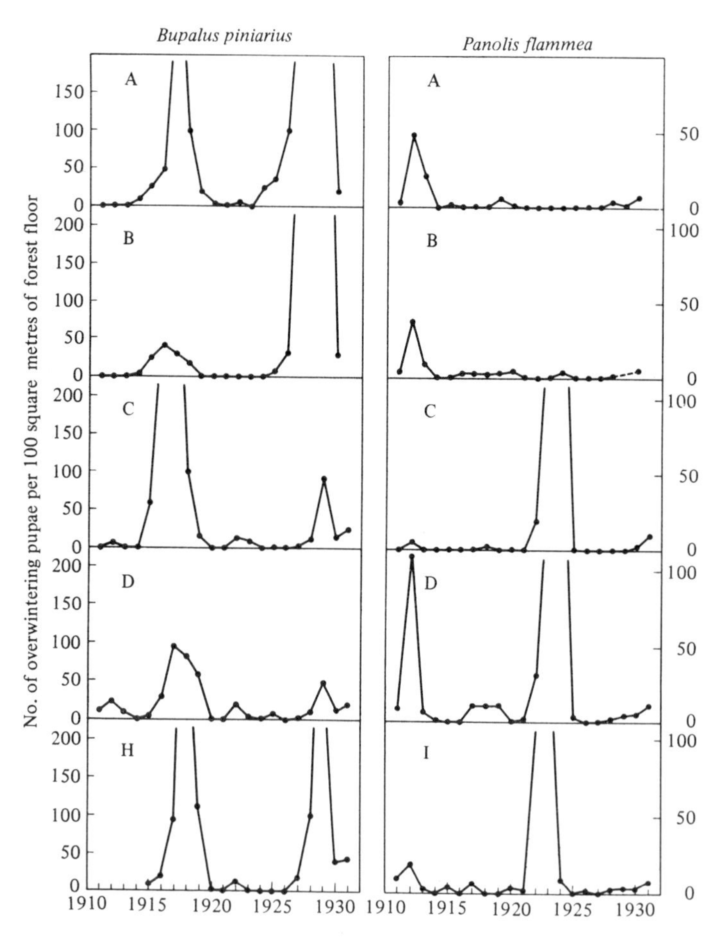

gain clear-cut results. According to him (1932), at the time of the outbreaks of *Bupalus* in 1893–94, 1900 and 1928 the temperatures from July to September (from egg stage to mid-instar larvae) and winter temperatures were higher than in normal years. On the other hand, decline occurred in the years of low winter temperatures. There was no obvious effect of monthly rainfall. The common factor for the increasing phase in the three outbreaks was the higher-than-normal temperatures, in August and September and throughout the larval stages over two to three years. This was also true for *Panolis* (Schwerdtfeger, 1935a). That is, there is evidence that the forest in which this species often irrupts has higher air temperatures from May to September than the forest in which it does not. According to this paper the areas of outbreaks of *Dendrolimus* is limited to the region of high summer temperatures. Friedrichs (1935) also found that *Panolis* in northern Germany outbreaks only in those years in which the continental climate is replaced by the oceanic climate. Holland enjoys an oceanic climate, thus it is warmer than Germany, and this might be responsible for the relatively high minimum densities of *Bupalus* in Holland. Generally speaking the population fluctuation is greater in less favourable habitats.

In Fig. 3.22 one can see that the outbreaks of *Bupalus* and *Panolis* in different forests of central Germany occurred mostly in the same years. As these forests are not continuous only the climatic factors can explain the synchronization. But as Schwerdtfeger (1941) recognized climate does not always cause outbreaks. From 1890 to 1893 winter temperatures were lower than normal but the number of *Panolis* kept increasing. Also the expected outbreaks of *Bupalus* or *Panolis* in warm, dry years, clearly favourable to these species, sometimes did not occur – perhaps because of the activities of natural enemies. Once the gradation has started the increase is not halted by low temperatures. Conversely, no recovery is seen in the decreasing phase in the years of high temperatures.

This reminds us of Voûte's (1946) escape theory explained on p. 109. Once an outbreak is triggered by high temperatures and drought, these lepidopteran populations escape from the pressure of all natural enemies and keep increasing until an epidemic disease spreads or they eat out pine needles and starve. And once they begin decreasing the course may not be reversed by the advent of a favourable climate.

This is considered to be the mechanism of gradation.

(2) The process of gradation starting from the 'escape' is seen, though small in scale, in the fall webworm *Hyphantria cunea* which invaded

Japan from North America after the war and became a pest. The fluctuation in the number of nest-webs found in the cherry trees of avenues in Fuchū City in Tokyo and in the *Platanus* trees of Nishigahara avenues, Kita Ward in the metropolitan area, some 50 km away from Fuchū, showed gradation (Fig. 3.23). The fall webworm usually lays one egg mass per female and the larvae hatched from the same egg mass form a nest-web in aggregation. This species in Canada is also known to fluctuate in number and has gradation with peaks occurring once in about every eight years. It has one generation a year in Canada.

That the patterns of fluctuations in the two areas in Tokyo are very alike suggests that one of the causes for increase or decrease is the climate (see below). As shown in p. 88 the survivorship curve of *Hyphantria* is unique among the moths. Many hatched young die before making the nestweb, but after the nestweb is made mortality is very low until they grow to the fourth instar. Between the fifth and the seventh instar the number is decreased drastically. From experiments using net cover, the main mortality factor in this period was found to be predation by birds for the first generation (additional predation by *Polistes* wasps towards the end) and predation by *Polistes* wasps for the second generation. Itô *et al.* (1969) divided the life cycle into three stages and plotted the number of individuals at the end of each stage against the number of individuals at the beginning of that stage (Fig. 3.24A–C). The relations between the egg stage and the third instar and between the pre-pupal stage and the adult are represented by the regression lines of $b \simeq 1$, but the relation between the third instar and the pre-pupal stage is given

Fig. 3.23. Fluctuations in the number of the fall webworm *Hyphantria cunea* around Tokyo (maximum–minimum ratio in parentheses). After Itô (1977).

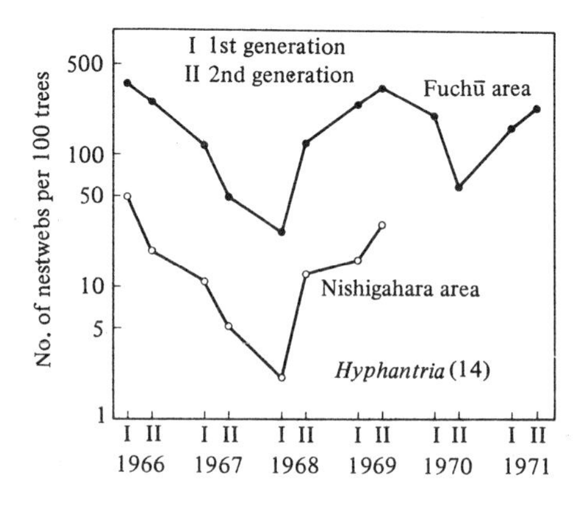

by the line $b \simeq 2$. If mortality occurred at random regardless of the density (that is, if mean mortality did not change significantly with change in density), the angle of the regression line should be 45°, or $b \simeq 1$. [In this treatment there are many statistical problems not mentioned here. These do not concern the discussion of this particular example. For details see Itô & Murai (1977).] The fact that $b \gg 1$ in the period from the third instar to the pre-pupal stage means that the greater the number of third instar larvae the lower (not higher) the mortality of the larvae until the pre-pupal stage. Birds and *Polistes* wasps form the main natural enemies in this period, but they cannot increase in the short time in response to the increased prey density because of the slow speed of reproduction and territorial restrictions. Thus the fall webworm population 'escapes' the limitation by predation.

Therefore, the fall webworm, once it starts to increase over the threshold, increases even faster because of the high survival-rate and, upon reaching a certain level in the downward trend, it decreases even faster. The reason for the gradation is found here. However, the trend must be reversed sooner or later. One of the factors responsible for this reversal is the relation between the density and the number of eggs laid, shown in Fig. 3.24D. [In this diagram the numbers of nestwebs for Nishigahara are multiplied by 10, see Fig. 3.23 and the original paper.]

Fig. 3.24. The relationship between density and survival in each of three developmental stages (A–C) and the density effect on the egg number (D) in the fall webworm in Tokyo district (drawn from Itô *et al.*, 1969, in Itô, 1977). Black circles, Fuchū data; white circles, Nishigahara data; the first and second generations are combined. The data for A–C were taken at permanent stations. The data for D were collected from avenue trees in a wider area.

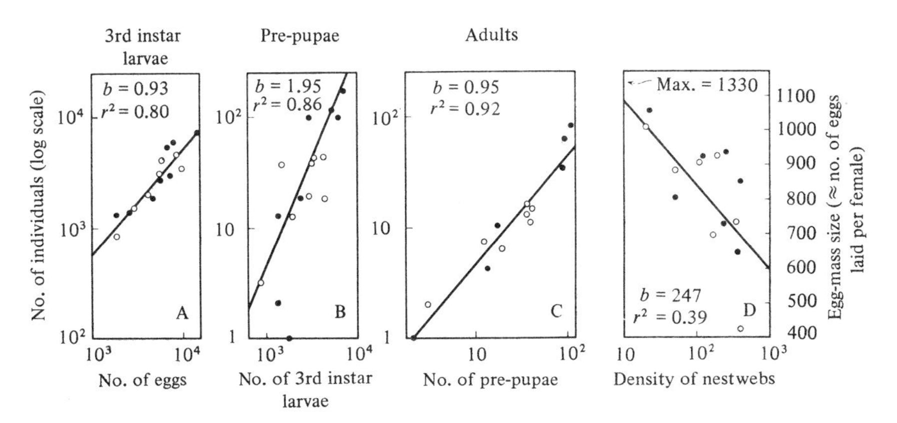

As can be seen from the diagram the number of eggs laid per female (this is usually the same as the mean number of eggs per egg mass) is halved when the density approaches the peak level of Fig. 3.23.

However, this is not sufficient to explain the reversal of the population trend. During the summer this insect suffers high mortality if the temperatures in July and August are high. The outbreak usually begins in a year when the summer temperature is low. Such climatic conditions, in combination with changes in the pest control effort of man, may determine the actual time of trend reversal from increase to decrease or from decrease to increase.

An astounding example of gradation has been studied in Switzerland by Bovey (1966), Baltensweiler (1971), Auer (1968) and others. This is of the larch budmoth *Zeiraphera diniana*, a small moth that attacks larch needles. As can be seen in Fig. 3.25 the amplitude of population fluctuation in this species reaches 35 000-fold and, besides, such large fluctuations occur cyclically eight to nine years apart (8.4 ± 0.4 years). The lower diagram shows the past outbreaks deduced by Baltensweiler from the damage statistics and isolated reports. These fluctuations are

Fig. 3.25. Fluctuations in the number of the larch budmoth *Zeiraphera diniana*. Upper graph, population fluctuation in Zuoz district (drawn from Baltensweiler, 1971) (maximum–minimum ratio, 35 000); lower graph, population fluctuation (lines) for the entire Upper Engadin district and the damage to larches (histogram) (drawn from Baltensweiler, 1964). The broken part of the line is not based on accurate data.

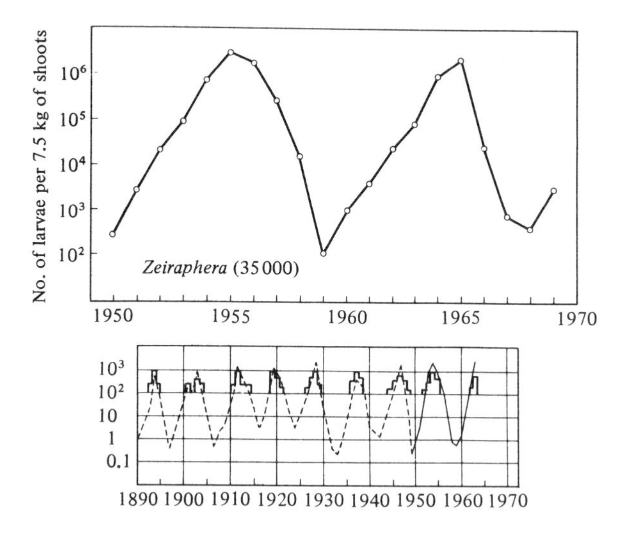

unique among the insects both in the scale of their amplitude and in the accuracy of their periodicity.

From the studies of Auer (1968) and Baltensweiler (1971, 1977), it has become clear that the periodic fluctuation of this species population rises out of the larch–budmoth system itself in the Swiss larch forest. There are neither parasitoids nor predators that can influence the population of this species during five or six years of its increasing phase. The densities of hymenopterous parasitoids do increase with an outbreak of the budmoth but exert no decisive influence. In this species there are two phases in the larvae: the dark-coloured phase and the pale-coloured phase. During population increase the survival-rate is high and hardly any selection operates with respect to the larval phase. The genotype of the dark-coloured larvae with fast metabolic rate and growth increases in the population though the vigour of this phase is low under unfavourable conditions. Increase of this genotype accelerates the population growth. When the larch trees are defoliated by the insects the dark phase is selected against; it dies faster than the light-coloured phase during food shortage. By the time the population reaches its trough, at a density less than 1/10 000 of the peak density, the larch has started its recovery. Such severe infestation would normally kill evergreen conifers but larches are deciduous and more resistant to infestation than the evergreen species. Thus the larch budmoth forms a special system with the host in the extensive larch forests of Switzerland and repeats its periodic increase and decrease in population size. This unique evolution of the system is similar to the relation between the lemming and tundra vegetation (see below). In this sense the nature of the population fluctuation in the larch budmoth, though large in amplitude, is different from that of other pest insects. The peak density of the larch budmoth is a product of the normal population processes of this species. Baltensweiler compared two population cycles in three districts of Switzerland (comparison of density among six peaks and six troughs) and found that, although differences among the minimum densities were in the order of ten-fold, differences among the maximum densities were only about three-fold. (2–6 million per 7.5 kg of foliage). Such densities might form the upper limit that the population can reach without destroying the larch forest.

Other examples of outbreaks that are part of a normal mode of life for the species concerned and are not the result of 'abnormal' emergence are found in the migratory locusts. As mentioned in Chapter 1 the migratory locusts belong to the same species as the ordinary grasshoppers that do not perform group flight. The Asian migratory locust

Locusta migratoria manilensis, which performs group flight over China and the Philippines, is nothing but an ordinary, familiar grasshopper in Japan, where it remains in the solitary phase (phase *solitaria*) and does not march or fly in groups. The migratory locust of Central Asia is another race of this species, which is also an ordinary large grasshopper in Europe, and used to be called *Locusta danica*. When the density of these grasshoppers becomes very high, the interference among individuals causes a change in the body colour from green to dark brown. The nymphs form dense groups and adults aggregate and lay eggs in concentration. As a result the aggregation becomes denser and the interference of individuals is intensified in the following generation. The grasshoppers eventually change into what is called the gregarious phase (phase *gregaria*) (Uvarov, 1921). As mentioned earlier the adults of the gregarious phase have a strong tendency to aggregate and the gonads do not mature unless the locusts fly for a certain period. Thus swarms of tens to hundreds of millions of locusts are formed and migrate to faraway countries, eating all of the green vegetation on their way. What is the cause of the increase in the solitary phase? For a long time it has been said that drought triggered this increase. Ma (1958) and Ma *et al.* (1965) examined the records of outbreaks and meteorological data in China and confirmed that outbreaks and phase variations of the migratory locust on the Chinese continent occur when wet grassland on the delta of large rivers, such as the Hwang Ho, has dried up as a result of prolonged drought and the survival-rate of eggs laid in the soil increases. In the past outbreaks of this species were considered to have periodicity but Ma was not able to demonstrate it. The outbreak disperses gregarious individuals from the devastated grassland to unfavourable land, where the number gradually decreases and the outbreak comes to an end.

Fig. 3.26 gives the flight appearances of the gregarious phase of the famous African species, the desert locust *Schistocerca gregaria*, studied by Waloff (1976). This graph shows the number of territorial units attacked by the locust and not the number of locusts though this may be reflected in the figure. In this graph there is no periodicity of outbreaks. However, the outbreaks are not occurring at random; they occur at some intervals and once an outbreak starts it continues for several years. According to Waloff, this species, like the Asian counterpart, breeds in the delta of large rivers and its increase and phase variation are caused by drought. However, for the formation of exceptionally large groups of more than a hundred million the existence of particular winds that bring together such enormous numbers at one place is considered necessary.

From measurements by aerial photography, etc. the swarms of the gregarious phase are known to contain several thousand million to several billion locusts. Such a magnitude is probably achieved also by other African species such as the red locust *Nomadacris septemfasciata* and the African migratory locust *Locusta migratioria migratorioides*. The number of individuals of the solitary phase in the breeding ground is known for the red locust – about ten million (Symmons, 1966; Stortenbeker, 1967). From this we can obtain the ratio of the density of the gregarious phase to that of the solitary phase. This is in hundreds and is reached after several generations. Considering the fecundity of this insect (not very high) this period is reasonable. The fluctuation of numbers during the solitary phase was studied by Farrow (1975) on the African migratory locust and was found to be about 4.6 in the maximum–minimum ratio of adult density per ha over four years (minimum 2350, maximum 10 900). The rate of yearly fluctuation N_{t+1}/N_t (t = year) was 0.3–3.8. The population of grasshoppers that does not change phase is also generally stable. The amplitude of fluctuation in the ratio of the maximum to the minimum was found to be 3.2 for *Chorthippus brunneus* and 5.5 for *C. parallelus*, both studied in England over five years by Richards & Waloff (1954), and 9.3 for a Japanese species, *Parapleurus alliaceus*, studied over five years by Nakamura *et al.* (1975.)

Our last example of gradation is taken from the Hymenoptera. This is the European spruce sawfly *Diprion herciniae*, which invaded Canada in the 1920s. There is a detailed report on this species by Neilson & Morris (1964), but Fig. 3.27 gives the data up to 1969, which have been made

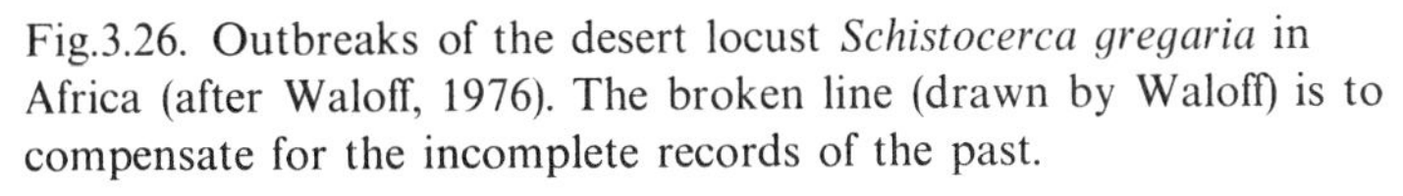
Fig.3.26. Outbreaks of the desert locust *Schistocerca gregaria* in Africa (after Waloff, 1976). The broken line (drawn by Waloff) is to compensate for the incomplete records of the past.

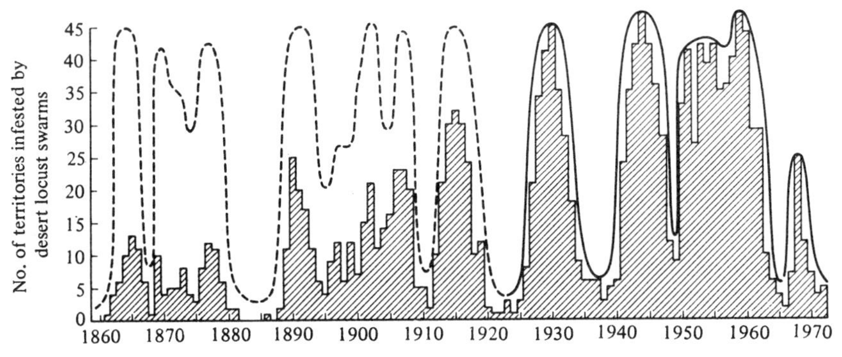

available by Dr M. M. Neilson and Mr D. Elgee. The fluctuation in numbers of this species is finer than in other examples discussed but it is still a gradation type and the amplitude, including the period of DDT application and immediately after, is 180 in the maximum–minimum ratio for the first generation and 360 for the second generation. The figure also shows the percentage of individuals killed by diseases and parasites. The particularly high death-rate around 1940 was caused by the polyhedrosis virus of this species introduced accidentally from Europe and spread among the host, which had a high density at the time. This disease terminated the outbreak of the late 1930s and played a role in bringing down the number in the outbreak of 1945. A hymenopterous parasitoid and a parasitic fly were introduced from the native country. After their establishment around 1945 the parasites began to increase before the sawfly reached its peak in the outbreak and showed a high rate of parasitism during the decreasing phase of the sawfly population.

Fig. 3.27. Population fluctuation of the European spruce sawfly *Diprion herciniae* and the percentages of those killed by diseases and parasites in New Brunswick, Canada (Maximum–minimum ratios are given in parentheses). (Data from plots 1 and 2, by courtesy of Dr M. M. Neilson and Mr. D. Elgee, Forest Research Division, Department of Environment, Canada).

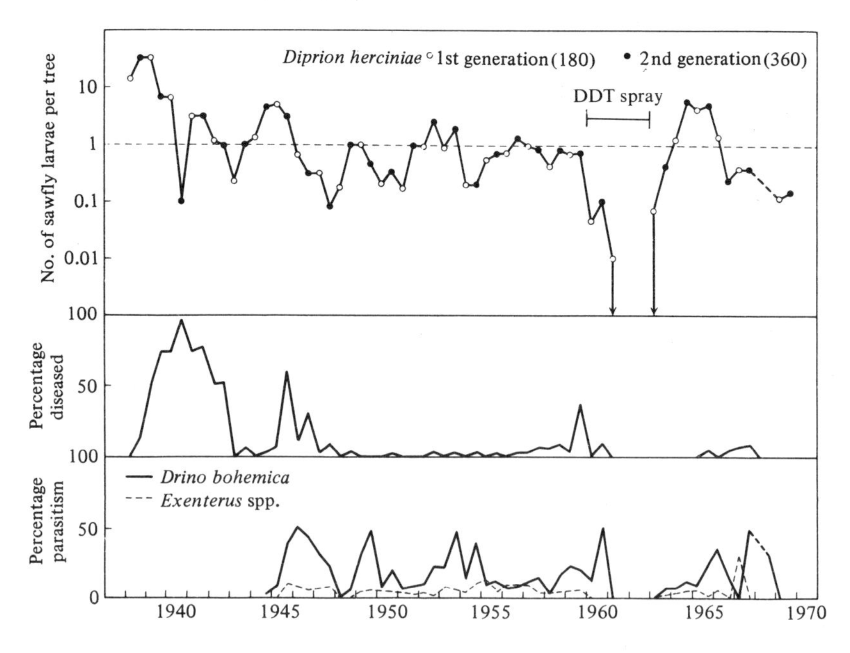

Thus both the parasites and the disease formed density-dependent mortality factors.

Apart from the above examples, an example of gradation is seen in the black-headed budworm *Acleris variana* which has been investigated by Miller (1966). This moth had two periods of gradation between 1947 and 1964 and their maximum–minimum ratios were 220 and 100 respectively. The population increase appeared to be triggered by high temperatures during the growing season in several successive years, but the decrease was caused by an increase in the number of parasitic wasps.

(3) In Japan the number of moths in the rice paddy, particularly the rice-stem borer *Chilo suppressalis* and the paddy borer *Trypolyza incertulas*, has been studied using light traps (forecast lamps) since the Meiji period. Ecological analysis of the forecast lamp data has been made by Miyashita (1955, 1963). The population fluctuation is plotted in Fig. 3.28 for *Chilo* and in Fig. 3.29 for *Trypolyza*. [In *Chilo suppressalis* (called Nikameiga in Japanese, meaning the moth with two emergences) the first emergence of adults is derived from the overwintered larvae of the second generation of the previous year and the second emergence of adults (the first generation adults) may have entirely different densities (outbreaks often caused by one or the other) and if these were plotted in actual numbers on the same graph the fluctuation would be comb-shaped. To avoid this, Miyashita (1955) expressed the fluctuations in the deviation from the mean density of adults killed at the light traps for each emergence group. In Fig. 3.28 this method was adopted. Since the Meiji period new types of lamp have been devised for attracting moths but for the forecast work the same white electric bulb has been used for comparison.] The numbers killed at the light traps are considered to express the relative densities of the two species in the rice paddy fairly accurately. For example, in Fig. 3.28 the result for Okayama Prefecture shows a peak density of the second emergence adults in 1931, which corresponds to the peak year of larvae obtained independently in the same Prefecture (see Fig. 3.30). [In these pest insects there is a big loss to the population each year due to transplanting and harvesting of the host plants even without the control measures being applied. Thus they differ from the forest pest insects.] In both species the number is fluctuating more finely than in other species examined so far. In the rice-stem borer outbreaks occasionally resemble a gradation in having few successive generations of increase or decrease but in the paddy borer the outbreaks are normally completed in one or two generations. I would call these out-

Fig. 3.28. Annual fluctuation of the moth (adults of the rice-stem borer) *Chilo suppressalis* killed at light traps in different Prefectures in Japan (from Miyashita, 1955, and additional information). White circles, the first emergence adults; black circles, the second emergence adults.

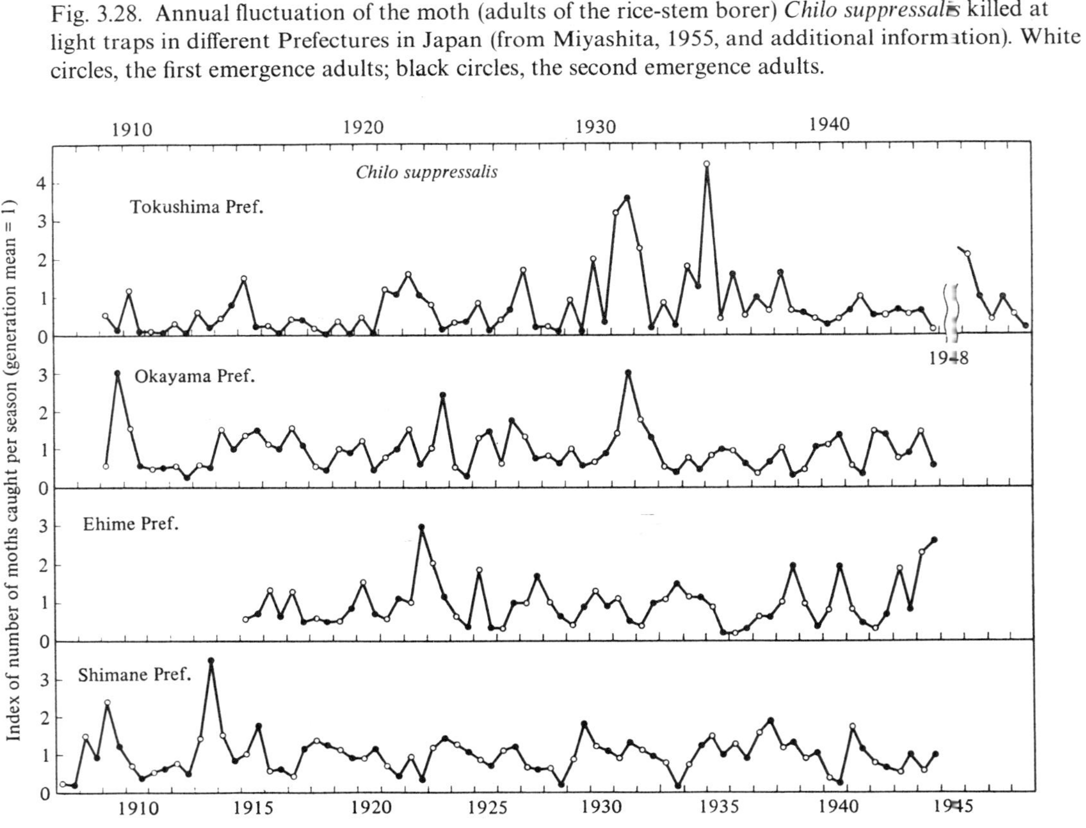

Fig. 3.29. Annual fluctuation of the moth (adults of the paddy borer) *Trypolyza incertulas* killed at light traps in different Prefectures in Japan (data mainly from Ishikura & Nakatsuka, 1955). White circles, the first emergence adults; black circles, the second emergence adults; circles with diagonal line, the third emergence adults.

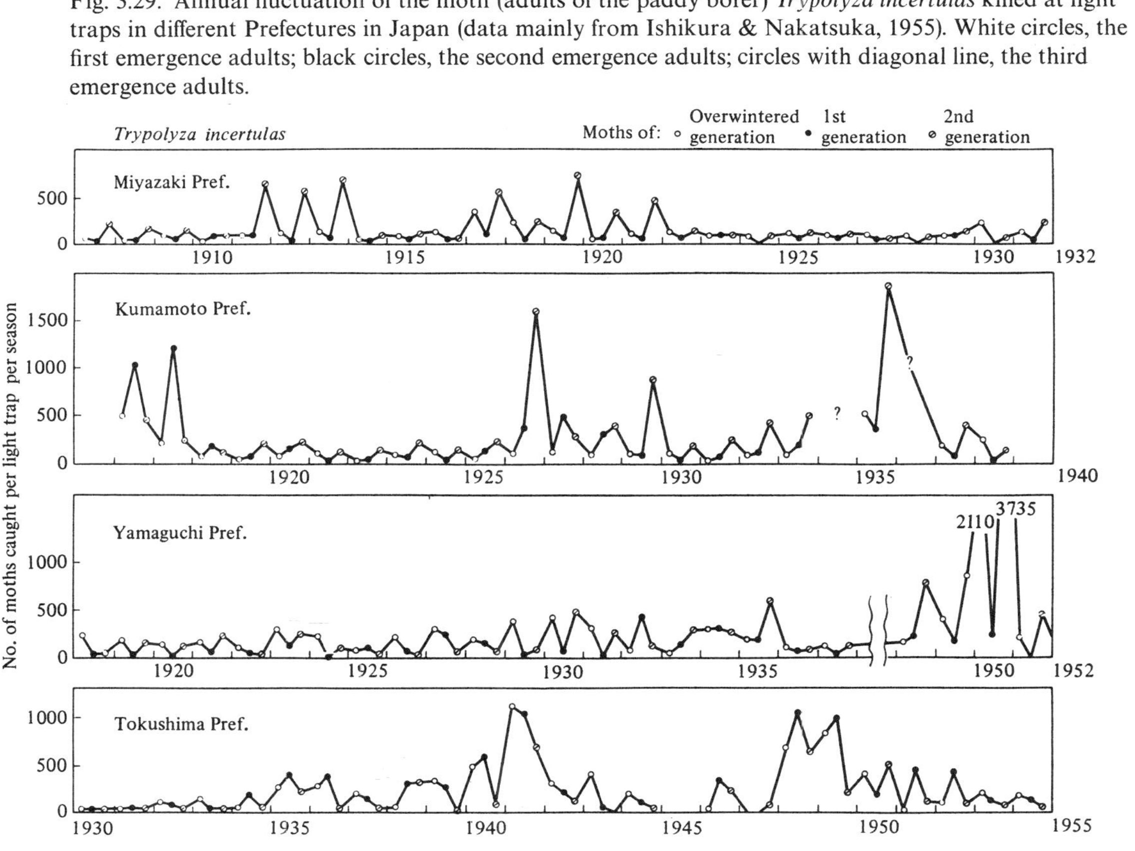

breaks the sporadic type of outbreaks. It is difficult to read the amplitude of fluctuations directly from the graph of the rice-stem borer, but in the case of the second emergence adults in Tokushima Prefecture the amplitude would be about 150 in the ratio of maximum to minimum while in Okayama Prefecture where it was small the ratio would be about seven. Fukaya & Nakatsuka (1956) tabled a record of the number of moths killed at the forecast lamps. Table 3.2 gives the maximum–minimum ratios of moth numbers calculated from their record which covered a period of at least ten years before the end of the war. In this table the amplitude in the maximum-minimum ratio for the rice-stem borer ranges from 3.7 to 83, being generally low (in Aomori Prefecture, not shown here, there was a record of fluctuation over nine years with an amplitude of 217). On the other hand, the amplitude of fluctuation in the paddy borer calculated from the data of Ishikura & Nakatsuka (1955) is large, reaching 111 for the first generation in Kumamoto Prefecture and 260 for the second generation in Miyazaki Prefecture. This seems to indicate that fluctuations are of the sporadic type and that the paddy borer is more strongly affected by climatic factors than the rice-stem borer, for it is generally accepted that the population fluctuation on the fringe of distribution where conditions are unfavourable to the species concerned is greater than in the centre of distribution (Richards & Southwood, 1968; Watt, 1968) and Japan lies at the northern limit of distribution for the paddy borer.

In the past egg masses were collected in the paddy as a measure of control. There is a valuable record of the number of egg masses collected over ten years in a district of Fukuoka Prefecture, which has been cited by Ishikura & Nakatsuka (1955). According to this, the amplitude of fluctuation for the egg mass of the rice-stem borer was 7.4 in ratio whereas that for the paddy borer, excluding the period of four years in which the number was reduced as a result of change in the rice transplanting season, was 40.

In the past the causes of outbreaks in the rice-stem borer and the paddy borer have been explained in terms of weather. Irregularity of irruption and very high early mortality in both species suggest the importance of climatic factors. In fact, *the cause of the beginning* of the outbreaks seems to be explicable generally in terms of climatic conditions.

The outbreak of the rice-stem borer often occurs when June and July temperatures are low. In the examples of Fig. 3.28 the outbreaks in Okayama Prefecture in 1923 and 1930–31, in Ehime Prefecture in 1922–

Table 3.2. *Fluctuation in the number of moths caught by light-traps (forecast lamps) in pre-war Japan*

(a) Rice-stem borer, *Chilo suppressalis*

Locality[a] (Prefecture)	Ibaraki		Gumma		Nara[b]	Shimane[b]	Okayama		Nagasaki		Miyazaki	
Generation	2	1	2	1	1	2	2	1	2	1	2	1
Max. no. (A)	3955	17 129	3668	981	454	156	9746	1103	5211	474	6979	500
Min. no. (B)	337	965	933	37	13	42	785	130	824	98	382	6
Ratio (A/B)	12	18	3.9	27	35	3.7	13	8.5	6.3	4.8	18	83

(b) Paddy borer, *Trypolyza incertulas*

Locality[a] (Prefecture)	Kumamoto			Saga			Miyazaki			Yamaguchi		
Generation	3	1	2	3	1	2	3	1	2	3	1	2
Max. no. (A)	562	1219	1806	281	369	1383	415	129	1303	425	433	582
Min. no. (B)	14	11	72	15	22	181	12	(2)[c]	5	43	9	9
Ratio (A/B)	40	111	25	19	17	7.6	35	65	260	9.9	48	65

Calculated from the light-trap records cited in Fukaya & Nakatsuka (1956) and Ishikura & Nakatsuka (1955).

[a]Light-traps were set at the Prefectural Agricultural Experimental Station or its vicinity.

[b]Records for the 2nd generation of Nara and the 1st generation of Shimane are missing.

[c]This excludes zero value obtained in one year. The next lowest value of two was used in the calculation of the ratio.

23 and in Shimane Prefecture in 1908–09 and 1931 occurred in the years when June or July or both had temperatures lower than normal (Miyashita, 1955). Also, the outbreaks recorded in Aichi Prefecture (around Anjō) in 1948–49 and in Kagawa Prefecture in 1946–47 were associated with low temperatures in May, June and July in the first year. An unusually low temperature for early summer leading to an outbreak can be explained from the life cycle and physiology of this species. That is, the low temperatures in May–June will retard emergence and will later increase the number of eggs laid in the paddy after the transplantation of rice seedlings (this usually takes place in May–June). Thus the enormous loss of eggs and larvae that would normally accompany rice transplanting may be avoided. Also, the larval mortality usually increases in July with increase in the temperature of rice fields. The low temperature in this month may raise the survival-rate of the larvae, resulting in an increase of first-generation adults. According to Fukaya & Nakatsuka (1956) the extensive outbreak of 1931 over the whole of Japan and the outbreak of 1947–48 in western Japan were accompanied by a prolonged wet season and low temperatures in July.

However, such low temperatures during early summer in the years following an outbreak did not cause another outbreak; e.g. Aichi Prefecture 1950, Kagawa Prefecture 1948–49. Considering this fact Miyashita (1955) cast doubt about the idea of weather determining the amount of emergence. For example, when the second-emergence adults (the first-generation adults) are abundant or at a peak in abundance there is a tendency for the first-emergence adults of the following year (derived from the abundant second-emergence adults of the previous year) to be greater than the first-emergence adults of the previous year. That is, once the first generation irrupts by getting favourable conditions this tends to carry over to the next generation.

As far as the *increase* of the population is concerned, abundance can be explained in terms of climate, but the fact that climatic factors fail to account for the decrease of the population or for the low density suggests the importance of density-dependent, biotic factors acting on the abundance of this species.

Harukawa *et al.* (1934) conducted a detailed survey of densities of the rice-stem borer and the rate of parasitism by its larval parasitoid *Temelucha biguttula* in Okayama Prefecture. There was an outbreak of this pest during the study period and a conspicuous increase in parasitism was found during the outbreak (Fig. 3.30). This was confirmed by Miyashita (1955) who found a positive correlation between the rate of

Fig. 3.30. Fluctuation (including an outbreak) of *Chilo suppressalis* and percentage parasitism (bars) by an ichneumonid larval parasitoid, *Temelucha biguttula* (from Harukawa *et al.*, 1934, in Itô & Miyashita, 1956).

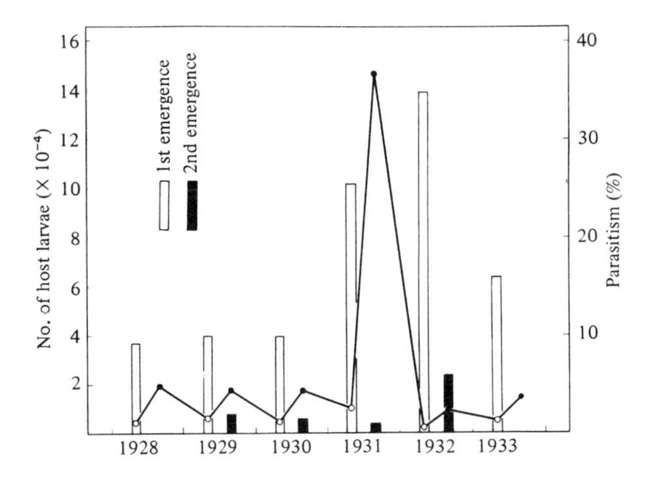

Fig. 3.31. Fluctuations in the number of *Trypolyza incertulas* in three areas of Ehime Prefecture (drawn from light trap records of Ehime Pref. Agr. Exp. Sta.).

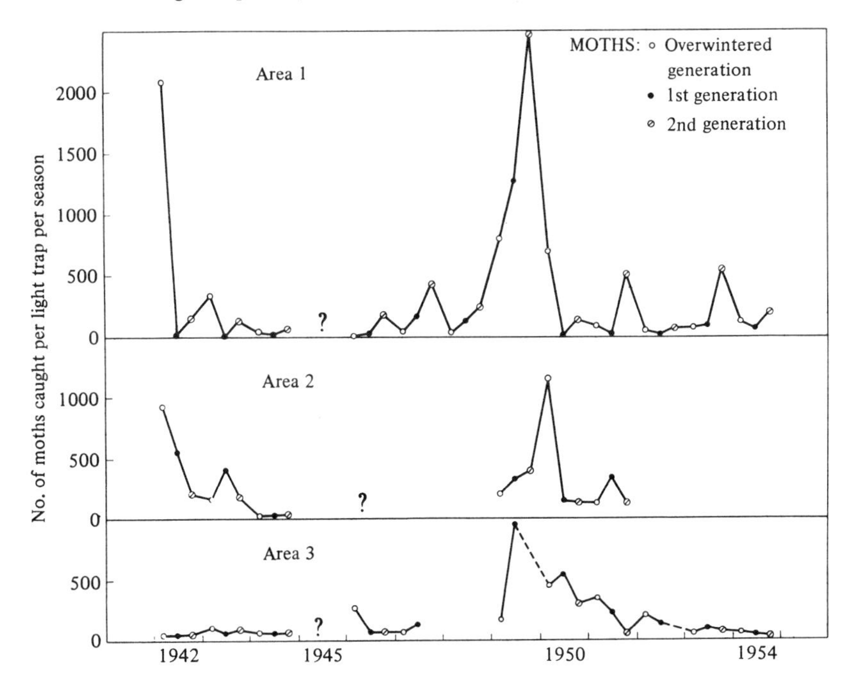

parasitism and the number of adults belonging to the corresponding generation killed at light traps in a study of egg parasitism over ten years at the Agricultural Experimental Station of Ōita Prefecture. Miyashita (1963) also reported a case of reduction in the number of eggs in the ovaries from 200 to 50 (down to one-quarter) when the number of adults killed at the light traps increased from 0.05 to 1.7 (thirty-five times) in the deviation from the mean.

The cause of outbreaks in the paddy borer has been found mainly in weather and the conditions of rice cultivation. If the temperature is low during the overwintering period then the number of first-emergence adults coming to the light traps decreases. Fig. 3.31 shows synchronization of fluctuations of this species in different parts of Ehime Prefecture, suggesting that the outbreaks were caused by climatic factors.

(4) Continuing with pest insects of rice plants, Fig. 3.32 is a graph constructed by Miyashita (1963), showing fluctuations of the whiteback planthopper *Sogatella furcifera* in Ōita Prefecture. This species arrives in rice fields in spring and produces three to four generations in the paddies. In summer females of the brachypterous form appear and breed while in autumn the proportion of the macropterous females increases, only to disappear when the rice is harvested. As it is, in reality, difficult to separate the adults of each generation in the forecast lamp data, Miyashita grouped the numbers of moths killed into the 'summer

Fig. 3.32. Fluctuation in the number of the whiteback planthopper *Sogatella furcifera* killed at forecast lamps in Ōita Prefecture (after Miyashita, 1963). While circles, 'summer generation'; black circles, 'autumn generation'. The index of abundance is given by the ratio of the number killed to the mean number (designated as 1) for each generation. (Maximum–minimum ratio given in parentheses.)

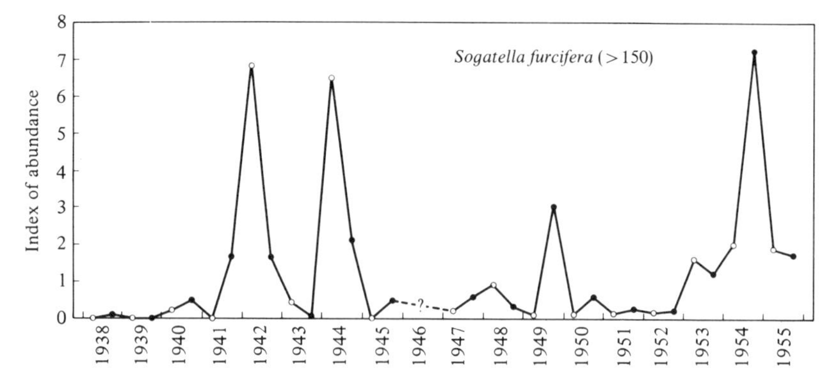

generation' and the 'autumn generation' and expressed population fluctuation in terms of the deviation from each mean. The result showed an amplitude of fluctuation of more than 150 in the maximum–minimum ratio. Although some outbreaks lasted for two to three generations a very high density was also observed in a single generation. The same trend was obtained in the data collected in Kumamoto Prefecture and Ehime Prefecture, which were cited by Miyashita. His analysis showed that the 'summer generation' outbreaks often occurred in the years in which sunny days were few and temperatures were low in April and May while the 'autumn generation' outbreaks often occurred in the years in which June and July temperatures were low. Since early days planthoppers have been known to irrupt in the years of low temperatures, but this cannot be a general rule because there have been outbreaks in a few high-temperature years as well.

Fig. 3.33. Population fluctuations of the brown planthopper *Nilaparvata lugens* (black circles) and the green rice leafhopper *Nephotettix cincticeps* (white circles) in the rice paddy of Kyūshū Agricultural Experimental Station (drawn from Kuno, 1968, and unpublished data of Kyūshū Agr. Exp. St.). Patterns of increase from the arrival (0 generation) to the peak generation are shown in log scale in inserts. (Maximum–minimum ratios are given in parentheses.)

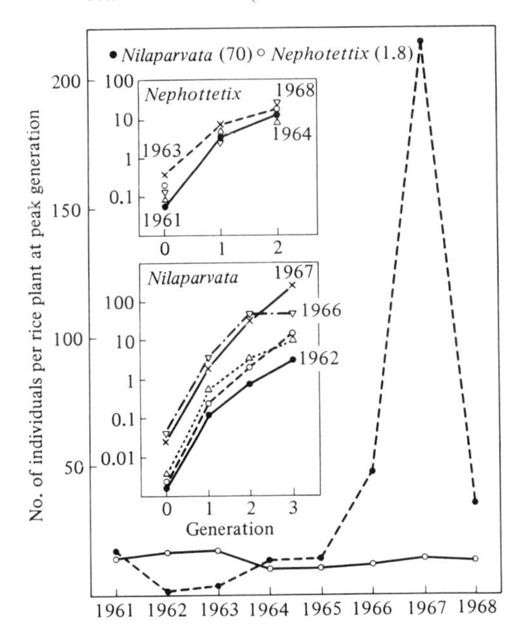

Violent fluctuations of numbers as seen in *Sogatella furcifera* are also found in the brown planthopper *Nilaparvata lugens* (Fig. 3.33). Kuno (1968, and unpublished) investigated this species, using a sucking machine to collect planthoppers from single rice plants. Different generations were separately treated in the analysis, thus the result was more accurate than the forecast lamp data for depicting population fluctuations. In Fig. 3.33 the peak generation (the third generation in rice paddies) of this planthopper fluctuated seventy-fold in eight years of study.

Another small plant-sucking insect in the rice paddy is the green rice leafhopper *Nephotettix cincticeps*, whose population fluctuation is entirely different. The peak generation of this species (the second generation in the rice paddy) is almost constant from year to year and in the eight years of study the maximum–minimum ratio was only 1.8 (Fig. 3.33). This fact confirms the intuitive impression of field workers in different districts that the density of this leafhopper at the heading season of rice plants is remarkably constant from year to year.

How, then, does such a difference arise between the planthopper and the leafhopper? In the insertions of Fig. 3.33 the population growth following the arrival of adults in the rice paddies in spring (generation zero) is plotted in log scale. The abundance of the generation zero varies from year to year, but in the case of the leafhopper if the number arriving in spring is small its reproductive rate is high whereas if this number is large its reproductive rate is low. Thus the variation of the second-generation population becomes much smaller than that of the number arriving. On the other hand, in the case of the brown planthopper if the number arriving in spring is large the number in the autumn is also large and if it is small the autumn population is also small. The brown planthopper shows little regulation (only in 1966 is such a trend noticeable), and the number fluctuated independently of density, largely determined by the number arriving in spring. In the rice leafhopper, it appears that some regulatory mechanism is operating within the population to normalize its density each year (this species overwinters in fallow fields etc.).

What is the mechanism of regulation in this species? In spite of the detailed field studies conducted by Kuno & Hokyo (1970) and Kiritani *et al.* (1970), this is still not clear. What we can say is that at least regulation is not likely to be achieved by parasitic wasps or diseases. Some form of interference among individuals within the species is suspected to be the major controlling factor.

As mentioned in Chapter 1, the Japanese populations of the brown planthopper originate in the Chinese continent. They fly across the sea to Japan in early summer and if this number is great an outbreak is imminent. Lack of intraspecific density regulatory mechanisms may be tied to such a high-fecundity strategy. It is known that among the planthoppers and leafhoppers that damage rice plants in Japan the brown planthopper lays the largest number of eggs.

The whiteback planthopper has also been suspected to arrive from China by Kisimoto (1971) though this species may overwinter in Japan as well. The years of outbreaks shown in Fig. 3.32 may have been those in which large numbers of these insects have flown across to Japan (two of the major outbreaks coincided with outbreaks of the brown planthopper). If this was the case, the fact that major outbreaks followed low temperatures and many cloudy days in early summer might indicate a greater frequency or strength of the humid air-masses that would bring the planthoppers from the continent in these years.

Kiritani and his collaborators (Kiritani, 1964, 1971; Kiritani & Hokyo, 1962; Kiritani *et al.*, 1963, 1967) studied the population dynamics of another rice plant pest, the southern green stinkbug *Nezara viridula* in the warmer region of Japan. They studied a population in Wakayama Prefecture at the northern limit of its distribution, where most adults with no habit of diapause died in cold winters. Thus the fluctuation in the number of adults immediately after overwintering was great, the maximum being thirty-six times the minimum in six years. Yet the amplitude of fluctuation in the number of third-generation adults arriv-

Fig. 3.34. Population fluctuation of the whitefly *Aleurotrachelus jelinekii* (maximum–minimum ratios are given in parentheses). Drawn from Bush C data of Southwood & Reader (1976).

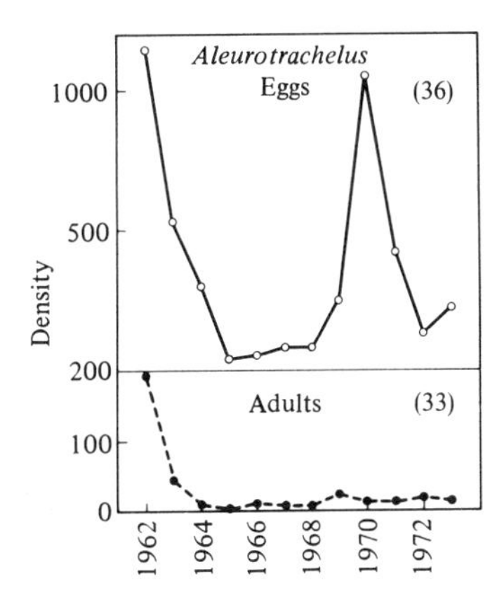

ing at the wintering sites in autumn was only five in the maximum–minimum ratio. This must be a result of density regulation taking place from spring to autumn, but as in other species the mechanism of regulation in this species is still unknown.

Among the plant-sucking insects studied in other countries is the whitefly *Aleurotrachelus jelinekii* which parasitizes a bush, *Viburnum tinus*. This insect has one generation a year and fluctuates in number (Fig. 3.34). According to Southwood & Reader (1976), who studied this species, no density-dependence was discernible in the fluctuation. Increase in number was brought about by changes in climatic factors and decrease was caused partly by squirrels which damaged the vegetation of parasitized bushes.

(5) On several islands of Okinawa the sugar cane cicada *Mogannia minuta* (= *M. iwasakii*) attained a very high density in the canefield after the war. This cicada used to live in tall grassland (*Miscanthus* field), where the density was not particularly high before the war. They began irrupting in the canefield since about 1965 on Ishigaki Island and since 1967 at Chinen-mura in the southern part of Okinawa Island. The outbreaks of this species are somewhat different from those of other species in that the amplitude of population fluctuation is small from year to year. For example, the following are the numbers of mature nymphs appearing per square metre of surface in an area of irruption at Chinen-mura. The maximum minimum ratio is only 2.5.

1971	1972	1973	1974	1975	1976
350	358	230	144	239	152

There are other characteristics in the outbreaks of the sugar cane cicada. At Chinen-mura this cicada is restricted in distribution and within the range the density is very high everywhere while outside the irruption areas there are no cicadas to be found though soil and vegetation are the same as in the irruption area. Thus there is no gentle slope of normal distribution of densities along a section of its geographical distribution. Instead, the form is like a rectangle or a trapezium with very steep slopes (Fig. 3.35). Itô & Nagamine (1974, 1976a, b) thought that one of the reasons for this peculiar distribution was a result of the 'escape' by the population. The important natural enemies of this species are the ants that inhabit the canefield in very large numbers (often several thousands per cane). However, in the irruption area the density of cicadas is so high (some tens of thousands of the first-instar nymphs hatch on a single plant and fall on the ground like snow flakes) that the ants cannot cope with them all. Any offspring of the cicada that venture out of this area, however, will be eaten in no time. Dispersal of adult cicadas is probably limited by the mutual attraction of calls, which might also be contributing to this peculiar distribution. This, however, does not explain the temporal stability of the number nor the constancy of density within the irruption area. M. Nagamine (personal communication) considers that one of the causes might be interference ('fighting') among the nymphs underground and another might be the prolongation of the nymphal stage (normally two years) associated with high density.

The number of cicadas is thought to be stable also in other natural populations. Fig. 3.36 shows the result of censuses of the cicada

Fig. 3.35. A section through an outbreak area of the sugar cane cicada *Mogannia minuta*, plotting the number of exuviae found on the ground and canes (after Itô & Nagamine, 1976a). Black circles indicate the sites with no exuviae. The centre of irruption is at Seifâ-Utaki, Chinen-mura, southern part of Okinawa Island.

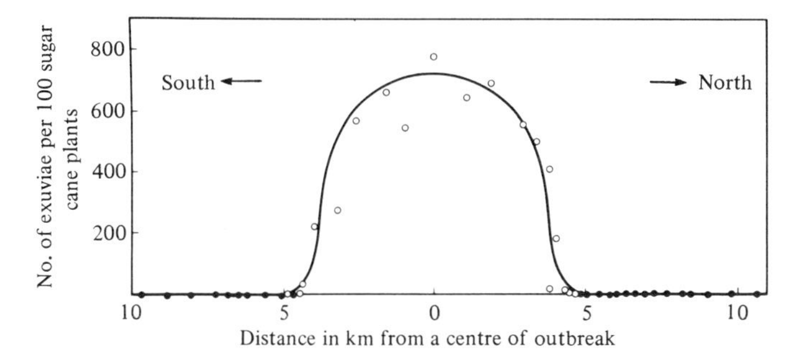

Fig. 3.36. Fluctuation in the abundance of the cicada *Platypleura kaempferi* at a garden in Okayama Prefecture (maximum–minimum ratios are given in parentheses). From Hirose (1977).

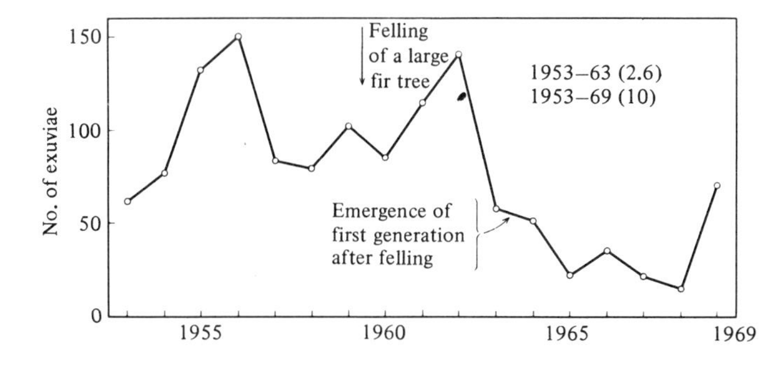

Platypleura kaempferi emerging in a garden in Okayama Prefecture, conducted by Hirose (1977). The garden was surrounded by cultivated fields so that interchange of cicadas between suitable areas was considered to be very limited. In 1959 a large fir tree was felled, which had provided a favourable habitat for cicadas living and ovipositing on it. As the life cycle of this cicada takes five years to be completed the effect of this felling was expected to appear from 1964 onwards. The maximum–minimum ratio of numbers observed over eleven years from 1953 to 1963 was 2.6, which indicated stability of the population. Since these cicadas lay a large number of eggs and suffer high early mortality, the stability of the population may be a result of the low mortality of the young during their long subterranean life.

In contrast the amplitudes of fluctuations in the populations of periodical cicadas in America are almost infinite. The populations of the seventeen-year cicadas *Magicicada septendecim*, *M. cassini* and *M. septendecula* irrupt every seventeenth year while the populations of the thirteen-year cicadas of the south, *M. tredecim*, *M. tredecassini* and *M. tredecula*, irrupt every thirteenth year. In the first three species the number emerging every seventeenth year reaches several hundreds per square metre and they defoliate new canopy leaves on every tree. The numbers emerging in the sixteenth and eighteenth year are negligibly small and from the nineteenth year of one cycle to the fifteenth year of the next no individuals emerge. Of course the nymphs waiting to emerge in the seventeenth year are abundant in the ground. Lloyd & Dybas (1966) who

Fig. 3.37. Fluctuation in abundance of mature nymphs in two species of Zygoptera, *Pyrrhosoma nymphula* and *Enallagma cyathigerum* (after Macan, 1974). Most pond weeds were washed out by the floods of 1966. (Maximum–minimum ratios are given in parentheses.)

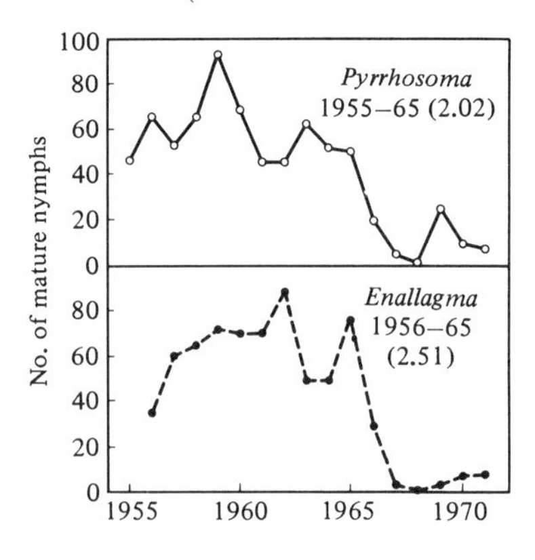

studied these cicadas give the reason for the maintenance of exact periodicity in spite of some variation in the life cycle as that the high densities in the seventeenth (or the thirteenth) year have 'escaped' predation pressure, bringing about the synchronization of outbreaks among different species.

The adult populations of dragonflies which, like cicadas, are Hemimetabola and characterized by high early mortality and low nymphal mortality, are relatively stable. Fig. 3.37 gives fluctuations of mature nymphal populations for two species of Zygoptera (damselflies), *Pyrrhosoma nymphula* and *Enallagma cyathigerum*, found in English ponds. Until 1966, when the pond weeds used as oviposition sites were mostly washed out by floods, the amplitude of fluctuation over ten years had been 2–2.5 in the maximum–minimum ratio. The populations of Anisoptera (dragonflies) are also considered to be relatively stable. The occasional large congregations seen in *Pantala flavescens*, *Sympetrum* spp. and *Anax parthenope* are probably due to winds which bring these migratory dragonflies together. Territoriality shown by many dragonflies (p. 203) plays a role in stabilizing the population.

Varley & Gradwell (1963) investigated populations of many lepidopterans for thirteen years in an oak wood near Oxford. As a rule every species had one generation a year so that the mature larvae falling to the ground to pupate could be counted accurately by catching them on traps in autumn (for the oak tortrix the mature larvae were counted in trees). According to this study the maximum–minimum ratios of densities in the order of abundance were as follows:

Species	mean no per m^2	ratio
Tortrix viridana	100	14
Operophtera brumata	90	64
Eucosma insertana	30	5
Tortricodes tortoricella	10	9
Erannis leucophaearia	10	32
Cosmia trapezina	2	14
Erannis defoliaria	1	26

The predatory beetle of these lepidopteran larvae, *Philonthus decorus*, was studied for nine years from 1957 in the same wood and its maximum–minimum ratio was found to be ten. Thus the amplitude of fluctuation of the predator population was lower than that of most of the moth populations which fluctuated greatly even in natural woodland.

In contrast to the moths, many of which have become pest insects of forest trees and agricultural crops, the butterflies (except the skippers belonging to a very special family Hesperiidae) do not normally constitute a pest. The number of species against which insecticides are applied is perhaps less than twenty in the world. This makes us suspect that butterflies have developed some density regulatory mechanisms. Unfortunately, the population fluctuation of this popular group of animals has not been studied to date in spite of many ardent amateurs in this field.

Our last examples of insects with stable populations come from the social insects which have a very well developed intraspecific regulatory mechanisms (ecological homeostasis). Long-term studies on the population fluctuation of this group are unfortunately few; one of these was reported by Pickles (1940) on three species of ant in England (Fig. 3.38). The amplitude of population fluctuation over five years was 3.5 in the maximum–minimum ratio for *Formica rufa*, 2.8 for *Myrmica ruginodis* and 4.3 for *Lasius flavus*. The cause of the large amplitude in the last species was said to be the variation in mortality during the investigation; the number was affected by the digging out because of their peculiar mode of life. In fact, the ratio in the fluctuation of colony numbers over a period of four years was 2.5 in comparison with 1.8 in *F. rufa* and 2.0 in *M. ruginodes*.

A colony of ants sometimes lasts many years and the population size remains stable. What about the wasps that have yearly colonies? Matsuura (1977) studied fluctuations in thc number of nests over eleven years in the paper wasp *Polistes fadwigae* (Fig. 3.39). The amplitude of

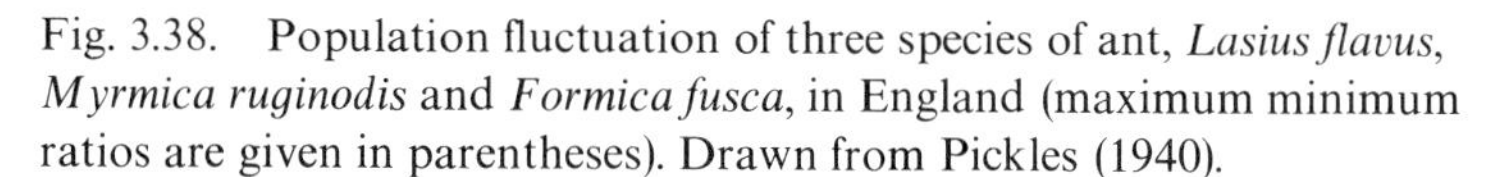
Fig. 3.38. Population fluctuation of three species of ant, *Lasius flavus*, *Myrmica ruginodis* and *Formica fusca*, in England (maximum minimum ratios are given in parentheses). Drawn from Pickles (1940).

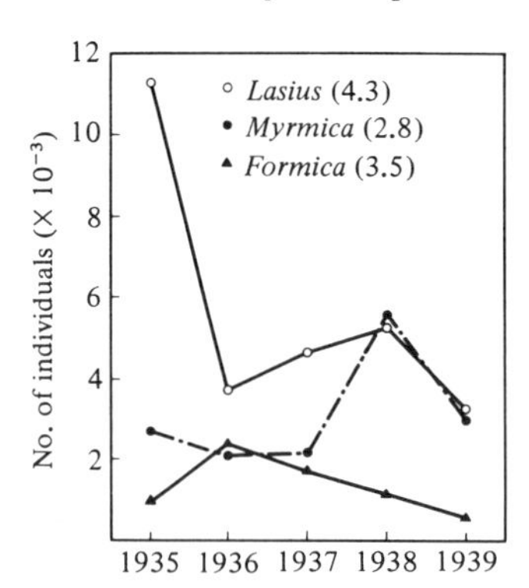

fluctuation was very small for an insect, suggesting that this wasp has a powerful density regulatory mechanism. Matsuura also recorded the number of nursery cells containing larvae in each nest. This showed very little variation from year to year, the minimum being about 140 and the maximum about 200, in the case of the village population. The main regulatory factors for *Polistes* were the attacks by hornets and 'cannibalism' from summer to autumn. The latter occurs when the new adults increase in number; the workers pull out larvae and pupae from the cells and feed adults and other larvae with them. This seems to contribute greatly to the stabilization of the population.

Population fluctuations of birds and mammals

(1) There are two very different types of population fluctuation found in birds and mammals. One extreme is seen in a group of animals inhabiting tundra and boreal coniferous forests and fluctuating in number enormously and cyclically. The other extreme is shown by the group with very stable populations. These are the results of two different reproductive strategies, which may even be found among closely related species. Naturally there are many species with intermediate strategies. Let us look at examples of cyclic fluctuations first.

The idea of periodicity in the fluctuation of animal numbers has been put forward repeatedly since the last century. Most of the cycles presented, however, were not testable in the strict statistical sense or did not survive the test; at present those that can be confirmed by means of correlogram used in statistics are restricted to the population fluc-

Fig. 3.39. Fluctuations in the number of nests of a paper wasp, *Polistes fadwigae* (maximum minimum ratios are given in parentheses). From Matsuura (1977).

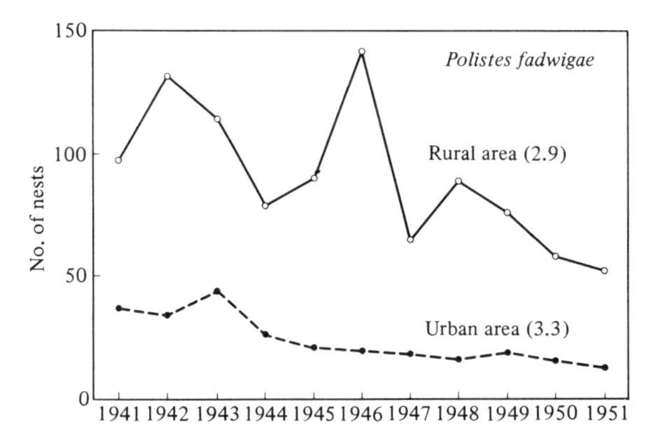

tuations of some birds and mammals with the exception of the adult members of the periodic cicadas. The presumed correlation between the number of locusts and the sunspot cycle (mean cycle, 11.125 years); with outbreaks and group flights occurring coincidentally with the low sunspot years, has been rejected by Ma (1958) who analysed the outbreak records over 1000 years in China. Waloff's (1976) analysis of data over 100 years in Africa was also negative. Only the larch budmoth in Switzerland (Fig. 3.25) can be considered almost certainly to have a cycle of 8–9 years, but the population data are unfortunately limited to only twenty recent years.

Coming to the homeothermal animals we find for the first time examples of cycles that can survive statistical tests. These are the population fluctuations of arctic animals which Seton (1911) so vividly described in *The Arctic Prairies.*

From the middle of the eighteenth century to the nineteenth century there emerged enterprises with large capital backing to deal in fur in the New World. Amongst them was the Hudson's Bay Company which became the sole dealer to buy fur from almost all of Canada and which accumulated complete hunting statistics over many years. Charles Elton used this material since 1924 to establish a theory of cyclic fluctuation which has been accepted to the present day.

In his first paper Elton (1924) investigated animal population statistics of the world and concluded that the following eight groups had cyclic fluctuations:

Cycle $3\frac{1}{2}$ years

1. Lemmings: *Lemmus lemmus* (northern Europe) and *Dicrostonyx groenlandicus* (Canada)
2. Arctic fox: *Alopex lagopus* (tundra)
3. Red fox: *Vulpes fulva* (tundra)

Cycle 10–11 *years*

4. Varying hare: *Lepus americanus* (North America)
5. Lynx: *Lynx canadensis* (North America)
6. Fisher: *Martes pennanti* (North America)
7. Red fox: *Vulpes fulva* (North America – taiga)
8. Sand grouse: *Syrrhaptes paradoxus* (Central Asia and Mongolian deserts)

Of the above, 1–3 are animals of the tundra ecosystem of the polar region while 4–7 are animals of the boreal coniferous forest (taiga). Of the latter, the fisher's cycle, though of about ten years, is far from being significant statistically. Also the record of the sand grouse is too short to be tested for its periodicity. However, the periodicity of population fluctuation for other species is clear as can be seen in Figs. 3.40–3.44.

Elton's discovery of two cyclic series in the arctic and subarctic regions was of superior merit but he sought their cause in the periodic change in climate associated with the sunspot fluctuations. [The periodic sunspot fluctuations change the amount of solar radiation falling on the earth, but the meteorological conditions on the surface of the earth are

Fig. 3.40. Population fluctuations seen in the number of skins collected of the arctic fox *Alopex lagopus* (white circles) and the red fox *Vulpes fulva* (black circles) (drawn from Elton, 1924; MacLulich, 1937; Lack, 1954b). Arrows indicate irruption years of the varying lemming *Dicrostonyx groenlandicus* (no record before 1900). Adjustments made for delay in fur collection.

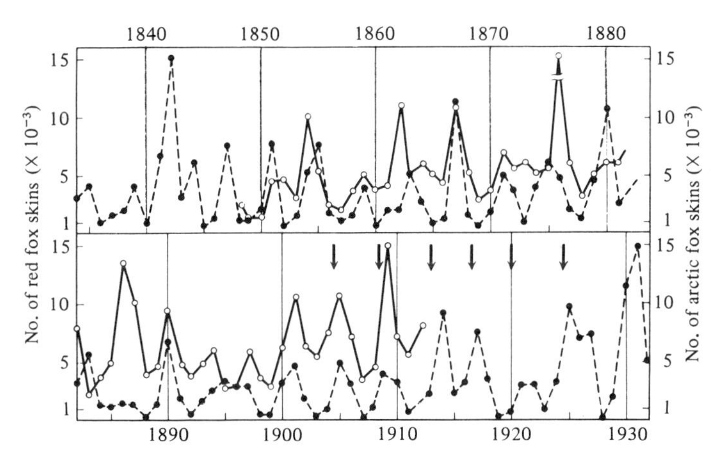

Fig. 3.41. Population fluctuation seen in the combined number of red and arctic foxes collected in Norway (redrawn from Elton, 1942, with additional data from Lack, 1954b). Arrows indicate the lemming years and the bars indicate the irruption years of willow ptarmigans.

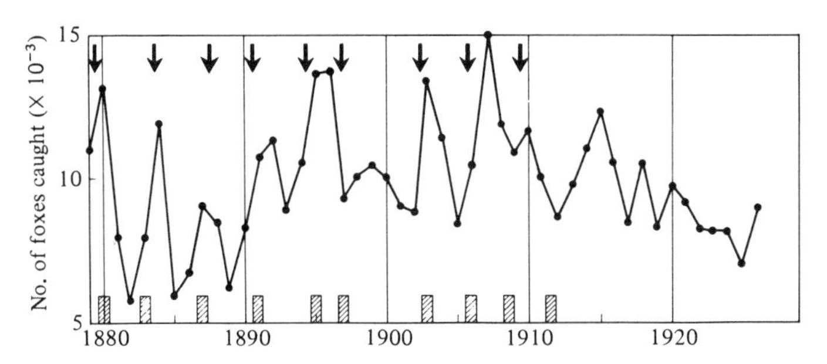

influenced by so many complex factors that the effect of this fluctuation is cancelled out. The only parameter, recognized today, that has the corresponding eleven-year cycle is the amount of rainfall in the tropics.] MacLulich (1937) showed that the peaks of population fluctuation in the lynx and the varying hare coincided with the peaks of the sunspots only for a period, after which the correspondence was lost gradually until it began to coincide with the troughs of the sunspots. The cycle in the taiga was not 10–11 years as Elton classified, but 9–10 years in the cor-

Fig. 3.42. Fluctuations in abundance of the varying hare *Lepus americanus* (white circles) and the lynx *Lynx canadensis* (black circles) collected in coniferous forests of Canada (drawn from MacLulich, 1937).

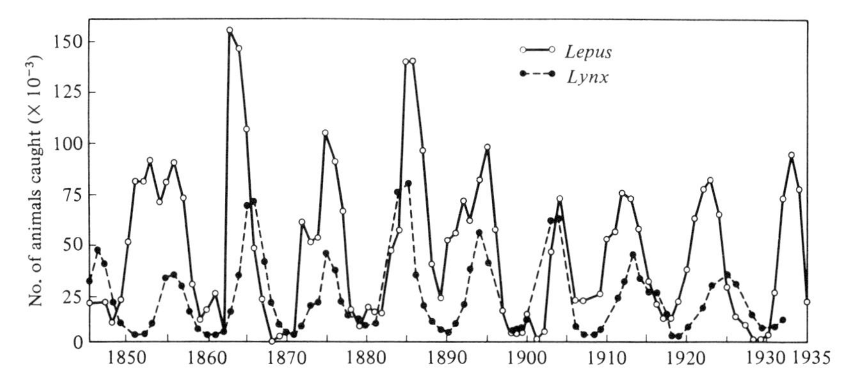

Fig. 3.43. Population fluctuations of the red fox *Vulpes fulva* and the fisher *Martes pennanti* as seen in the statistics of the Hudson's Bay Company (after Elton, 1924).

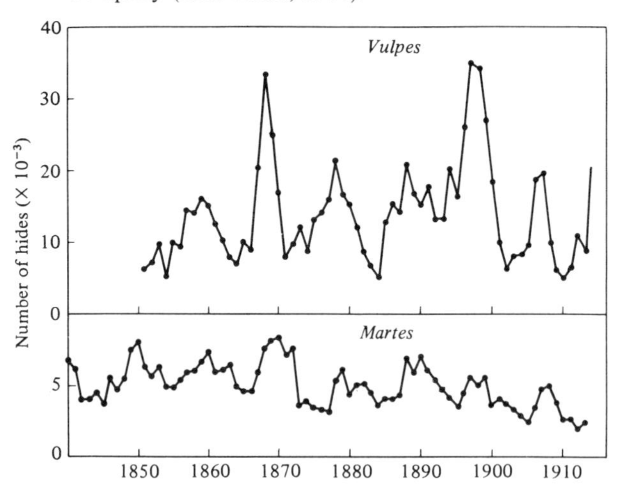

relogram analysis. The 11-year cycle of the sunspots may have influenced Elton's idea of 11-year cycles for animals. As Lack (1954b) and Moran (1949) stated, to verify interrelations of two time series with long peak intervals we need data over a very long time.

As shown in Fig. 3.40 the cycle of 3–4 years for the red fox and the arctic fox populations occurs in agreement with outbreaks of lemmings on which they feed. The lemmings were not important fur-bearing animals and hunting statistics for them were not available. In recent years, however, long-term ecological studies have accumulated more reliable population data for lemmings than the hunting statistics of other animals. Fig. 3.45 is one of these, depicting the fluctuation in the number of the brown lemming *Lemmus trimucronatus* in Barrow District, Alaska (Pitelka, 1973). The peak interval is not four years but the amplitude of fluctuation is great, the maximum–minimum ratio being 612. Other data of Pitelka (1957a) give a ratio of 370 in six years and those of Krebs (1964) give 105 in four years. Pitelka suggested that in the past the amplitude was perhaps over 1000-fold but it decreased in recent years as a result of rapid development of the region and that the absence of an

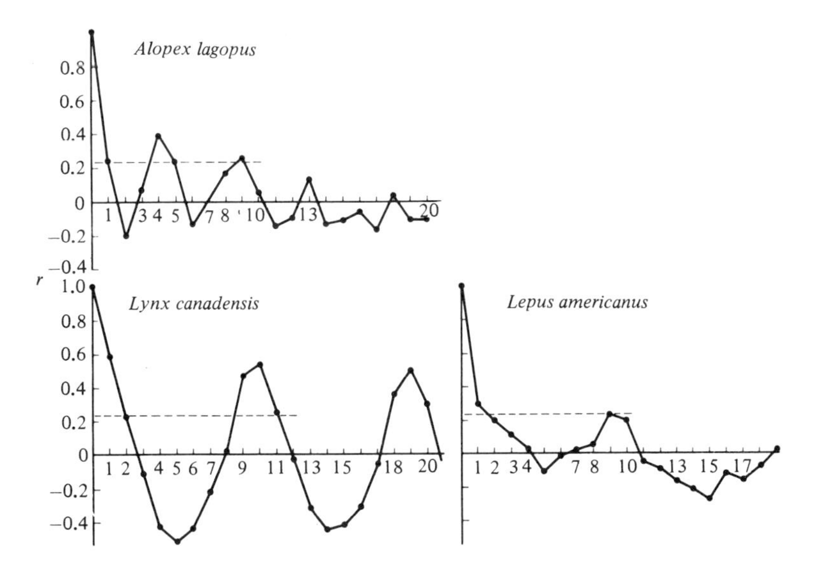

Fig. 3.44. Correlograms of population cycles in the arctic fox *Alopex lagopus* and lemmings in tundra and the lynx *Lynx canadensis* and the varying hare *Lepus americanus* in taiga (calculated from Elton, 1927, 1942; Elton & Nicholson, 1942). Broken lines indicate the 5% level of significance.

expected peak in 1968–69 was due to an abnormal phenomenon associated with this development.

It will be clear from Fig. 3.42 that the population fluctuation of the lynx follows that of the varying hare on which it preys. Thus tundra and taiga each has an established predator–prey system. The best evidence for this is that the red fox population has a cycle of four years in tundra and ten years in taiga. Butler (1951) studied the red fox in an intermediate zone in Quebec and found that the peaks of fluctuations in the two habitats did not correspond with each other, indicating that the two cycles are independent.

The most noticeable feature of population fluctuation in tundra and taiga is the scale of the fluctuations. For example, the number of red foxes caught in a district of Labrador between 1834 and 1925 fluctuated between 1 and 1495 (Elton, 1942). Excluding the year 1919 when only one was caught the range is 27–1495, a ratio of 55. Also, if the numbers of red foxes and arctic foxes in Norway are added together then the maximum–minimum ratio is found to be 59 for the northern districts, 10 for the central areas and 4.6 for the southern districts. [The reason for the decrease in amplitude in southern districts may be an influence of forest-dwelling foxes. However, ten out of twelve peaks coincided in all these districts.] For the arctic fox the ratio taken from Fig. 3.40 is 24 and for the lynx the data in Elton & Nicholson (1942) come to 179. The varying hare had ratios of 75 and 15 in Minnesota and as high as 2250 in Ontario (Keith, 1963). The large amplitude of lemming numbers has already been mentioned.

Fig. 3.45. Fluctuations in the number of brown lemmings caught in the Barrow District, Alaska. The numbers represent June captures from permanent traplines. (Maximum–minimum ratio given in parentheses.) Drawn from Pitelka (1973).

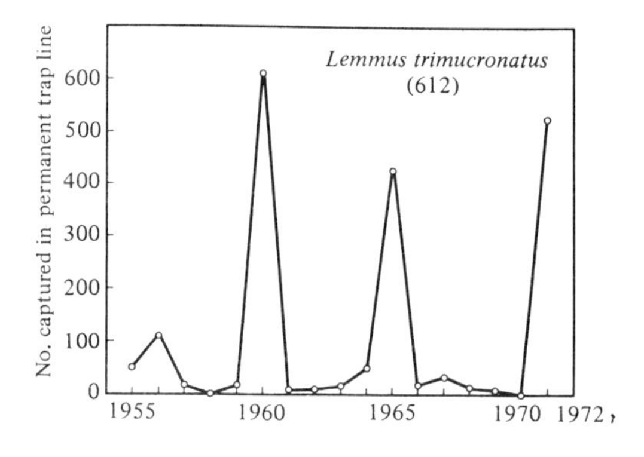

(2) The fact that different combinations of predator and prey populations fluctuate cyclically in tundra and taiga makes us suspect that some relationships, such as expressed in the differential equation models of Lotka (1956) and Volterra (1926), exist within the system. But this idea is not entertained much today. This is because, as Lack (1954b) points out, at the peak density of lemmings or varying hares the predators, though approaching their peak density, are not abundant enough to cause a crash of the prey population. In Krebs's (1964) study of mortality there is no evidence that *most* individuals are killed by predators. Although predation is an important aspect of population control, what is more important is the cause of death of those that have escaped predation.

The basic theory accepted by many current workers is that the cycle of vegetation – its destruction and regeneration – influences lemmings. This was suggested by Lack (1954b), and Pitelka (1957a, b) provided evidence. In addition, induction of the stress syndrome (intensification of fighting, increased susceptibility to diseases) and recovery from it (C. J. Krebs, 1964, and others) have been combined with this idea. Foxes certainly are a factor in the decrease of lemming numbers but they are not the only cause of population decline nor does a decrease in fox numbers necessarily result in an increase in lemmings. Foxes are thought of as being dependent on the cycle produced in the vegetation–lemming system.

According to Pitelka the fragile vegetation of tundra is completely destroyed by the impact of lemmings as they increase several hundred-fold in two to three years. From about the third year of the lemming cycle plants begin to die in patches and in the fourth year lemmings run around in the exposed space in search of food. Being starved they are also active in daylight. They suffer great mortality then from predation since predators are attracted by their large numbers and increased activity. The population 'crash' begins with such predation and death caused by the stress of overcrowding; they disappear through the winter, and by the following summer it will be difficult to find a lemming in the field. The lemmings then are distributed in surviving patches of vegetation. If the crash is severe the density remains low for two years and the following peak does not come till the fourth year, but if the crash is mild the peak comes in the third year. They begin increasing in the second or the third year. The new individuals survive well since the vegetation has recovered and predators have decreased by then. By the autumn of that year lemmings may be found everywhere and the vegetation begins to show damage again. During the winter they eat out vegetation under the snow and approach the next crash.

The above is the model presented by Pitelka. There are other theories such as Schultz's (1964), which attribute the population decline to the reduction in the amount of phosphorus in plant tissues at the height of the outbreak. This is not very convincing as the main governing factor of population cycle. Krebs (1964) questioned Pitelka's theory on the ground that at the time of the observed crash vegetation was not much destroyed.

Although vegetation may be completely destroyed by the irruption of lemmings, at times some vegetation remains green when the lemmings disappear. The first thing thought of as the cause of the crash, then, was disease (Elton, 1925, 1931).

After a population crash many lemmings may be found dead in the field. Also, the behaviour of lemmings during migration is said to be very abnormal (Clough, 1965). According to Elton, however, the causes of death are varied, sometimes it is tuberculosis, sometimes it is bacterial infection (*Bacilus pestis-lemmi*) and sometimes it is neurosis symptoms without pathogens.

In 1938 Green & Larson of Minnesota University published a paper (1938a, b) in which they claimed that the cause of the population crash in the varying hare – occurring with a ten-year cycle – was the 'shock' disease. This was the year in which H. Selye began his study of adaptive syndromes using rats. With the concept of shock disease or stress one can explain easily the variety of diseases that animals suffer at the time of a crash. The irrupted lemmings (and foxes), suffering from hunger and social pressure, are in the fourth stage of the adaptive syndrome of Selye. From the winter to spring at the height of the outbreak the population maintains an extremely high density under conditions of (1) food shortage, (2) shortage of shelter from cold, (3) exhaustion from foraging effort, (4) intraspecific competiton and (5) general stress of increased exposure to cold, etc. as a result of the above (Christian, 1950). When daylight begins to increase and secretion of reproductive hormone is stimulated, animals face this stimulus under a high degree of stress overtaxing the pituitary. Many individuals which have endured the winter die in spring and the few young that are born from these animals have low viability, reflecting the poor conditions to which they have been subjected during the embryonic stage. This is the theory of Christian (1950), who applied the stress theory to the lemming cycle.

Krebs (1964) who studied the population crash of lemmings is opposed to the application of the stress theory to the lemming situation because in lemmings the enlargement of the adrenal cortex, which is considered a necessary component of the stress theory, is not demonstrable. He proposed that increased mortality at the time of outbreaks selects for aggressive individuals, resulting in even more severe fighting among them (they are aggressive animals to start with, see Frank, 1962) and leading to physiological weakening and a population crash. To prove this hypothesis in *Microtus* Krebs and co-workers conducted electrophoresis of the blood of marked individuals in an enclosure and demonstrated changes in the ratio of some isozymes over a few generations (Myers & Krebs, 1974).

But the hypothesis of Krebs *et al.* is not sufficient to explain the cyclic crash. For example there is little evidence that specific isozymes are associated with the genes carrying the traits of agonistic behaviour, and the fluctuation of isozymes themselves is not all that conspicuous. Although the narrow definition of the adaptive syndrome with enlargement of the adrenal cortex does not apply to the lemmings there may be other physiological pathways, as yet undiscovered, which bring about death in a broad sense of stress disease (see Itô & Kiritani, 1971), since it is apparent that the animals do behave abnormally at the time of an outbreak.

Perhaps the mechanism of cyclic fluctuation in lemming populations is not simple. As shown in Fig. 3.46, predation, hunger and social stress all operate one after another and cause the crash. The basic relationship in this is between vegetation and lemmings, and foxes increase or decrease

Fig. 3.46. A schematic representation of the population cycle in the lemming in relation to vegetation and predators. Arrows indicate negative (solid line) or positive (broken line) factors acting on the population, D, disease; A, aggression; M, migration; R, recovery of health.

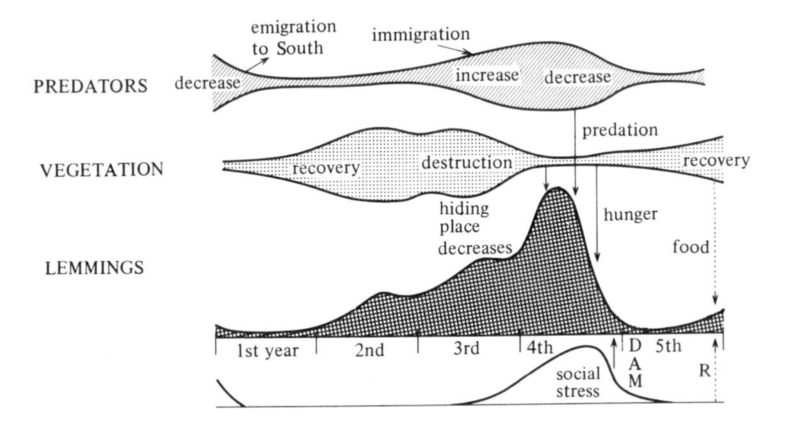

Table 3.3. *Four-year cycles of rodents and predators in North America*

Peak years of rodents	Years of invasion			Peak years of arctic fox
	Snowy owl	Great grey shrike	Rough-legged buzzard	
—	1889			1890
—	1892			1893
—	1896			1897
—	1901	1900		1901
1904, 1905	1905	1905		1905
1908, 1909	1909	1909		1909
1913	1912	1913		1913
1916, 1917	1917	1917	1917	1917, 1918
1920	1921	1921	—	1921, 1922
1924, 1925	1926	1926	1926	1926, 1927
1929	1930	1930	1930	1930, 1931
1933, 1934	1934	1934	1934	1934
1936, 1937	1937	—	1937	1938
1940	1941	1939, 1940	—	1941, 1942
1943	1945	1945	—	1946

From Lack (1954b).

in number depending on this relationship without affecting it. The cyclic fluctuation of foxes has been studied in detail by Tschirkova (1955) in U.S.S.R. and Macpherson (1969) in Canada. From their work it seems certain that the crash of the lemming population induces the failure of reproduction and mass mortality in foxes.

There are other animals, such as predatory birds, which fluctuate in number depending one-sidedly on the lemming population. For many years it has been known that many snowy owls *Nyctea scandiaca* appear periodically in southern cities. Lack (1954b) constructed a table listing the known years of abundance of some predatory birds in the south (Table 3.3). Such arrivals of northern predators are believed to occur as a result of food shortage in the tundra where they have increased. All three species appear in the same years, coinciding with the peak year of lemmings or a year after. Lack also described the peak years of abundance for the rough-legged buzzard *Buteo lagopus* and the goshawk *Accipiter gentilis*, which occurred about every four years, coinciding with the peak abundance of foxes and six months, on average, after the outbreak of the Norway lemming *Lemmus lemmus*. According to Lack the willow ptarmigan *Lagopus lagopus* has similar cyclic fluctuations in Norway and its peak years are about six months behind those of the lemming. The willow ptarmigan increases in number when foxes (its predators) decrease and decreases when the predators turn to ptarmigans after the decrease in the lemming population.

Not many quantitative studies have been published on the ten-year cycles. The peaks of the lynx population are known to occur, on average, two years after the peak of the varying hare population. It is almost certain that an increase in the number of varying hares (prey) leads to an increase in the number of lynx (predator) and the decrease of the prey population is due to the high mortality that follows the peak abundance. According to Keith (1963) the mortality of young hares till the spring in Alberta, Canada, was 9% in one year and 4% in another during the increasing phase of the population but increased to 92% in the year of the crash. A large part of this mortality was due to predation by the lynx, but hunger and perhaps the stress disease in a broad sense may also have been responsible. [The reason why Green & Larson (1938a, b) proposed the concept of the shock disease was that they found a low level of blood sugar (later recognized as one of the characteristics of the adaptive syndrome) in the varying hare found dead during a population crash, but Chitty (1959) criticized their work saying that the use of dead hares made it impossible to interpret the low level of blood sugar as a *cause* of death or a *result* of death. But as the abnormal behaviour of hares during the

Table 3.4. *Ten-year cycles of prey and predators in North America*

Peak numbers trapped in		Invasion of Toronto		Peak numbers shot in Great Lake region
Hudson Bay or Ontario Varying hare	Hudson's Bay Co. Northern Dept. Lynx	Horned owl	Goshawk	Ruffed grouse
1857	1857	—	—	1857
1865	1866	—	—	1866
1876	1876	—	—	1877
1887	1885, 1886	1887	1886	1887
1896	1895	1897	1896	1898
1905	1905	1907	1906	1905
1914	1913, 1914	1916	(1916)	1914
1924	1925	1927	1926	1923
1934	1934	1936	1935	1933
1943	—	—	—	1942

From Lack (1954b).

crash has been reported many times by different people, it is possible that social stress does exist during this period.] It is not known if the recovery of the varying hare population is due to a decrease in lynx abundance or interaction with forest vegetation, as in the case of the lemmings.

In taiga as in tundra there are dependent cycles of predatory and other birds. These are ten-year cycles dependent on the cycles of the varying hare and the lynx. As can be seen in Table 3.4 the southern movement of the horned owl and the goshawk occurs one or two years after the peak of the varying hare population. Though not significant in correlogram analysis the fisher population (predator) in Fig. 3.43 fluctuates in a cycle of about ten years, being influenced by the abundance of the varying hare. The ruffed grouse *Bonasa umbellus*, a forest-edge species in taiga, fluctuates in abundance in agreement with the varying hare just as the willow ptarmigan does in tundra. According to the detailed study of Williams (1954) on population fluctuations of grouse in Canada, the ruffed grouse population fluctuates in a cycle of 9–10 years along a large wave in the graph. An interesting point is that the willow ptarmigan population in taiga seems to fluctuate in a cycle of about ten years (see below) as opposed to four years in the tundra population.

In summary, Fig. 3.47 gives a simple model of interactions among cyclic populations of homeothermal animals in tundra and taiga.

(3) So far we have examined species that show cyclic or violent fluctuations in abundance. But most birds and mammals have only small fluctuations in number. Of course those short-lived animals with a relatively large clutch or litter size, such as rats, mice, rabbits, hares and grouse, show larger amplitudes of population fluctuation than others, but

Fig. 3.47. A model of population interactions in tundra and taiga.

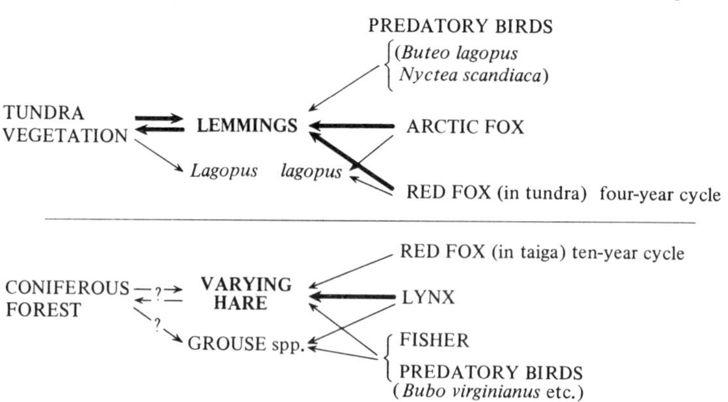

even in these species the maximum–minimum ratio is not in the high order of several hundreds. For example, population fluctuations of the wood mouse *Apodemus sylvaticus* and the bank vole *Clethrionomys glareolus* studied over twenty years in Wytham woods near Oxford had maximum–minimum ratios of 8.7 and 13.9 respectively (Fig. 3.48). In many species of rodents the population fluctuation is irregular and is affected by a variety of factors including food conditions and weather. For example, in dry areas the amount of grass seed strongly influences

Fig. 3.48. Fluctuations of December populations of the wood mouse *Apodemus sylvaticus* and the bank vole *Clethrionomys glareolus* in Wytham Woods near Oxford (maximum–minimum ratios are given in parentheses). Drawn from Southern (1970) and Lack (1966).

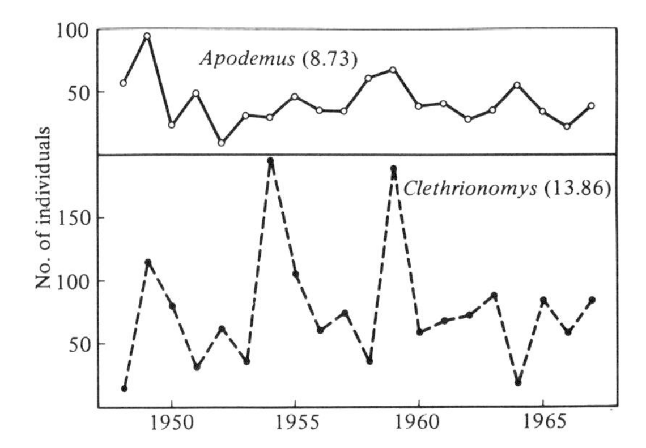

Fig. 3.49. Population fluctuation of the common hare *Lepus europaeus* on a small island off Denmark (after Abildgård *et al.*, 1972). The insert shows the relationship between the adult density in spring and the ratio of young to adults in autumn. (Maximum–minimum ratio given in parentheses.)

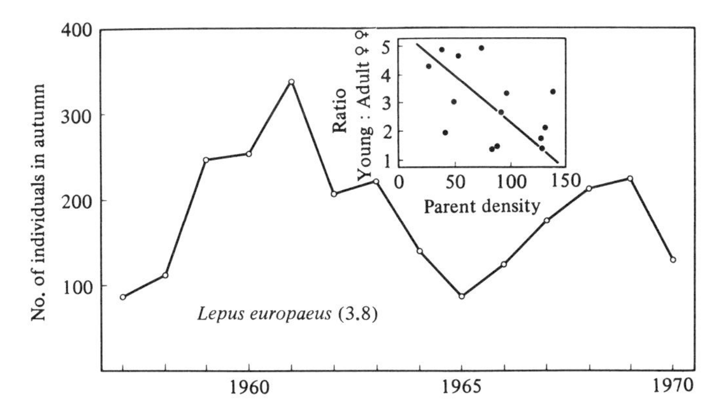

the population density, which increases in the years of much rainfall (Lidicker, 1973). In spite of such trends, Lidicker found it impossible to explain the whole of the population fluctuation in terms of one or a few factors. It seems certain, though, that the upper limit of the rodent population is determined ultimately by social stress in a broad sense. Krebs & DeLong (1965) and Krebs (1966) found through their field experiments that a *Microtus* population that was being fed increased rapidly at first, then stopped increasing and declined to a level below the density in the control area. In another experiment Krebs (1966) found that when a few individuals from an increasing population were introduced into a decreasing population the latter showed a very rapid increase. There is an enormous amount of literature on the rodents of the temperate regions, but there is no example of predation being the major cause of their population decrease.

Fig. 3.49 shows population fluctuation of the common hare *Lepus europaeus* on a small island off Denmark. In this case the maximum–minimum ratio was almost four. Abildgård *et al.* (1972) who reported on this fluctuation did not suggest causes of the fluctuation. As shown in the insertion in the figure, if the hare was abundant in spring the ratio of young to adults in autumn decreased markedly, suggesting either a decrease in litter size or an increase in early mortality in a density-dependent manner.

The domesticated rabbit, *Oryctolagus cuniculus*, which was introduced

Fig. 3.50. Fluctuation of warren numbers of feral rabbits *Oryctolagus cuniculus* per 55 sq. miles in southeastern Australia (drawn from Myers, 1971). Numbers in the figure indicate mean numbers of holes per warren (maximum–minimum ratios in parentheses).

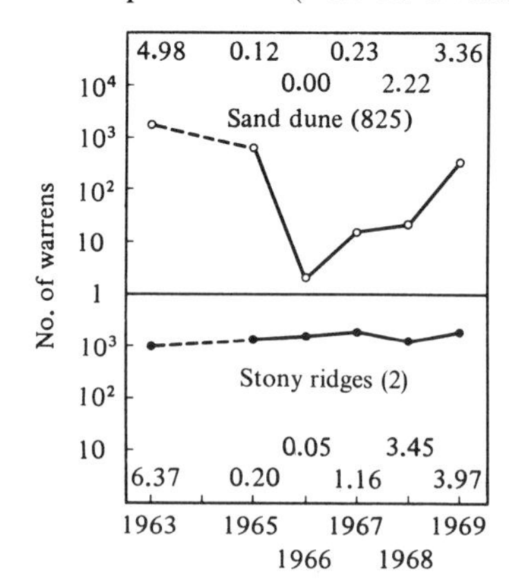

into Australia, has increased there in the wild and created problems of pasture protection and conservation of marsupials. Fig. 3.50 shows the fluctuation of the number of warrens per 55 square miles in southeastern Australia. The number fluctuated little on stony ridges but greatly on semi-desert dunes. According to Myers (1971) rabbits can seldom breed in sand dunes where usually food and water are limited and sand storms often bury their warren. When the rainfall is high (e.g. 1963), however, annuals grow everywhere and the rabbit density increases to a level higher than that on the stony ridge. The habitat of the stony ridge contains perennial grasses, herbs and shrubs which can support the rabbit population during the dry season, while rainfall affects plant production little in this area. Thus the rabbit population is fairly stable. The number of individuals is of course not only proportional to the number of warrens but also to the number of nest holes per warren, thus the amplitude of the population fluctuation is greater than that of the warren-number fluctuation. The maximum–minimum ratio is estimated to be larger than 825 for the sand dune population which became almost extinct locally in 1966. On the stony ridge, also, the population of 1966 would have been very small. The dune population of 1967 is derived from the ridge population. This shows how a high-fecundity strategist maintains its population by utilizing a heterogeneous environment.

Among birds we have already discussed cyclic fluctuations of polar

Fig. 3.51. Fluctuations in the total number before hunting (white circles), the number of young birds in spring (black circles) and the number of adults in autumn (triangles) of the red grouse *Lagopus lagopus scoticus* in Scottish mountains (maximum–minimum ratios are given in parentheses). Drawn from Jenkins *et al.* (1963) and Watson (1971).

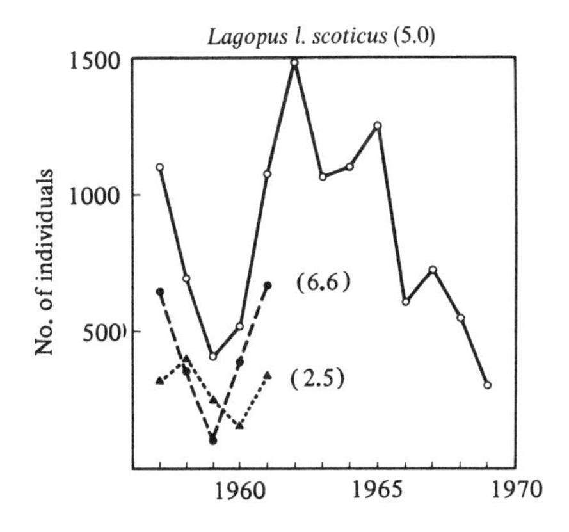

species. The amplitude of fluctuation of the ruffed grouse population (Keith, 1963) was greater than ten in the maximum–minimum ratio. This is characteristic of taiga and much larger than the amplitude of fluctuation recorded for the red grouse *Lagopus lagopus scoticus* in the mountains of Scotland by Jenkins *et al.* (1963; and others) and Watson (1971; and others). The red grouse population had a ratio of five over thirteen years (Fig. 3.51). Jenkins *et al.* (1963) presented detailed data for the first five years of study, in which the number of adults (two years and older) fluctuated with a ratio of 2.5 while the number of young hatched in spring fluctuated with a ratio of 6.6. Hunting of this population is permitted (mortality due to hunting is less than 10%), but the strongest influence on the population size is exerted by winter mortality and the rate of reproductive success. This depends on the state of the heath. What influences winter mortality is not the cold itself but the number of individuals that have acquired territories in autumn (those without a territory most certainly die during the winter). The number of territory holders changes from year to year, depending on the state of the heath and the physiological condition of the birds themselves. In all important mortality factors, 'delayed density-dependence' has been observed and this is considered to bring about population fluctuation.

Weeden & Theberge (1972) investigated a population of the rock ptarmigan *Lagopus mutus* under no hunting pressure in Alaska (Fig. 3.52). The amplitude of its fluctuation was small; 2.8 for the young and 2.6 for the breeding females in the maximum–minimum ratio. Winter mortality and predation of eggs by the weasel *Mustela erminea* were related to population fluctuation and their reduction caused the sharp

Fig. 3.52. Population fluctuation of the rock ptarmigan *Lagopus mutus* in Alaska (maximum–minimum ratios are given in parentheses). Drawn from Weeden & Theberge (1972) in Dempster (1975).

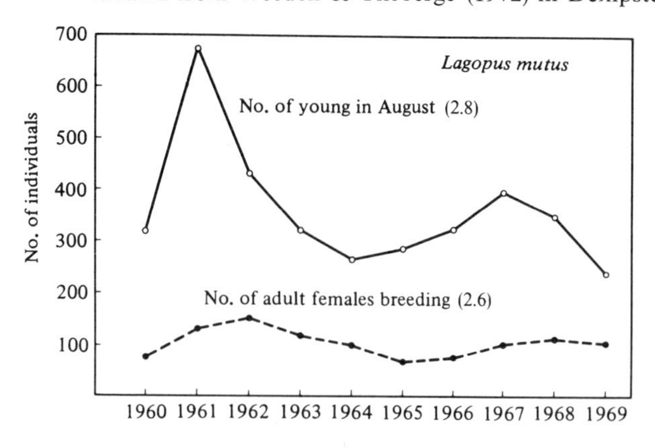

increase in 1961. As in the red grouse winter mortality acts on non-territorial individuals, but why the number of territories changes from year to year is not known. The winter of 1962–63 was the coldest and mortality was indeed high, but looking at the whole period of the study the amount of snow-fall was not considered important.

The interval between peaks of population fluctuation seems to be shorter in the red grouse than in the rock ptarmigan in the temperate zone. Hunting statistics kept over scores of years in Scotland show the peak interval to be five years on average and the maximum–minimum ratio to be 20–30 (Mackenzie, 1952). The fluctuation is more gentle in the ptarmigan population than in the red grouse population. This may be a result of strong territoriality in ptarmigan which live on the higher parts of mountains.

Among other birds of the Galliformes with high fecundity, hunting statistics are available for the partridge *Perdix perdix* in England. As shown in Fig. 3.53 its fluctuation is irregular and the amplitude is large (34 in ratio).

The geese also have high fecundity but their life-span is longer than that of gallinaceous birds. Fig. 3.54 shows the population dynamics of the Canada goose *Branta canadensis* breeding on twenty-one islands of the Columbia River in Canada. In this graph the number of hatched eggs fluctuated with a ratio of three while the number of breeding pairs

Fig. 3.53. Fluctuations in the number of hunted partridges, *Perdix perdix*, in Norfolk (drawn from Middleton, 1934).

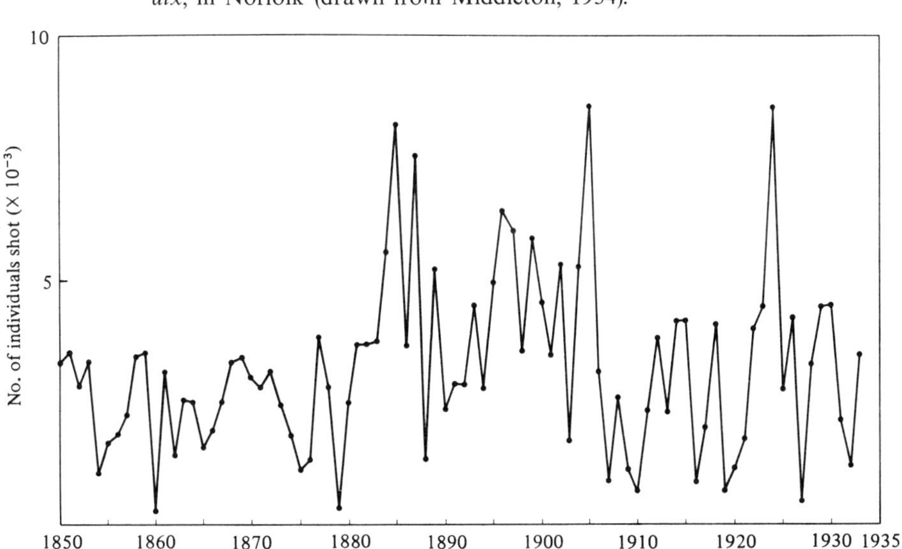

fluctuated little. Non-breeding adults were found on the fringe of the colony or left the area. Thus similar numbers of nests were constructed from year to year. The variation of hatchability was mainly due to predation by the coyote (predation of eggs or parents or desertion of eggs by escaping parents). Since the geese shifted their nesting area away from the coyote's influence a stable population has been maintained.

Thus in birds and mammals the effect of climatic changes on a population is not great though in the temperate regions there are cases where a severe winter produced a significant change in population size. Fig. 3.55 shows the population fluctuation of the wood pigeon *Columba*

Fig. 3.54. Population fluctuation of Canada geese on twenty-one islands in the Columbia River, Canada (maximum–minimum ratios are given in parentheses). Drawn from Hanson & Eberhardt (1971).

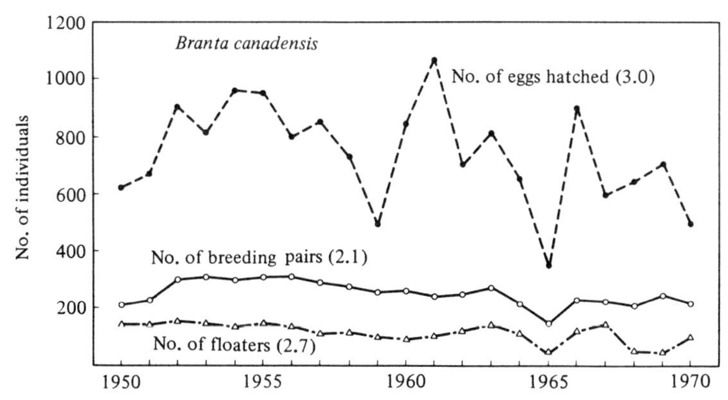

Fig. 3.55. Population fluctuation of the wood pigeon *Columba palumbus* in England (maximum–minimum ratios are given in parentheses). Drawn from Murton *et al.* (1974) in Dempster (1975).

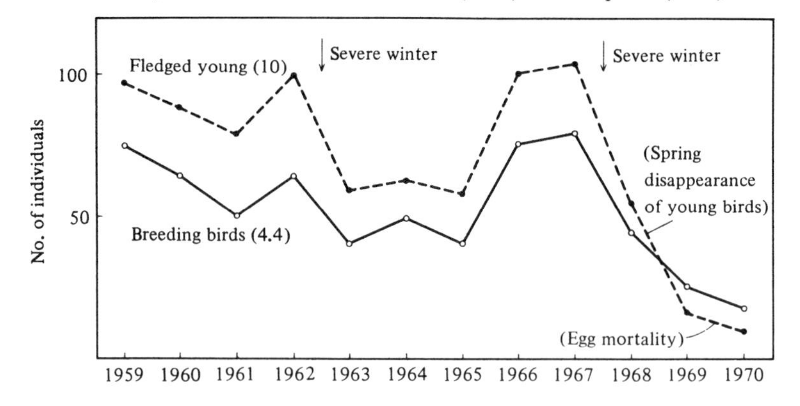

palumbus; the number of breeding birds decreased as a result of a severe winter on two separate occasions. Desertion of many nests in 1969 and 1970 was due to an elimination experiment. Among the examples showing the effect of a severe winter on bird populations the case of the grey heron *Ardea cinerea* is famous (Lack, 1966). In this case also, the amplitude was less than two in ratio over nineteen years.

Fig. 3.56 (upper graph) gives the population fluctuation of the great tit *Parus major* and the blue tit *P. caeruleus* in Marley Wood. The populations of both species increasing or decreasing together show the fluctuation of climate as the main cause of their fluctuations (according to Lack (1966) the years of population decrease or increase were shared by other populations in nearby areas within Great Britain). Fig. 3.56 (lower graph) shows the fluctuation in number of the pied flycatcher in the Forest of Dean. This area is not far from Marley Wood, but the pattern of fluctuation is different from those of *Parus*. As the mode of life of this species is quite different the effect of weather on the population would be different. The amplitude of fluctuation for the pied flycatcher was 1.9 in ratio while that for the tits is essentially 2–3 (the population has been increasing gradually with changes of the environment over many years and the ratio was not calculated). In any case the population fluctuation of these species is small. Of course the number of breeding

Fig. 3.56. Fluctuations in the number of breeding pairs of the great tit *Parus major* and the blue tit *P. caeruleus* in Marley Wood (above) and of the pied flycatcher *Muscicapa (Ficedula) hypoleuca* in the Forest of Dean (below) near Oxford (maximum–minimum ratio in parentheses). Drawn from Lack (1966).

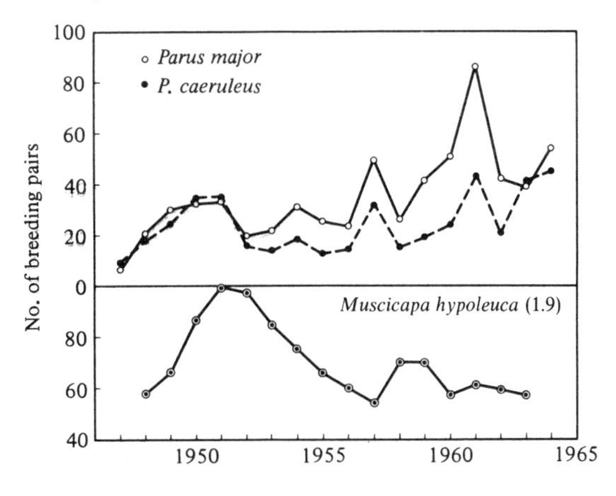

pairs almost equals the number of territories established as non-territorial birds are forced out of the area.

Kluyver (1971) presented data in Holland, showing this moving-out of great tits to less favourable areas. The deciduous broadleaf forest with oak is rich in prey insects, in both the number of species and abundance, and had thirty-three pairs of great tits nesting per 10 ha. On the other hand, the pine forest is poor in prey insects and had on average only four pairs. When the density was low in the oak forest it was also low in the pine forest, but when the density increased beyond a certain level in the oak forest the density in the pine forest increased rapidly. Thus the amplitude of fluctuation in the oak forest is smaller (2.6 in ratio) than in the pine forest (4.6) (Fig. 3.57).

Britain has the best census data of birds in the world, particularly large is an accumulation of data on small birds. Ginn (1969) summarized the results of ringing and nest records on many species. Of these the ringing results of nine species are given in Table 3.5. In this table the wren (a small number) and the goldfinch have large amplitudes of fluctuation, but for the remainder the maximum–minimum ratio is only 2–3 over sixteen years, showing the stability of passerine populations in general. Parental care and associated territoriality may buffer the effect of weather and reduce the amplitude of fluctuations.

It is expected that as the strategy of low fecundity/parental protection becomes further advanced then the smaller will be the amplitude, and the more gentle the slope, of population fluctuation. A typical example may be seen in the tawny owl *Strix aluco* studied by Southern (1970). In Fig.

Fig. 3.57. Fluctuations in the number of breeding pairs of the great tit in 10 ha of oak forest (mean thirty-three pairs) and pine forest (mean four pairs) in Holland (drawn from Kluyver, 1971). Log scale is used to permit comparison of the amplitude of fluctuation between the two areas. (Maximum–minimum ratios are given in parentheses.)

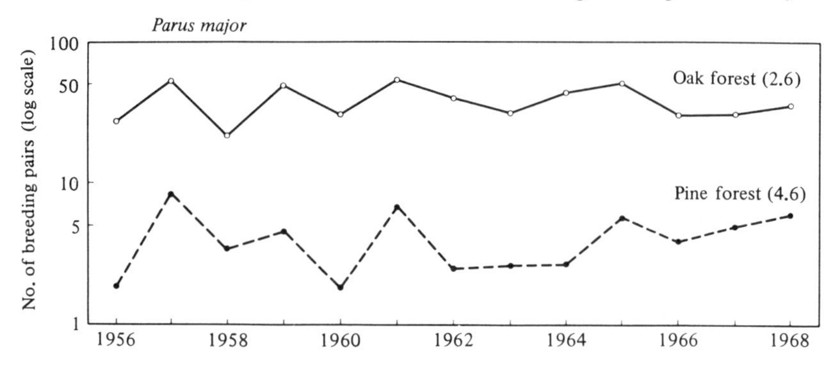

3.58 the maximum–minimum ratio is 1.88, but if the fact that the population increased gradually over the period of study (with the growth of vegetation and associated increase in other organisms) is taken into consideration the breeding population of this species may be considered

Fig. 3.58. Change in the number of established territories of the tawny owl *Strix aluco* and the relation between the number of fledged young and the density of rodents (*Apodemus sylvaticus* and *Clethrionomys glareolus*) in spring (maximum–minimum ratio in parentheses). After Southern (1970).

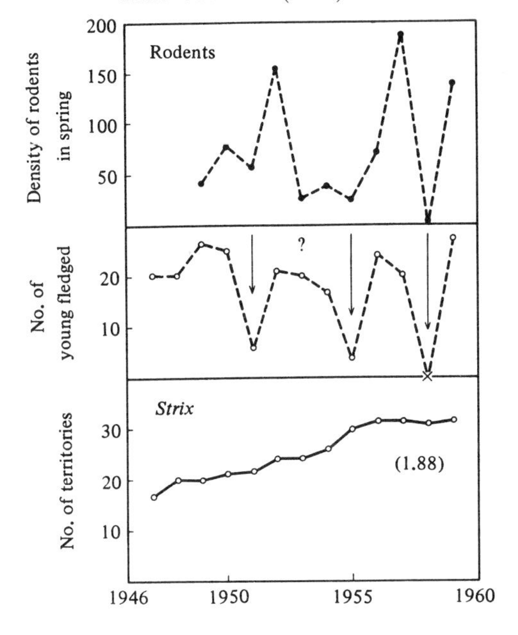

Table 3.5. *Fluctuation in the number of nestlings of nine species ringed in Great Britain from* 1953 *to* 1968

Common name	Scientific name	Max.	Min.	Ratio
Goldfinch	*Carduelis carduelis*	31	189	6.1
Greenfinch	*Chloris chloris*	254	796	3.1
Song thrush	*Turdus philomelos*	1214	3575	2.9
Blackbird	*T. merula*	2463	4863	2.0
Wren	*Troglodytes troglodytes*	12	88	7.3
Skylark	*Alauda arvensis*	183	447	2.4
Pied wagtail	*Motacilla alba yarrelli*	266	725	2.7
Robin	*Erithacus rubecula*	527	964	1.8
Chaffinch	*Fringilla coelebs*	281	590	2.1

Data from Ginn (1969).

as practically not fluctuating at all. The change in number was conspicuous, however, of the fledgelings (excluding the year marked × in the figure, when no bird fledged, the amplitude was 7 in ratio), which were few when the number of rodents (top figure) was particularly small (why the number fledged was large in 1953 when the rodent population was low is not known). If the rodent number is small in spring, not only do birds starve but the incubating females are also not fed by the males. Consequently they leave their nests to hunt and the eggs die of cold. In spite of such fluctuations in the number of fledgelings, the population is very stable because the adults live long and territoriality severely limits the number of breeding pairs in the forest.

Othe birds with relatively low fecundity include gulls and gannets. They also seem to fluctuate little in number. Here let us examine a population of penguins which exhibits the low-fecundity/parental-care strategy instead of the high-fecundity strategy expected to be found in the severe environment. Fig. 3.59 shows part of the result of a detailed investigation of the life history made by Richdale (1957) on the yellow-eyed penguin *Megadyptes antipodes*, which nests near the southern end of New Zealand (cool temperate). Here the number of residents (there was a small number of individuals joining from other rookeries) varied between 90 and 150, which gave a ratio of only 1.6 in amplitude. Fluctuation in the number of reared young was also small, being 2.7 in ratio. From this it can be seen that the small amplitude of fluctuation is

Fig. 3.59. Fluctuations in the number of residents (number of individuals older than one staying in the breeding ground) and of the young reared till Christmas in the yellow-eyed penguin *Megadyptes antipodes* (maximum–minimum ratios are given in parentheses). Drawn from Richdale (1957).

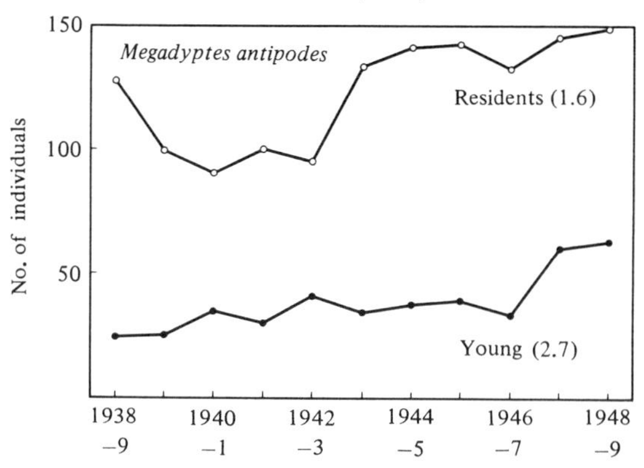

due not only to the long life-span but also to the small fluctuation in the number of young produced.

The yellow-eyed penguin lays 2–3 eggs in the cool temperate environment, thus it is not a typical species of penguin, but the small amplitude of its population fluctuation suggests the process by which the emperor penguin and the king penguin have successfully colonized the extreme environment of the antarctic, breeding in blizzards, with a clutch of one egg in their low-fecundity/parental-care strategy.

(4) The rodents discussed earlier in this section belong to the group of high-fecundity species with type B or C survivorship curves. They have no territoriality or complex social communication systems. There are, however, other types of rodents with different modes of life, such as territorial ground squirrels, communal beavers, prairie dogs and other group-living species with complex communication systems.

Carl (1971) studied the ecology of the arctic ground squirrel *Spermophilus undulatus* on the coast near Point Barrow in Alaska. The fact that the lemmings show violent population fluctuations in the tundra nearby has already been mentioned. This ground squirrel had a very stable population (at least in the three years of study). This is because this species, unlike lemmings, limits its number by territoriality (the vegetation in which this species is found is different from the vegetation in which lemmings occur).

Slade & Balph (1974) studied another species of ground squirrel,

Fig. 3.60. Population fluctuation of the Unita ground squirel *Spermophilus armatus* in the experimental forest of Utah State University (drawn from Slade & Balph, 1974). Black circles indicate numbers during the period of elimination experiments. (Maximum–minimum ratio in parentheses.)

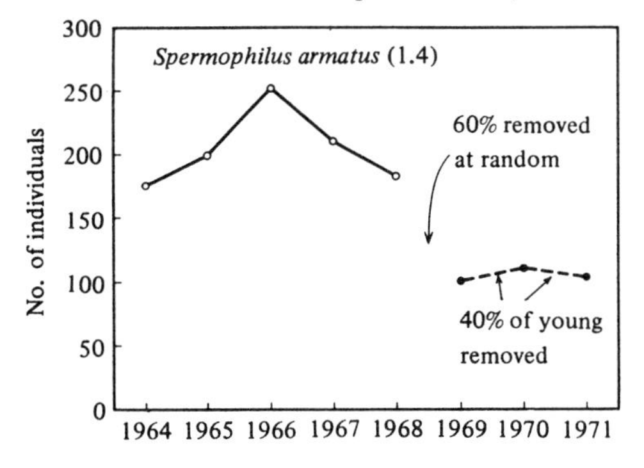

Spermophilus armatus, in the State of Utah (Fig. 3.60). This study lasted only for five years and the amplitude of population fluctuation during this period was 1.4 in ratio. A field experiment has eliminated about half the population since the summer of 1968, but the density, though lower than before the experiment, appeared to be stabilized. The habitat of this species can be divided into lawn and a mixture of shrubs and grass, of which the lawn is best suited for the species. Before the elimination experiment more than 80% of the females living in the lawn succeeded in reproduction while in other areas the reproductive success was 60% for those older than one year and less than 50% for the first-year animals. After the elimination, however, the remaining individuals in the latter habitat reproduced as well as those in the lawn. The interpretation of the results is that this species breeds in the lawn and the surplus individuals invade other habitats where suitable areas for reproduction are limited and many animals fail to reproduce.

Seton (1911) presented graphs of hunting statistics of furred animals kept at the Hudson's Bay Company. Figs. 3.40–3.43 were based on the same statistics. Here, Fig. 3.61 shows fluctuations in the numbers shot of the American beaver *Castor canadensis* and the brown bear *Ursus arctos* from Seton. The population fluctuation of either species is non-cyclic and its amplitude (apart from a swell in the graph) is small. Elton (1925, 1927) also noted the stability of the beaver population among the rodents. The stability is achieved probably because this species builds dams to regulate the water level – i.e. it has the ability to stabilize the environment – and is protected from predators by its strong nest. The litter size of beavers is 1–6, the fecundity being lower than in most rodents except arboreal species.

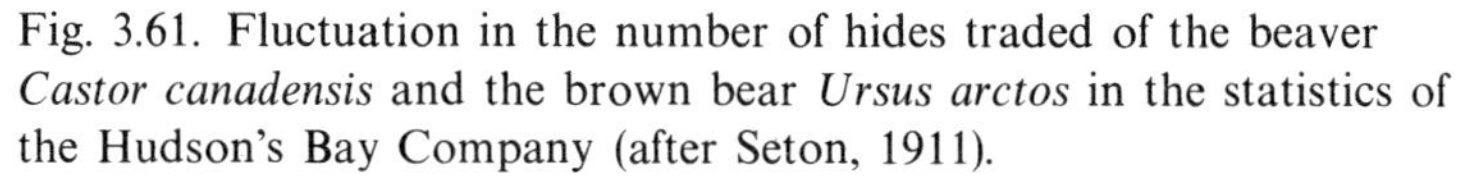
Fig. 3.61. Fluctuation in the number of hides traded of the beaver *Castor canadensis* and the brown bear *Ursus arctos* in the statistics of the Hudson's Bay Company (after Seton, 1911).

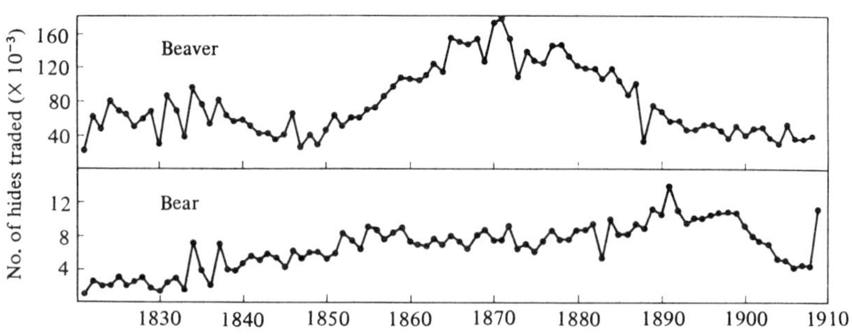

Population data for large carnivores are unfortunately scarce (fish-eating sea mammals, such as fur seal and ringed seal, have been counted on islands where they come to breed; the populations of these animals have hardly any fluctuation but exhibit a gentle trend of increase or stability). Buechner (1960) in his paper on the bighorn sheep presented the figure for pumas (cougars) *Felis concolor* shot in the States of Arizona and California. This is shown in Fig. 3.62. A bounty was placed on the carcass but the hunters sometimes left the nursing female alone, thus the number does not necessarily reflect population fluctuation in the field. It is almost certain, however, that the amplitude of the fluctuation is small in this species.

Fig. 3.62. Fluctuations in the number of the puma *Felis concolor* killed by individual hunters in California and in Arizona (maximum–minimum ratios are given in parentheses). Drawn from Buechner (1960).

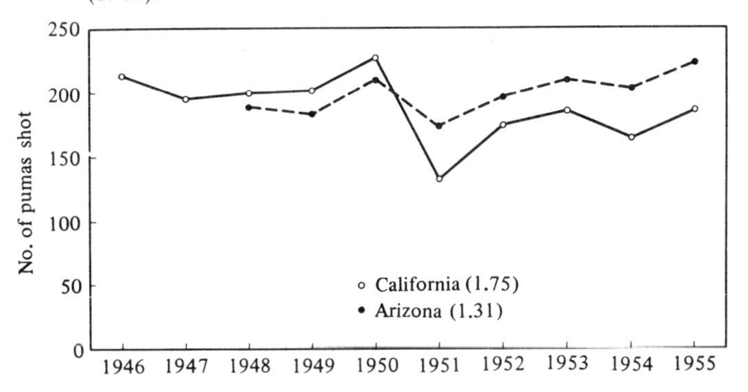

Fig. 3.63. Population fluctuation of the bighorn sheep *Ovis canadensis* in Yellowstone National Park (drawn from Buechner, 1960). Broken line indicates absence of data. (Maximum–minimum ratio in parentheses.)

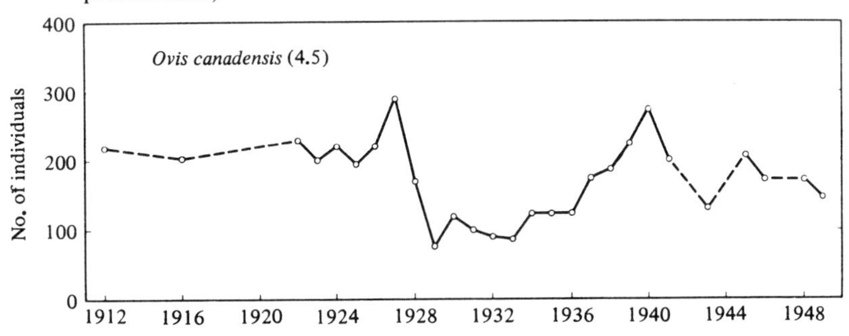

The large ungulates have a litter of 1–2 at a time and their number is very stable. Fig. 3.63 shows the population fluctuation of the bighorn sheep in the Yellowstone National Park, drawn from a table in the monograph of Buechner (1960). The statistics cover a period of thirty-six years, yet the maximum–minimum ratio of numbers is only 4.5. Normally the puma and the wolf are the only predators of this species. Their large body size, long life-span and group life probably minimize death caused by chance fluctuations of environmental factors and help keep the population stable.

A surprising fact is that even such a large animal can suffer great mortality as it did when a disease spread. Buechner investigated population fluctuation of this species in Pikespeak, Colorado, where the maximum–minimum ratio of numbers was less than 1.5 over the period 1949 to 1952, yet in 1953 the number fell from 160 to a mere 10 (Fig. 3.64). According to Buechner this was due to an infestation of a lungworm. The population has since shown recovery. Another noticeable feature of this graph is that only 5–10 young were born from 50 to 100 females. Some sort of social regulation is probably operating on the population.

Fig. 3.65 shows the population fluctuation of the wapiti (American elk) *Cervus canadensis* which fluctuated in number with a small maximum–minimum ratio of 1.8 over twenty years. Apart from these examples, Bergerud (1971) also presented data showing a gentle fluctuation in a population of caribou, *Rangifer arcticus*, in Newfoundland.

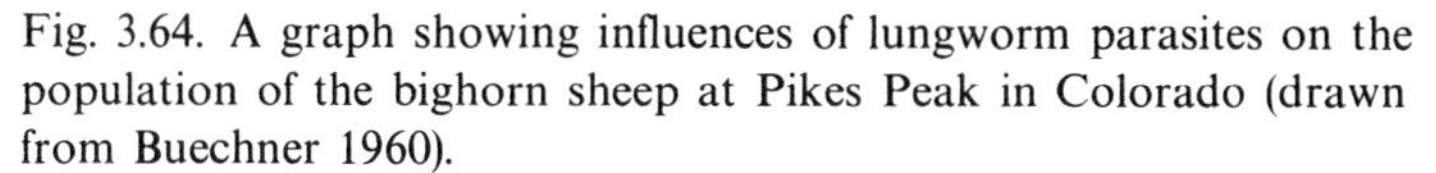

Fig. 3.64. A graph showing influences of lungworm parasites on the population of the bighorn sheep at Pikes Peak in Colorado (drawn from Buechner 1960).

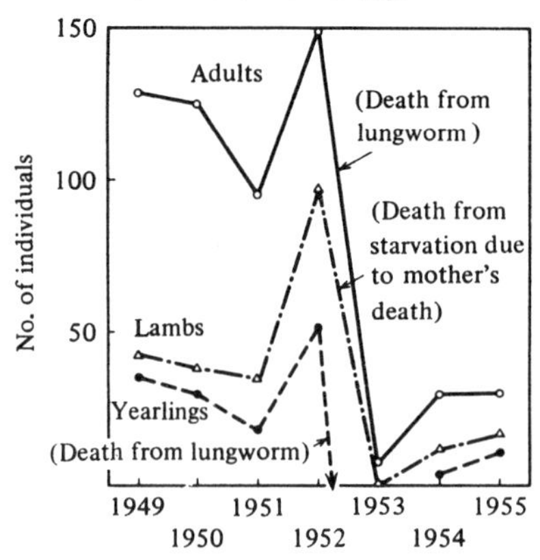

As yet long-term population data of primates are not available (there are many data on troops of 'provisionized' monkeys but they live an unnatural life). The study of Suzuki *et al.* (1975) of the Japanese macaque *Macaca fuscata* in the Yokoyugawa basin, Shiga-kōgen, includes observations of two natural troops (B_2 and C) whose population fluctuations are shown in Fig. 3.66. The fluctuations are very mild and judging from the number of infants in the troops the number born was also considered to vary little (generally it is less than half the number of females in the troop). The number is not the total for the area but is believed to represent the density, for the home range of each troop did not change in spite of budding off of new troops and shifting of individuals to other

Fig. 3.65. Population fluctuation (winter) of the wapiti (elk) *Cervus canadensis* in Gullatin Canyon, Wyoming (maximum–minimum ratio in parentheses). After Peek *et al.* (1967).

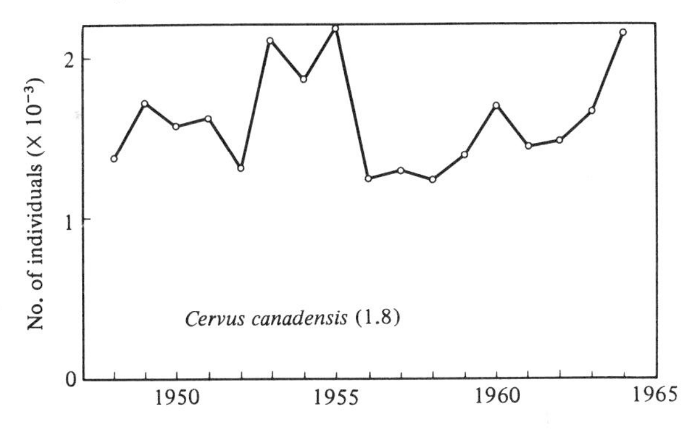

Fig. 3.66. Population fluctuations of two troops (B_2 and C) of the Japanese macaque *Macaca fuscata* at Yokoyugawa, Shiga Plateau, Japan (maximum–minimum ratios are given in parentheses). After Suzuki *et al.* (1975).

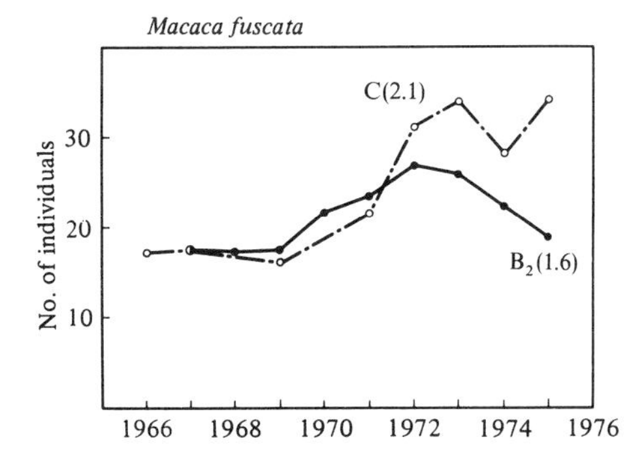

troops. In other words, this population of the Japanese macaque is controlled by splitting of troops and departure of individuals from the troop.

In summary, the higher animals with low fecundity and parental care are little influenced by the caprice of physical environment and have long life-spans, resulting in only mild fluctuations of numbers unless they are hunted by man or their habitat has changed drastically by man's activities.

Reproductive strategies and population fluctuation

(1) In Chapter 1, I stated that many a time organisms had faced the choice of two strategies – the high-fecundity strategy and the low-fecundity/parental-protection strategy – and that high fecundity developed in the environment in which it is relatively easy for the young to procure food and low fecundity with parental protection developed in the opposite environment. In Chapter 2, I have shown that the differences between these two reproductive strategies have influenced the way mortality occurs in the population. That is, high early mortality has been selected for in the high-fecundity group, and low early mortality and concentrated mortality either about the time of attaining the reproductive age or thereafter in the low-fecundity/parental-protection group. Such differences in the strategy of life are also thought to influence the way thc population fluctuatcs. If so, wc should be able to predict types of fluctuation from the strategies adopted by the species and determine the method of collecting data for the study of population dynamics. For this we do not have to wait until the life-table has been constructed and analysed for all species as some Russian and European workers suggest. I put forward this methodological idea in 1959 in the first edition of this book. At that stage, however, the choice between the two reproductive strategies was not considered to have been repeatedly faced by organisms at various stages of evolution. The phylogenetically older groups were thought to have high fecundity and the so-called 'higher' animals were thought to have low fecundity and parental care. Also, the important difference between the irregular fluctuation and the cyclic fluctuation of populations was not fully appreciated. Only the amplitude of the fluctuation was treated. For this reason it was not possible to explain, for example, why the oceanic fishes while laying enormous numbers of eggs had relatively stable adult populations or why the lynx, a large carnivore with well-developed parental care, fluctuated several

hundred-fold in number. Here I shall reflect on these points and examine the relationships between reproductive strategies and population fluctuation.

Increase of micro-organisms is governed in almost all cases by external factors. This is a reasonable deduction from the low level of their ecological homeostasis. Thus the fluctuation of numbers is irregular and its amplitude is often very great. Decrease in number may be induced by changes in the external conditions, but because these organisms have a high reproductive potential they often deplete nutrients rapidly and decline in number as a consequence (a typical example is 'red tide').

Population fluctuations of many marine invertebrates which lay large numbers of eggs (except for sessile species) are brought about basically by external factors and are irregular and violent. The amplitude of fluctuation is great, usually the maximum–minimum ratio is 50–60 even for a population at a certain stage of maturity, but sometimes it reaches as high as 1000. Also, the longer the period of planktonic life the greater the magnitude of fluctuation as shown by Thorson (1950).

However, it is not always the high-fecundity strategy that older phylogenetic groups – morphologically simpler groups – have chosen. For example, some freshwater planarians produce only a few young which at the time of birth are as big as 20–30% of the adult in body length. Their early mortality is low and the population is stable. They produce venom to protect themselves against predators but their environment is often poor in nutrients. They have adopted the low-fecundity/parental-protection strategy to colonize small bodies of water and mountain streams. The population fluctuation of such species is naturally very different from that of marine invertebrates with planktonic larvae.

With respect to the high-fecundity strategy the marine fishes occupy a unique position among all animals. The fluctuation of catches is generally very stable and many of them fluctuate less than ten-fold. However, this is a trend of the adult population which consists of fishes of many age (year) groups, each year group differing in number reflecting the independent survival-rate of that year group during the fry stage but *in toto* masking such differences and damping down yearly fluctuations. In fact, the few statistics that deal with young fish show a much greater fluctuation of fry or juvenile populations compared with adults, and such fluctuations have been attributed mainly to external factors such as sea temperature, current, influx of freshwater and so on. I shall discuss this point further on fishes in the section on 'metamorphosis'.

(2) In insects, species with the high-fecundity strategy, such as noctuid moths, pyralid moths, budworms and planthoppers, generally show irregular and great fluctuations of numbers. In contrast, species with relatively low egg numbers, such as grasshoppers and butterflies, or social insects with the low-fecundity/parental-protection strategy (only few quantitative data are available), fluctuate little in abundance (for dragonflies and cicades, see below).

At the same time, there is a special group among the insects. This is a group of pest insects whose outbreaks constitute a gradation. Typical examples include migratory locusts and the larch budworm. In the case of migratory locusts, the population increase is generally triggered by changes in the rainfall pattern, but once it breaks through the threshold density the phase variation appears. When the population is further multiplied a thousand-fold or more, the gregarious phase takes over and the population departs for migration, leaving the site of the irruption now devoid of vegetation. In the case of the larch budworm the population reaches its peak density every 8–9 years at a level tens-of-thousands of times as great as the low-level density. At the peak density the needles of larch trees, the insect's exclusive food, are eaten out over a very large area and the population begins to decline as individuals starve and suffer from diseases. From the value of r in its increasing phase the larch budworm may be classified as an r-strategist, but this species is not merely trying to overcome the difficulty of surviving in a fluctuating environment with a large r. In fact that is not the case; the influence of the external environment (climate) is hardly felt by the population. The larch budworm is forming a system with the larch that is deciduous by nature and is not killed by insect-induced defoliation. The large cyclic fluctuation which this insect population undergoes is a result of the special strategy adopted by the budworm in this simple ecosystem of the Swiss Alps (at altitudes of 1300–1800 m) where the trees are covered with snow for several months of the year. The low species diversity in this ecosystem, the strong ability of larches to recover from defoliation and the high fecundity of this insect are combined to evolve such a strategy.

Such dynamic equilibrium with a kind of stable, rather than unstable, fluctuation has also developed in the lemmings. The lemmings, though belonging to the high-fecundity group (litter size of 5–15 and three litters a year in the increasing phase) among the mammals, are far less fecund than many fishes and insects. In spite of this the maximum–minimum ratio reaches several hundreds or more. This is a strategy developed in a simple ecosystem of the harsh polar environment. In the increasing phase these herbivores are not checked by any major mortality factor, but hunger, social stress and concentrated predators all take a toll and reproduction decreases before the vegetation is completely destroyed. This strategy evolved through the adaptation of lemmings to widely fluctuating polar conditions. Initially, this was probably adopted to counteract local extinction, to meet the need to increase rapidly from small pockets and then to disperse. Stability of the system with large cyclic fluctuations was probably achieved by the process of co-evolution among plants and animals including the foxes.

In Chapter 2, I also stated that the arctic fox was able to adapt itself to the severe environment of the tundra (where prey species fluctuate in number) by the high-fecundity strategy (the arctic fox is the most fecund among the North American foxes). In the lemming year additional foxes move into the tundra from forests so that the fluctuation of the number appearing in the hunting statistics is exaggerated. Nevertheless, it is certain that foxes are fluctuating violently in number and much more conspicuously than other mammals of similar size, and this too is an adaptation to the cyclic fluctuation of prey populations. Initially, it may have begun with the high fecundity strategy to cope with the unstable environment.

At present the relationship between the varying hare with a ten-year population cycle and the vegetation is not clear. The relation between the varying hare and the lynx might be a rare example of prey–predator interaction which produces population oscillation in nature. If this is accepted we may interpret the large magnitude of several hundred-fold in population fluctuation in this case not as representing instability but as a dynamic equilibrium of the populations established through co-evolution.

These examples of cyclic fluctuations in tundra and taiga are exceptional among the warm-blooded animals. Among the birds and mammals, most of those with the high-fecundity strategy (in the relative sense) characteristically show large and irregular fluctuations of numbers whereas others with the low-fecundity/parental-protection strategy produce fluctuations of small amplitudes. The first group is represented by rats, mice, hares and many ground birds and the second group by forest birds, island birds, many large ungulates, carnivores and primates (though more quantitative data are required to substantiate this).

Of these, the large mammals and primates mostly have long life-spans, and yearly fluctuation of new arrivals may be masked as we have seen in marine fishes. Various data indicate, however, that in these animals the young population does not fluctuate as in rats or hares. Territorial forest birds, particularly the predatory species with low fecundity (though only one species has been studied in detail), have very small amplitudes in population fluctuation. Another fact worth noting is that in this group of animals the population suffers little effect from external factors such as the weather.

The above discussion does not necessarily mean that the high-fecundity strategist is primitive in its response to environmental changes and the low-fecundity/parental-protection strategist is advanced. The truth is that by adopting different strategies the animals have solved the problem differently. For example, the high-fecundity strategist was able to adapt itself to a wildly fluctuating environment by adopting this strategy (even when environmental instability causes high mortality). Thus black rats and Norway rats were able to conquer the world while rabbits became rampant in Australia.

The adaptations of organisms are extremely varied. Many animals try to colonize polar regions by adopting the high-fecundity strategy. Yet, some penguins, by adopting their special mode of life – small clutch size and communal caring for young – succeeded in maintaining the population in the antarctic blizzards, in the absence of predators.

The choice between these two strategies is a problem confronting species at all times.

The origin of metamorphosis

(1) In the first edition of this book I stated that the population fluctuation of small-egg/high-fecundity species had a large amplitude and speculated that in these species the fluctuation was governed mainly by the change in external conditions.

But many marine fishes with a large number of small eggs have relatively stable adult populations. Besides, in many species the density effect is seen in the retardation of growth-rates, reduced fecundity and increased mortality of adults. Why should this be so?

Here, we arrive at the question of the ecological significance of *metamorphosis* in the life cycle. In all marine fishes with small eggs and high fecundity the young hatching out of these eggs are very small planktonic organisms. Most of them die during the planktonic stage which lasts for a set period of time according to the species. The few that survive change their morphological features (become adult-like) and enter the adult population. Once they reach this stage their life expectancy increases, which means that their survivorship curve, though called type C, is in fact a combination of two lines broken at the age of transformation. It is this fact that explains the apparent contradiction that these fishes, while laying a large number of unprotected eggs and suffering from high mortality during the early stage of life, can maintain fairly stable adult populations.

The small egg size and high fecundity were the superior strategy for reproduction in the unstable environment of a vast ocean. The alternative was to become a large predator with a viviparous tendency (such as sharks). But for the mode of life with high fecundity and a 'compensating' pre-reproductive mortality of 99.99% or greater, if this high mortality occurred just before reproduction it would be a tremendous waste of resources; consumption of all that energy for nothing. Nature does not permit such superfluous expenditure! The type C survivorship curve with a colossal mortality of small young before their food consumption becomes wasteful is also an economic strategy under the premise of high fecundity. Individuals leaving the planktonic life change their body shape, size and food habits. Many of them enter the life of large predators whose survival and continued growth are governed by the food supply. In other words, marine fishes have adopted a special strategy of transforming from the *r*-strategist's life to the *K*-strategist's life by metamorphosis. They have competed successfully among the adult fishes in the ocean. [My view, unlike the theory of *r*-*K* strategies, is that interspecific competition is severer in the environment in which the high-fecundity strategy has been selected. See p. 42, Chapter 1.]

At the other extreme we find elasmobranchs which originated before the evolution of the teleosts. Their survival till the present day may have been due to the low-fecundity/parental-protection strategy that they have adopted. The young born from the mother (ovoviviparous) or a strong egg-case are endowed with strong powers of locomotion and the ability to escape from predators. It seems that this mode of life could only evolve when they already possessed a powerful predatory ability. The largest sharks, such as *Rhincodon typus*, are plankton feeders, but this is considered a secondary adaptation, like that of baleen whales, which occurred after they had attained the large size as predators. This change to plankton feeding assured further enlargement of the body size.

It is noticeable that the mode of life in the ocean has evolved either in the direction of production of enormous numbers of eggs or towards low fecundity with viviparity. There appears to be no intermediate way of life.

(2) There are other animals with a wildly fluctuating juvenile population and a very stable adult population. The population of fiddler crabs is stable on the seashore but the larval population (zoea stage) in the sea is thought to fluctuate enormously (Takeo Yamaguchi, personal communication, 1976). Yamaguchi considers that this is typical of those intertidal animals whose larvae live a planktonic life and whose adults become sessile or sedentary as benthos.

In Chapter 2 mention was made that the survivorship curve of a freshwater turtle was characterized by high early mortality and low adult mortality. Here too, the curve is broken into two parts. This tendency is probably more conspicuous in the sea-turtles because they are the most fecund group among the reptiles. Turtles do not metamorphose, but they too have a drastic change of life. This is another adaptive strategy combining high fecundity with a long life-span.

Morisita (1976) drew attention to the fact that the crocodiles, which protect eggs and young, lay many more eggs than many lizards which show less parental care. This was explained in terms of the protection achieved by lizards burying their eggs in the ground. Crocodiles also bury their eggs in the ground but here it means there is a transformation of life style from land to water. This transformation probably induced crocodiles to adopt a combined strategy of high fecundity and long life-span.

(3) Was the metamorphosis of insects beneficial in a similar switch-over of the mode of life? Here, I do not mean the distinction between the Hemimetabola and the Holometabola made in entomology textbooks. These two are separated according to whether or not they have a pupal stage. Among the insects with incomplete metamorphosis (Hemimetabola) there are two groups: cockroaches, grasshoppers, pentatomid bugs, planthoppers and others, in which nymphs and adults are essentially similar in shape and food habits, the only difference being that the adults have functional wings and can reproduce; and mayflies, dragonflies, damselflies, cicadas and others, in which nymphs and adults are ecologically different, living in entirely different environments and eating different food items. There are very few insects with complete metamorphosis (Holometabola) in which larvae and adults use the same methods of feeding (e.g. ladybeetles, chrysomelid beetles). The egg number of the Holometabola, excluding species with parental protection, is mostly several hundreds to over a thousand, but the pentatomid bugs and grasshoppers that do not transform ecologically do not lay so many eggs.

Those with a long life-span can lay more than once but their longevity in the field is limited. In reality the number of eggs laid per female is fewer than 100. And the survivorship curves of the pentatomid bugs, grasshoppers and leafhoppers are intermediate between type A and type B.

On the other hand, cicadas and dragonflies, though they are classified in the Hemimetabola because of the absence of the pupal stage, have an entirely different nymphal life. The cicadas lay several hundred eggs and their survivorship curve shows very high early mortality unlike those of the leafhoppers or the planthoppers. In dragonflies there are species known to lay several thousand eggs and their survivorship curves would probably be characterized by high early mortality. However, the abundance of adults shows little fluctuation in cicadas and dragonflies – particularly in the territorial species. Among the Odonata those scattering eggs over temporary bodies of water lay more eggs (typical high-fecundity strategists) than those laying on pond weeds (e.g. damselflies). In spite of this the number of adults is regulated, showing the effect of the transformation of life history.

In summary, the strategies of high fecundity and low fecundity/parental protection are influencing the fluctuation of animal numbers in many known cases; generally, in the former the population fluctuates irregularly and drastically whereas in the latter the population fluctuation is small. At various stages of evolution animals repeatedly selected one or the other of the two strategies, and accordingly the type of population fluctuation was determined. In this process, some species evolved a complex type of population fluctuation, with fluctuating larval populations and comparatively stable adult populations, while a few others combined the high-fecundity strategy with cyclic population fluctuation and succeeded in occupying a special position in simple but harsh environments. Thus the types of population fluctuation are not passively derived characters of different species, dependent on external factors, but they are different adaptive strategies of the species.

4

Territoriality

The emergence of territories

The word territory is one of those zoological terms that has been used in many different ways. Here I follow Noble's (1939a) definition: 'any defended area'. In the 'defence' of an area I include not only direct attacks and threats but also use of vocal communication, pheromones and ritualized postures, which cause the intruder to withdraw.

Eliot Howard, who published a book entitled *Territory in Bird Life* in 1920 and diffused knowledge about the existence of territories in nature amongst the professional and amateur biologists of the English-speaking world, used the word territory as implying defensive acts of the territory owner or holder. However, among the mammalogists the word often meant a range of movement of an animal during its normal activities and sometimes was even referred to as the range of a species. Noble's definition has been accepted by most zoologists particularly since Burt (1940, 1943), a rodent ecologist, stated that we should distinguish between the home range, which encompasses the area occupied during the normal range of activities of an individual or a group of individuals, and the territory, which is that part of the home range which is defended against intruders or is advertized by means of calls, pheromones or other signs to prevent invasion. There are some recent proposals for a new definition including one by Wilson (1975), but these are all based on Noble's definition and give different phraseology in the modification of it. [Wilson's (1975) definition: 'an area occupied more or less exclusively by an animal or group of animals by means of repulsion through overt defense or advertisement'.]

Armstrong (1947) proposed that we should use the term only in relation to the phenomena seen between individuals of the same species but Hinde (1956) found this restriction inappropriate, for there were cases where different species could be driven out of the territory by mechanisms operating among the individuals of one species. However, considering the fact that territory serves a density-regulatory function and ordering of mating within the species concerned, it should be basically thought of as an area defended against other individuals or groups of individuals of the same species. [If the 'defensive acts' were expanded to include defence against other species, the distinction between territorial defence and prey-catching behaviour or predator-mobbing behaviour would create a problem.]

The word territoriality is used where some regulation of normal social life is maintained by animals having territories (see Imanishi, 1951). Under special circumstances, such as fish kept in an aquarium, the defence of an area observed will not constitute territoriality unless it has significance in the *normal* life of that species in the field.

In order to establish a territory it is necessary for an animal to have acquired certain behavioural abilities. Emergence and development of territoriality are thus made possible by a certain degree of evolution of individuality in animals. In the first edition of this book I wrote that 'at present the territory as defined by Noble has not been recognized outside of the arthropods and vertebrates – even among the arthropods it is limited to some crustaceans, spiders and a relatively small number of insects'. In spite of many publications on the subject during the following seventeen years – far more than the total number of such papers published till then – according to a review of Wilson (1975) the discovery of a territory outside these groups is reported only for a species of limpet, *Lottia gigantea* (Stimson, 1970). [Another gastropod example was reported in 1975 (see below).] Thus territoriality can be said to be a social system basically found only in the arthropods, fishes and higher animals. [We must note, however, that we have very little knowledge of tropical animals today. As tropical research progresses the emergence of this social behaviour may be found in lower animals.]

A primordial territory can be recognized in very small animals with limited behavioural ability; a phenomenon of dispersal, which occurs in a population of high density and by which a suitable density is maintained.

Itô (1952a,b, 1953, 1960a) found in the viviparous generations of the soybean aphid *Aphis glycines* and a few other species of aphid that, when the density on a plant leaf becomes very high as a result of reproduction, they disperse to other intact leaves of the host plant and the population equilibrium is maintained by dispersal. Similarly, aphids disperse from one plant to another as the density on the plant becomes too high. Miyashita (1954) reported that the upper limit of density reached by the green peach aphid *Myzus persicae* which attacks cabbage leaves is different depending on whether unparasitized host plants are available or not in the surroundings.

Morisita (1950) reported that if overwintering adults of the water-strider *Gerris lacustris* appearing on a pond increase in number beyond a certain density some individuals disperse to other, less favourable ponds. Such density-dependent dispersal has been established for various crabs (Harada & Kawanabe, 1955; Ono, 1957; Connell, 1963), frogs and toads (Martof, 1953a; Pearson, 1955; Kikuchi, 1958). In the apple maggot fly *Rhagoletis pomonella*, females produce a trail pheromone while laying in fruit and deter repeated oviposition by others (Prokopy, 1972). Similar use of pheromones is known for the cowpea weevil *Callosobruchus chinensis* and many parasitic wasps but in these cases only one piece of food for the young is marked. It is hardly an area defended though marking is a form of defence here. However, if such behaviour has a survival value it is possible for it to develop into defence of a group of food items and finally to evolve into defence of a specific area.

The definition of Noble does not imply that the defended area must be fixed. Thus the defence of sites not fixed may be referred to as a territory, for example water-striders attacking approaching individuals while moving slowly over the surface of water, but from the functional viewpoint the word territory should perhaps be reserved, as a rule, for the defence of fixed areas tied to residence (residentiality). [The reason for writing 'as a rule' is that, though we don't call the space surrounding a walking spider a territory, examples such as the defence by a male bitterling, *Rhodeus amarus*, of a moving mussel, *Anodonta*, into which eggs are laid (see p. 212) or the defended area around an adult male in a herd of wildebeest slowly migrating over the African continent (see p. 253), may be included in territoriality.] For the existence of territoriality the structure of the environment has to be such that animals can be conditioned to certain sites or can commit sites to their memory (in a broad sense). Outside the higher vertebrates well-developed territoriality is found mostly among the insects and fishes that have a two-dimensional life, for example ants rather than bees and benthic rather than pelagic fishes. This probably ties in with their limit to the power of topographic memory.

In a conflict over a territory the characteristic of territoriality is that, unlike social hierarchies (Chapter 6) seen within groups of non-territorial species, the territory holder wins the fight as a rule even if he is smaller than the intruder. This is called the effect of prior residence.

Classification of territories

Mayr (1935) classified territories into the following categories:

Type A, defence of an area in which mating, nesting and food gathering for young occur;

Type B, defence of an area in which mating and nesting, but not feeding, occur;

Type C, defence of a mating station only;

Type D, defence restricted to narrow surroundings of a nest.

Nice (1941) added the following two to the above:

Type E, winter territories;

Type F, roosting territories.

Among many subsequent proposals, Wilson (1975) gave the following classification:

Type A, a large defended area within which sheltering, courtship, mating, nesting and most food gathering occur;

Type B, a large defended area within which all breeding activities occur but which is not the primary source of food;

Type C, a small defended area around the nest;

Type D, pairing and/or mating territories;

Type E, roosting positions and shelters.

The types A and B of Wilson are almost identical to those of Mayr. The types C and D of Wilson are similar to the types D and C respectively of Mayr while the type E of Wilson corresponds to the type F of Nice. I would add the following category, Type F, to the Wilson's classification. This would include Type E of Nice.

Type F, a defended area in which food supply is guaranteed whether it is for reproduction or not.

In the following the six types of territory mentioned refer to this modified version of Wilson's classification.

Territories of marine invertebrates

Many gastropods of the rocky shore exhibit homing habits. They move with the tide but return to their home sites on rocks at low tide. Ohgushi (1955) confirmed such movements of *Siphonaria japonica* and *Patelloida saccharina* by marking the shells and their sites of attachment on rocks with enamel; the same individuals returned to the same places for over ten days. No defensive behaviour of these was observed by Ohgushi but it was reported for another species of limpet, *Lottia gigantea*, by Stimson (1970). *Lottia* has a territory of about 1000 cm^2 in size, in which it feeds on algae. The algae can grow only within the territory of *Lottia* because other grazing molluscs are kept out of the territory. Similar behaviour has been reported for *Patella* by Branch (1975). These two examples seem to be the only published accounts of territoriality among the molluscs.

In the annelids there is a report on the defensive behaviour of a polychaete, *Nereis caudata* (Evans, 1973). This was observed when another individual was introduced experimentally near the nest. Since the defence is limited to the nesting hole it is doubtful whether the observed behaviour constitutes territorial behaviour.

Connell (1963) discovered a territory in the amphipod *Erichthonius braziliensis*. This species lives on the surface of algae and builds a tube in which to settle. They feed on small algae attached to the tube, which was defended if other individuals tried to enter. This observation was made in an aquarium but in the experiment these small animals were housed in a large tank simulating field conditions (Fig. 4.1). The same spacing mechanism probably operates in the field. This territory is of Type F in the classification presented above.

Among other crustaceans, lobsters are known to defend an area round

Fig. 4.1. Distribution of tubes (rectangles) and surrounding territories defended by a species of marine amphipod, *Erichthonius braziliensis* (after Connell, 1963). The left figure shows territories formed under an over-crowded condition; the individuals that have disappeared within eight days are indicated by black rectangles. The right figure shows the distribution after twenty days when there were no longer overlaps of territories (thus no defence); the normal feeding ranges are indicated by the cocoon-shaped areas around the rectangles.

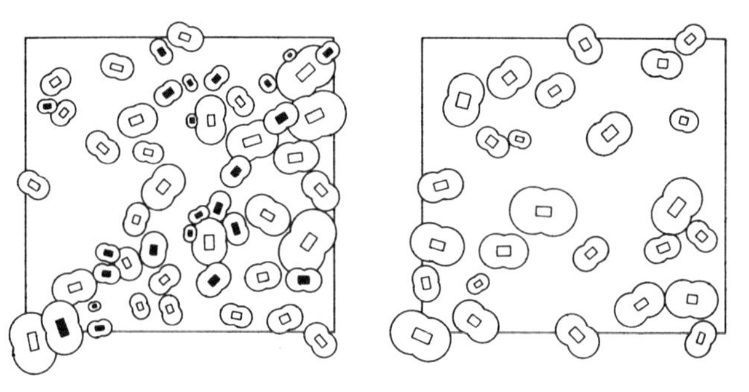

the shelter but little quantitative observation has been made in the field. In an aquarium study Fielder (1965) reported on the territorial defence of the spiny lobster *Jasus lalandei*. Since the defended area of lobsters is very small and not related to reproduction or feeding, their territoriality is of a primitive stage.

Among the crabs that have colonized the sandy beaches there are some estuarine species that defend the feeding area (a circular or fan-shaped area of sand round the nesting hole) as well as the nesting hole (Crane, 1941; Harada & Kawanabe, 1955; Ono, 1957). Harada & Kawanabe who studied *Scopimera globosa* reported that the nesting holes were always defended but the feeding area around it was not always defended. However, this species is known to exhibit density-effects on territoriality: Sugiyama (1961) and Yamaguchi & Tanaka (1974) have shown that when the density is increased the territory size diminishes and the crabs utilize deep layers of sand for feeding; if the density is further increased non-territorial wanderers appear. Thus in this species the type F territory may be considered established. According to Ono (1957) the fiddler crab *Illyoplax pusillus* has territorial defence among adult males for the nesting hole and the area with a radius of several centimetres round it. The females and young do not defend their nesting holes while the adult male tolerates the digging of their nesting hole near his own. In the territorial defence of adult males the territory holder usually wins the fight but occasionally the invader succeeds in taking over the nesting hole and the loser becomes a wanderer. Thus in the fiddler crab, males have territories while females and young have home ranges only. That behavioural differences exist among the crabs of different sexes and ages indicates that territoriality has become established as a social system in this group.

Among the isopods the territory is known for *Hemilepistus reaumuri* which lives in the semi-desert of Africa (Linsenmair & Linsenmair, 1971). This species lives in pairs in deep holes to escape the heat and defends the hole and the feeding area around the hole.

Territories of spiders and insects

(1) Many wandering spiders (e.g. *Atypus*, *Lycosa*, *Menemerus*, etc.) are known to have home ranges (Kuenzler, 1958; Mizuno & Funakawa, 1956). For example, the jumping spider *Menemerus confures* has a fixed range of movement, which can be shown by tracing movements of several individuals while they are catching prey on a wall. If one individual finds itself within 20 cm of another interference occurs and

one of them is driven away. If they are equally matched fighting ensues, but how this attack can be distinguished from prey-catching behaviour is not known.

(2) In insects territories are developed in very few groups: Orthoptera (crickets and some grasshoppers), Odonata (many species), Isoptera, Diptera (some tabanid flies), Hymenoptera (some Aculeata). Water-striders and butterflies show primitive forms of territory.

Here simple defence of food items is excluded. Defence of food is shown sporadically by a breeding pair of subsocial beetles. For example, the male of the burying beetle *Necrophorus vespillo* calls the female (probably by means of a pheromone) when he finds a lump of meat which he defends against other individuals. When the nest is built and parental care by the female begins the defence is terminated (Pukowski 1933; Milne & Milne 1976).

In the Orthoptera territorial behaviour is known for crickets (Alexander, 1961). The call plays a role of display in the crickets. Social hierarchy reported for crickets confined in a box may be concerned with modification of territorial behaviour rather than a hierarchy.

There are species of grasshopper which call aloud. Otte & Joern (1975) recognized territorial behaviour in a large grasshopper, *Ligurotettix coquilletti*. In this species each male exclusively occupies a bush in semi-desert and calls from the bush. It attacks any male entering the bush and usually wins the fight. As a result the individuals of this species have a very uniform distribution.

In the Lepidoptera some butterflies have been observed to chase others while showing site attachment. For example, Baker (1972) reported that males of the nymphalid butterflies *Aglais urticae* and *Inachus io* defended definable areas against intruding males and mated within the defended area. Suzuki (1976) considered, however, that in the small copper *Lycaena phlaeas daimio* the attachment to a site and indiscriminate chasing of other individuals (both sexes were attacked) were not advantageous for mating, and referred to this behaviour as 'so-called territorial behaviour'. At present, no butterflies are known to have clearly defined territories as seen in dragonflies.

In the Diptera there is a phenomenon known as 'hovering' among the Tabanidae and Syrphidae. These insects hover in a stationary position in the air above forest clearings or other fixed places where they chase approaching individuals and return to the same position in space. This is often called a territory (Toyoshima, 1955; and others) but this behaviour is merely indiscriminate chasing. Among the blood-sucking tabanid flies which attack domestic stock, there are species that chase conspecific males and mate with females during hovering. In these species mating occurs only at the site of hovering (Blickle, 1959; Anderson, 1971; H. Hayakawa, personal communication, 1974, reported at the general meeting of the Japan Society of Sanitary Zoology 1974; and others). Thus these flies have type D territories. Although there are many reported observations of hovering in tabanid flies, not many of them show mating only at the hovering site.

Apart from the social insects, territoriality is best developed in the Odonata [the first proof of this comes from D. Saint-Quentin (1934 cited by Moore, 1952) using a marking method]. There are very few territorial species in other groups of insects but in the Odonata reliable descriptions

Fig. 4.2. Distribution of territories held by the dragonfly *Orthetrum albistylum speciosum* (after Itô, 1960b).

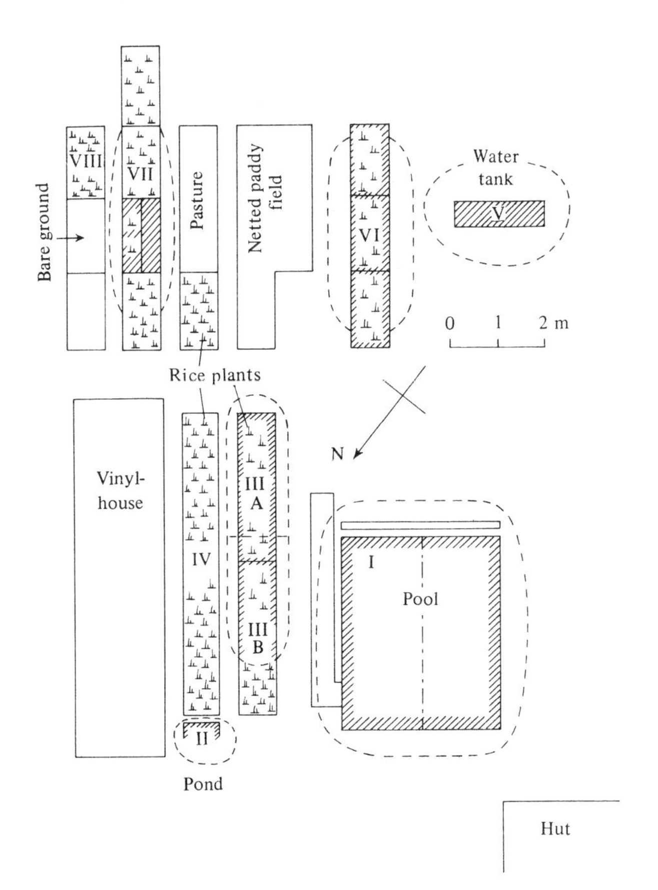

of territorial behaviour have been given for dozens of species. More than half the species of Anisoptera (dragonflies) and a considerable number of Zygoptera species (damselflies) are thought to have territoriality. No territory is known for the Anisozygoptera.

I studied territoriality of a common Japanese dragonfly, *Orthetrum albistylum speciosum* (Itô, 1960b). This species spends a few days in a bush after emerging. The male has white patches on the dorsal side of the abdomen and is called 'shiokara' (salted) while the abdomen of the female is mottled yellow and black and is called 'mugiwara' (straw). The subadult male is of 'mugiwara' type in colour. The emerged male stays in the bush until its abdomen becomes 'shiokara' type and then comes to a pond, where he secures a site containing some water surface and defends it against other males (Fig. 4.2). The dragonfly usually remains stationary at specific spots (not necessarily one spot) within his territory and if an intruder appears he takes off to attack and chase the intruder out of the territory. If a female enters his territory he catches her by the abdomen and they fly in tandem before copulating. Copulation does not last long and the female separates herself to start laying while the male hovers on guard above the female (Fig. 4.3). The male copulates many times as new females appear in the territory (there is also a case of one female copulating in two different territories). The territory holder usually wins the fight over an intruding conspecific but old individuals sometimes lose their territory to younger males and may shift to apparently less favourable sites such as a small pond covered with weeds (II in Fig. 4.2). At nightfall both sexes leave the water and spend the night in bushes but the same marked individuals tend to return to the same places on the following day. As can be seen in Table 4.1, No. 08 occupied one territory over six days including two rainy days of absence while No. 05 occupied another for four consecutive days. Such strong site-attachment (residentiality) may be the result of occupying small bodies of water. In the

Fig. 4.3. The male dragonfly *Orthetrum albistylum speciosum* (above) hovering on guard above the laying female (after Itô, 1960b).

Table 4.1. *Changes of marked (numerals) and unmarked (male signs) territory holders (all adult males) in the dragonfly* Orthetrum albistylum speciosum *(after Itô,* 1960*b*). *The location numbers correspond to Fig.* 4.2 *and Location III has been divided into two parts during the period of observation. A dash, an equal sign and a plus sign indicate, respectively, no observation, no territory holder and fighting severe with ownership undecided. The brackets indicate the territory holder being expelled during observation*

Date, Sept. 1959	Location						
	I	II	III		V	VI	VII
			a	b			
7 (p.m.)	01	—	—		—	—	—
8 (p.m.)	01	—	—		—	u	u
9 (a.m.)	03	01	05		04	—	—
9 (p.m.)	03	01	05		04	—	—
10 (a.m.)	=	01	05		04	07	—
10 (p.m.)	=	=	05		04	07	—
11 (a.m.)	—	—	05		=	07	08
11 (p.m.)	10	01	05		09	(04)	08
12 (a.m.)	♂	=	05		09	07	08
12 (p.m.)	12	=	(05)		09	11	08
13 (p.m.)	12	=	♂		=	13	08
14 (a.m.)	=	=	=		=	=	=
15 (a.m.)	=	=	10?		=	=	=
15 (p.m.)	17	=	♂		♂	18	(08)
16 (a.m.)	24	=	23		=	=	08
16 (p.m.)	26	—	—		—	08	—
	♂					♂	
17 (a.m.)	=	=	–		=	=	=
17 (p.m.)	(30)	=	23		♂		27
18 (a.m.)	♂	34	+		♂	28	—
18 (p.m.)	33	34	23	32	—	♂	16?
19 (a.m.)	33	=	23	32	34	=	—
19 (p.m.)	33	(36)	32	35	34	=	—
20 (noon)	♂	=	33	40	=	=	23
21 (a.m.)	42	=	41	40	23?	*	—
21 (p.m.)	42	40	=	44	23	43	39
22 (a.m.)	=	=	=	44	=	=	=

*Drought condition at the location.

large pond (I in Fig. 4.2) which lacks boundary markers site-attachment was not very strong (Table 4.1). A large dragonfly, *Anax parthenope julius*, appeared during the period of observation but territory holders showed no interest in it. Thus *Orthetrum albistylum* has typical territoriality in that it (i) chases only the intruding males of the same species, (ii) distinguishes males and females, and (iii) shows site-attachment. The territory has little to do with direct feeding, thus it belongs to type D. As Moore (see below) points out it serves to regulate the density.

Moore (1952, 1953) recognized territories in nine species of dragonfly in Britain. They show attachment to sites and perform mating and oviposition in their territory but in defence they cannot distinguish sexes very well. Thus the function of the territory was considered to be mainly for density regulation [Moore (1964) observed that when the density of one pond increased territorial fighting resulted in the dispersal of some individuals to other ponds].

Another view was expressed by Jacobs (1955) in America. She studied *Plathemis lydia* and *Perithemis tenera*, both with colourfully spotted wings. In both species new males acquire territories by displacing more mature males; the territory holder wins the fight against individuals of similar maturity. A male without a territory cannot acquire a female even if he is capable of copulating. This is probably true for the Japanese species reported above. The behavioural patterns used in territorial

Fig. 4.4. 'Duel flight' of male dragonflies, *Plathemis lydia* (drawn from Jacobs, 1955).

defence and in courtship are clearly different in both American species. In *Plathemis* the territorial fight is called the 'duel flight' in which dragonflies face each other and display the white powdery surface on the dorsal side of the abdomen (Fig. 4.4). Young males lack this coloration so that their display is not powerful. If the white part is painted over with enamel a mature male becomes weak in his display, thus he loses his territory and cannot acquire a female. The response to the appearance of a female is quite different; the male flies with the abdomen drooped. The two types of behaviour, one used in territorial defence and the other in courtship, are easily distinguishable also in *Perithemis*. Thus Jacobs considered that the function of the territory in dragonflies is the removal of interference in sexual activities – sexual isolation and sexual selection.

Since the above work territoriality has been discovered in many species of Odonata; *Nannophya pygmaea*, the smallest Japanese species (Tachikawa, 1957), *Jagoria pryeri* (Taketō, 1958), *Crocothemis servilia* (Higashi, 1969), *Hemicordulia ogasawarensis* (Sakagami *et al.*, 1974) and others in the Anisoptera and *Hetaerina americana* (Johnson, 1961), *Calopteryx maculatum* (Johnson, 1962) and others in the Zygoptera. *Anax parthenope julius* and *Macrodiplax cora* among the Anisoptera have not been seen to defend a fixed territory (Sakagami *et al.*, 1974). These species oviposit in tandem flight, thus there is no need for guarding during oviposition. They also perform group flight though aggressive encounters do occur among dispersed males. Most species of *Sympetrum*, in which oviposition occurs in tandem flight, and many small damselflies, which lay eggs in plant tissues while perched on the plant, are not known to hold territories. According to Sakagami *et al.* (1974), the females of *Rhinocypha ogasawarensis* and *Ischnura senegalensis* in Ogasawara lay unaccompanied by males and there is no territorial behaviour associated with mating and oviposition. Sakagami *et al.* claim that there is no anisopteran species in which a male does not attack other approaching males but I did not include aggressive behaviour as being territorial if it was not associated with a fixed site.

Recently, Ubukata (1975) made an interesting observation on the dragonfly *Cordulia aenea*. The mature males of this species range widely and do not hold territories while the density is low. When the density increases they occupy territories about 10 m long and 1–2 m wide, along the shore of a pond. If the density becomes very high the surplus males fly over the water without territories. Thus the territorial behaviour of dragonflies is still flexible.

Incidentally, very few members of the Anisoptera hold feeding territories (type F), as opposed to the mating territories discussed above. Sakagami *et al.* reported that in *Hemicordulia ogasawarensis* (Corduliidae) males showed territorial behaviour in the forest away from the mating and laying territories held at the pond. Other species of this family may also have similar behaviour when they are not engaged in reproductive activities.

(3) Among non-social members of the aculeate hymenopterans, type D territory is reported for *Sphecius speciosus* (Sphecidae) which hunts cicadas (Lin, 1963). The males of this species occupy territories near emerged holes of other individuals on the surface of the ground and chase males and other animals that arrive within a certain range of the hole. If a female enters the area she will be mated. Peckham *et al.* (1973) studied the ecology of five species of *Oxybelus*, which also belong to the Sphecidae but hunt dipterans, and observed in *O. subulatus* that the male positions himself at the entrance to the nest constructed by the female and attacks other insects including the conspecific males that appear near the nest while allowing the female to enter the nesting hole. The female, when she has carried enough food into the nest, drives the male out and seals the nest. Among the solitary bees, species of *Centris* (Anthophoridae) are reported to have mating territories defended by males (Raw, 1975). The male of *Centris decolorata* occupies a territory of 1–3 m^2 within a nesting colony and drives other males from it. Thus the nesting colony is divided amongst many males. When a female comes he follows her and they mate. This territory appears to be marked by a pheromone secreted from the mandibular gland. The same male is known to defend the same territory for many days.

In the three examples given above it was the male that defended the territory. As is well known, the males have no other functions than mating in the social hymenopterans. Thus it is interesting to note that when the Aculeata were evolving with the low-fecundity/parental-protection strategy the defence of nesting holes by males appeared. This habit will be found in many other species of wasp. However, instead of evolving into a monogamous family it has given way to the development of a matriarchal supraindividual system (Chapter 5).

In the social hunting wasps and bees defence of the nest (by females) is common. Yoshikawa (1973) reported that in the paper wasp, *Polistes fadwigae*, if two nests were artificially brought to within 5 cm of each other then severe fighting took place between the queens. These wasps often raid other nests located within a few metres and devour the larvae.

Honeybees and stingless bees also defend their nests. Kalmus (1941) found in the domesticated honeybee *Apis mellifera* that if a group of bees is trained with sugar water in a dish placed at a particular place the group defends the dish against bees of other strains. The degree of defence against bees from other colonies of the same strain, however, was found to be very weak. Thus the defence of territory in social wasps and bees is limited to an area around the nest, and defence of a large area as seen in ants has not been found. This is because ants are 'plane' dwellers and can occupy a fixed area behaviourally without difficulty, while wasps and bees need to visit scattered sources of food, distributed discontinuously and changing according to the season, be it nectar in flowers or live insects, which are not connected to the survival value in the defence of a fixed area.

Since the work of Elton (1932) the territories of ants have been studied in detail by many workers. The defence of a territory is made by the colony as a unit and tends to exclude not only the conspecific members of other colonies but also all other organisms except those insects which have special associations with the ants.

Morisita (1939) observed interspecific fighting among the ants *Lasius niger*, *Pristomyrmex pungens* and *Formica japonica* on a single tree. The first two species occupied large branches separately and collected honey from the aphids while defending the branches against other ants. The third species moved singly and stole honey from the aphids occupied by other ants instead of trying to occupy large branches exclusively.

Brian (1955) investigated territories and foraging behaviour of four species of ant by distributing many honey dishes in Scottish moors. The results showed that:

1. *Formica fusca* had a home range of 3.3 m in radius, of which 2.5 m radius was defended as a territory;
ii. *Myrmica rubra* had a home range and a territory of 1.3 m and 1 m radius respectively;
iii. *Myrmica scabrinodis* had a home range and a territory of 1.2 m and 1 m radius respectively.

The last two species occupy colonies of aphids exclusively (Fig. 4.5) as in Morisita's report.

Tsuneki & Adachi (1957) studied the territories of four species of ant by attracting them to sources of food as Brian did, except that they used dead insects instead of honey. According to this study, *Camponotus herculeanus japonicus* had a home range of 10 m in radius for large nests and 2 m radius for small nests. The ants defended most of their home range against conspecific ants belonging to other colonies (Fig. 4.6 left). The home range of *Formica japonica* is much larger than this, the frequently used part of it being greater than 10 m radius, and overlaps with home ranges of conspecific ants from neighbouring colonies. The third species, *Tetramorium caespitum jacoti*, formed a large colony with many nest holes but there was a home range around each hole which did not overlap with others. This species and the fourth species,

Fig. 4.5. An example of territorial arrangements in ants (after Brian, 1955). A, *Myrmica rubra*; B, *M. scabrinodis*; black circle, a plant on which *M. rubra* was rearing aphids; black triangle, a plant on which *M. scabrinodis* was rearing aphids; plant figure with no marking, a plant without aphids.

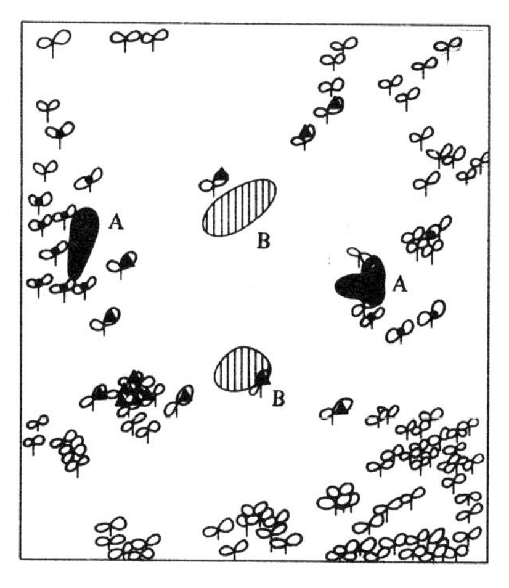

Fig. 4.6. Arrangements of home ranges in ants (after Tsuneki & Adachi, 1957). Left figure, *Camponotus herculeanus japonicus*; right figure unbroken lines, *Tetramorium caespitum jacoti*; right figure dotted lines, *Aphaenogaster famelica*.

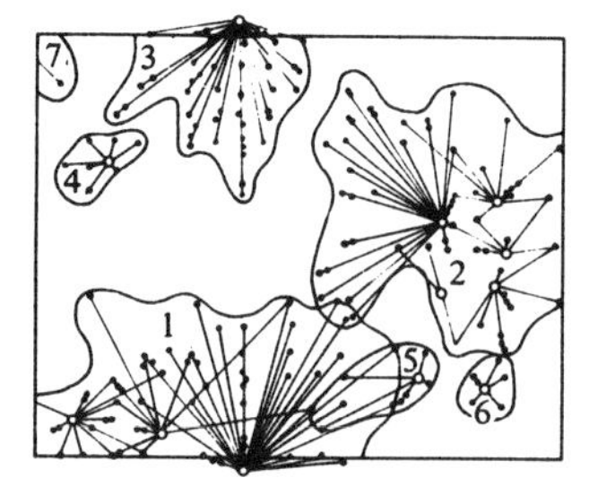

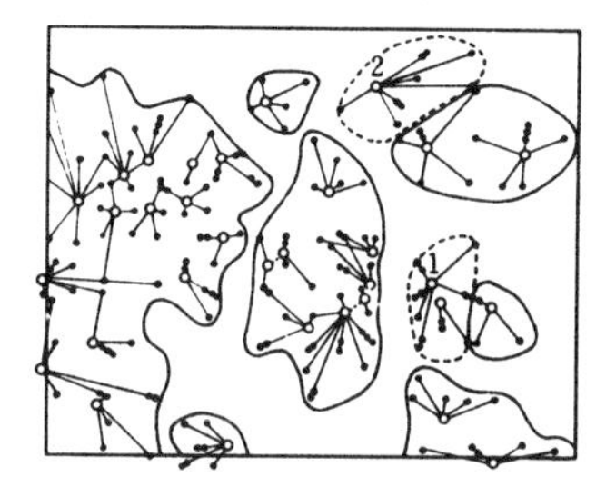

Aphaenogaster famelica, had separate home ranges, and it was predicted that they defended their territories against each other. However, these two species and the first two large species had completely overlapping home ranges.

Both Morisita and Tsuneki & Adachi observed that *Formica japonica* did not possess large territories and this was confirmed by the observations of Yasuno (1965). This species moves singly and does not defend a territory other than its nest. It may have evolved as an 'opportunist' (Wilson, 1971). Ants started with type C territories in which only the nest and its vicinity were defended. In many species territories developed as far as type B or mixed types C–F, in which aphid-infested trees were defended. Territories close to type A have arisen in a few species (e.g. *Camponotus herculeanus japonicus*) while territoriality seems to have been lost in a few others which have become opportunists and wander everywhere individually. The common factor in the defence of nests of all species is that it is effected mainly by the nest odour, which is characteristic for each nest, and the collective attack on intruders aroused by the secretion of alarm pheromones. The termites also defend the nest with a combination of nest odour and alarm pheromones.

Wilson (1975) infers that territoriality in ants may be related to the origin of 'slavery' in ants. If a few colonies of *Leptothorax* without 'slaves' are placed in a small area, territorial fights will develop and the queen and workers of small colonies will be killed. The pupae are transported back to the conqueror's colony and when the worker ants emerge they provide labour in the colony.

Recently, a very interesting phenomenon has been discovered in the coreid bug *Acanthocoris sordidus*. Fujisaki (1977) found that the females of this species are gregarious, up to ten of them sucking sap from a single stem of a host plant, while the male defends such a stem against other males and mates with one female after another over two months.

Territories of poikilothermal vertebrates

(1) The fishes that hold territories are a very small portion of the class Pisces. Most of them are freshwater fishes. Territories are known today only in the Teleostei; no territories are found among the Elasmobranchii (sharks and rays), the Holocephali (chimaeras), the Chondrostei (sturgeons) and the Dipneusti (Lungfish).

Apart from the type F territories held by *Plecoglossus altivelis* and possibly a few other medium- and large-sized fishes in rivers, territoriality of fishes is almost all connected to reproduction, e.g. nest building, protection of eggs and young. Except for the hymenopteran territories, those discussed so far are not related to reproductive strategy. In the phylogenetic positions reached by fishes and above we find the development of territoriality as a basis for adopting the low-fecundity/parental-protection strategy. Thus it is not surprising to find well-developed territories in fishes of mountain streams and oligotrophic springs. An interesting fact is that parental care in most fishes is performed by the male.

There are reports of schooling fishes forming territories under special circumstances. Kawabata (1954) found a minnow-like fish called 'medaka', *Orizias latipes*, forming territories unrelated to reproduction in an aquarium, while Mori (1956) observed young rudder fish *Girella punctata* defending territories in a tide pool. In both cases the fish would be schooling in its natural habitat and the territories observed under these conditions would not constitute territoriality.

The male of the bitterling *Rhodeus amarus* defends a freshwater mussel and its surroundings against other males. He attracts a female to the mussel and lets her lay eggs inside the mantle cavity of the mussel. Until young fish appear the male moves with the mussel and his defended area moves with it (Tinbergen, 1953).

The three-spined stickleback *Gasterosteus aculeatus* remains in a school outside the breeding season but males leave the school first to select territories at the beginning of the breeding season. If another fish, particularly a male, enters an established territory the owner threatens the intruder and attacks it. In this fighting the territory holder almost always wins. Within the territory the male digs a small pit on the bottom and builds a nest with algae. When the nest is complete, a female entering his territory is led to the nest by a special chain of behaviour. The female after spawning in the nest leaves the male's territory but the male remains at the nest to aerate the eggs by fanning and protects the nest from intruders (Tinbergen, 1953).

In Japan, since Hatta (1897) reported on the building and defence of the nest by a stickleback (species unknown) many similar accounts have been published. Horikawa (1921) among others deduced from the number of eggs in the nest and the egg number in the female that sticklebacks are polygynous. This is of course true and one sometimes finds several females laying in the same nest built by one male (see also Kobayashi, 1933).

Tinbergen (1953), in one of his many experiments, put a male stickleback that had established a territory into a glass tube and introduced it into the territory of another male which was also put in a glass tube in his own territory. The strange male tried to escape while the territory holder tried to attack the 'intruder' from inside the glass tube. The territorial defence of the stickleback is therefore based mainly on the visual stimulus.

Territoriality is best developed in benthic fishes. Winn (1958) discussed the evolution of territoriality among the darters, a group of benthic freshwater species found in the New World. Like the Japanese cottid complex, the darters are differentiated in many habitats from the torrent to the lake shore waters and exhibit different degrees of parental protection according to the species. Winn compared in detail the habitat, spawning habit and social behaviour among fourteen species. Table 4.2 summarizes his results. Different spawning methods particularly the variation in the angle of the body axis to the substrate, is illustrated in Fig. 4.7.

According to the information summarized, *Percina caprodes*, considered to be the most primitive morphologically, has no territory and spawns on sand or gravel. In *Etheostoma caeruleum* and *Hadropterus maculatus*, which bury their eggs, an area of 20–40 cm radius round the female is defended. However, the territory moves with the female and defence is not earnest. The more advanced are those laying among algae or moss (*Etheostoma blennioides*) and laying in crevices (two species belonging to the subgenus *Urocentra*), in which large fixed territories (50–100 cm in radius) are defended. Finally, in *Etheostoma maculatum* and *E. flabellare*, which lay adhesive eggs on the underside of rocks with co-ordinated actions of a pair fertilizing each egg separately, the territory is restricted to the 'nest' but the defence is strongest. In all these cases the establishment and defence of the territory are the role of the male. Here it is clearly shown that territoriality has developed along with the advancement of the low-fecundity/parental-protection strategy in association with colonization of oligotrophic environments. In the last two species the number of eggs laid is reduced though their adult body weight differs little from the other species.

Since many fish with parental care are smaller than related species without parental care, Williams (1959) wondered whether the reduced fecundity of some darters was due to their smaller size or associated with the evolution of parental care, and compared the ratio of ovary weight to body weight between those that protected eggs and those that did not among the darters. He naturally selected *Etheostoma nigrum* and *E. flabellare* for the species with egg protection and *E. caeruleum* and *E.*

Table 4.2. *Reproductive data on fourteen species of darter*

Species	Reproductive habitat	Sexual dimorphism and sex recognition	Territorial defence	Size of territory	Average no. of eggs laid[a]	Female spawnings with 1 male & eggs per spawning
Percina caprodes	Lake shores; stream riffles and raceways	Slight; none to weak	None, to weakly intraspecific around ♀	None to moderate	±2000	1–2; 10–20 or more eggs
Hadropterus maculatus	Stream pools & raceways sand or gravel	Slight to moderate; weak	Weakly intraspecific around ♀	Moderate	1758	?
Etheostoma caeruleum	Stream rubble or gravel riffles	Extreme; strong	Strongly intraspecific around ♀	Moderate	380	1 several; 3–7 eggs
E. spectabile	Small streams, fine gravel riffles	Extreme (colour); strong	Strongly intraspecific around ♀	Moderate	1254	1 several; 3–7 eggs
E. saxatile	Streams, gravel riffles	Extreme (colour); strong	Strongly intraspecific around ♀	Moderate		1 several; 3–7 eggs
E. exile	Lake shores or slow streams with organic debris	Extreme (colour); weak to strong	Strongly intraspecific, stationary by shore	Large	619	1–several 3–7 eggs
E. blennioides	Streams, rubble riffles with algae or moss	Extreme (colour); strong	Strongly intraspecific, interspecific tendency, stationary by shore	Large	784	Several; 4–7 eggs
E. microperca	Vegetated lake shores or stream pools	Moderate (colour); weak to strong	Weak to strongly intraspecific, stationary by plants	Moderate	358	Many; single eggs
Hadropterus copelandi	Lake shores, stream gravel raceways	Moderate; strong	Strongly intraspecific, stationary by rock	Large	721	1–several; 4–10 eggs

Etheostoma (*Ulocentra*) sp.	Stream pools, race-ways & shelves of limestone rock	Extreme (colour); strong	Strongly intraspecific, interspecific tendency, stationary by rock	Large	798	Many; single eggs
E. nigrum	Lakes or slow streams with rubble	Extreme (colour and fins); strong	Interspecific by rock nest	Small	1043	In clutches of 30–200 eggs laid singly
E. maculatum	Streams, rubble riffles or heads of riffles	Extreme (colour and fins); strong?	Interspecific by rock nest?	Small?	242	In clutches that average 65 eggs laid singly
E. flabellare	Stream rubble race-ways or slow riffles	Extreme (colour and fins); strong	Interspecific by rock nest	Small	449	In clutches that average 34 eggs laid singly

After Winn (1958).
[a]By two-year-old fish per season.

spectabile for the species with no (or little) egg protection. As shown in Fig. 4.8 the species with egg protection had a significantly greater ovary/body weight ratio than the species without egg protection. Thus the reduced fecundity is not related to the body weight of parents but is a reflection of the large-egg/low-fecundity/parental-protection strategy. It is noteworthy that the proportion of energy given to the ovary in relation to the body weight is greater in the species with parental care than in others. Tinkle (1969) called this ratio the 'reproductive effort'. Thus the adaptation to the environment of mountain streams resulted in increased reproductive effort. Also, the period of time that the male stays with the eggs is lengthened with the evolutionary change mentioned above, and in the last species (*E. maculatum* and *E. flabellare*) the male stays near the eggs until they hatch.

Fig. 4.7. Evolutionary changes suggested for the spawning types of some North American darters (Percidae) (redrawn from Winn, 1958). Arrows indicate presumed directions of evolution in the behavioural pattern, corresponding well with the phylogeny based on morphology.

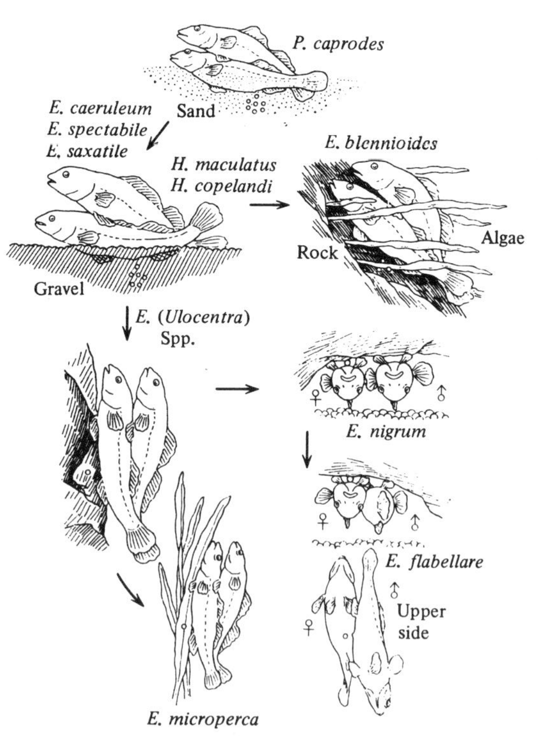

The Gobiidae and the Cottidae in Japan are adapted to benthic life, each independently from the darters. Very similar morphology and habits between the two families are considered to be a result of conversions (see Fig. 2.45, p. 96). Kitahara stated as early as 1897 that in *Cottus* the female lays eggs on the underside of a rock in the torrent and the male defends this site. Unlike the darters most gobiid and cottid species that protect eggs lay on the undersurface of an object with the female upside down in position. This might mean that these two families acquired benthic life early in the evolutionary history and some, retaining the benthic habit, colonized the eutrophic region of lower waters while others were even carried into the sea with the current and spent a long pelagic life there. The parrot goby *Sicyopterus japonicus* is one of those species with an apparently inconsistent life history. The female lays a large number (tens of thousands) of small eggs on the underside of a stone and the male protects them. But the young go down to the sea and spend a long period of pelagic, planktonic life – longer than gobies – before changing to the benthic life.

Next to the parrot goby, the largest number of eggs is laid by the spined sleeper *Eleotris oxycephala*, which lays probably the smallest eggs among the Gobiidae and which is the only species of Gobiidae that does not protect its eggs. On the other hand, the species whose young do not enter the sea or a lake but start benthic life soon or immediately after hatching include the dark sleeper *Odontobutis obscura*, the common freshwater goby *Rhinogobius brunneus* in Okinawa (with medium-sized

Fig. 4.8. Ratios of ovary weight to body weight in four species of darter; *Etheostoma nigrum* and *E. flabellare* with parental protection of eggs and *E. caeruleum* and *E. spectabile* without parental protection (redrawn from Williams, 1959). Each bar consists of a knob in the centre (mean value), black portion (standard deviation), white portion (standard deviation ×2) and the horizontal line (range of values).

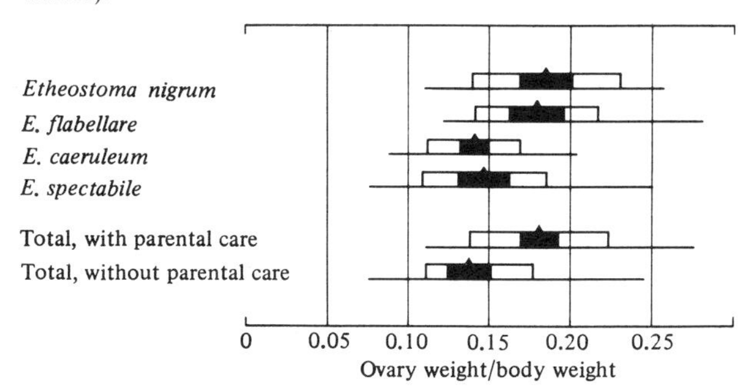

eggs), the lizard goby *R. flumineus* among the Gobiidae and wrinklehead sculpin *Cottus pollux* (the river form) and the Japanese sculpin *C. hilgendorfi* (the river form) among the Cottidae. In these species the eggs are defended most strongly, and feeding territories have also been reported for some of them (the dark sleeper and the common freshwater goby).

The defence by the male of the spawning site during egg laying is seen in salmonid fishes, e.g. rainbow trout *Salmo gairdnerii*, the char *Salvelinus leucomaenis pluvius*, the chum salmon *Oncorhynchus keta*. Other examples of the territories associated with reproduction are seen in the New World sunfish *Euponotis gibbosus* (Noble, 1938), the jewel fish *Hemichromis bimaculatus* (Noble & Curtis, 1939), the paradise-fish *Macropodus opercularis* (Sakaguchi, 1922) and many other tropical fishes (see Noble, 1939b; Aronson, 1957).

The territory of the ayu *Plecoglossus altivelis*, studied in detail by members of Kyoto University, is a typical example of the type F territory (Miyadi *et al.*, 1952; Mizuno & Kawanabe, 1957; Kawanabe, 1966; and others). This species migrates to sea during its life cycle of one year. The young spend the winter in the sea and come up the river in spring. They mature through the summer eating diatoms on the river bed and go down the river in autumn to spawn near the mouth of the river (there is no territory associated with reproduction). When the young reach a suitable grazing area in the stream they break up schools and establish territories around the rocks supporting growth of algae. The 'tomozuri' (decoy fishing) is a traditional method of rod fishing which has been developed to catch ayu by exploiting the aggressive behaviour of the territory holder which gets hooked while chasing a decoy fish. This feeding territory has no connection with reproduction and is individually held, irrespective of sex. As shown in Fig. 4.9 the size of the territory is about 1 m^2 in area though varying with density and the condition of food. If the density is low and the area is not packed with territories, they have a home range extending from the territory and covering an additional area of about 1 m in diameter. Some of them remain in schools, and the territory holders may abandon their territories to join schools when the density is raised above a certain level. According to Kawanabe (1966) the body weight of individual fish decreases with increase in density (related to reduction in territory size) but, if the density becomes very high and territories disappear, the weight of individual fish increases. Thus the type F territory of ayu secures more food than necessary for growth even when the area is fully packed with territories. Why should this be so? The productivity of algae in the river fluctuates from year to year, which is difficult for the fish to predict. Natural selection over a long period of time may have selected the size of territory large enough to support the territory holder in the years of low productivity. However, there is no mechanism to reduce the territory size below this level when the density becomes high temporarily in normal years. Thus territoriality collapses and school life appears, and otherwise surplus food is now effectively used. In the years of very low algal growth there would be no gain in body weight by shifting to school life.

In this connection the type A territory of birds usually contains more food than necessary to rear young (for this reason Lack thought that food was not the ultimate factor governing territory size but in the long run the function of the type A territory is also to secure the supply of food). According to Bustard (1970) the male of a gecko, *Gehyra variegata*, occupies a territory containing several shelters in which females are housed. The number of females is not determined by the amount of food but by social behaviour. There is always a surplus of food in the territory even with the largest number of females recorded in the territory. Here, too, considering the fluctuation of food production the territory size may be related to the securing of food in the year of lowest production.

Fig. 4.9. Territorial arrangements of the ayu *Plecoglossus altivelis* in the Ukawa, Kyoto (after Mizuno & Kawanabe, 1957). Dotted circles with numbers indicate positions of schooling fish with the number of individuals involved. Two areas circled by + signs indicate home ranges not defended. The classes of body length of fish: a = 15–17 cm, b = 12–14 cm, c = 9–11 cm, d = 6–8 cm.

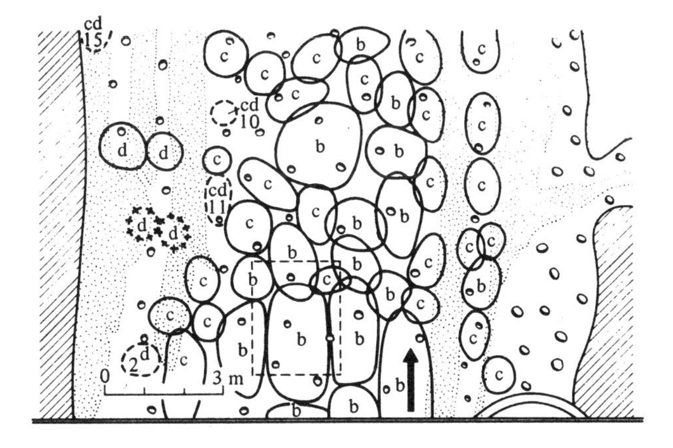

Why, then, did the strong type F territory develop only in the ayu (which is a high-fecundity strategist) among the many stream-dwelling fishes? This may be related to the fact that the algae (diatoms) growing on rocks in streams are relatively uniform and fixed in distribution and do not fluctuate much with time. As mentioned earlier, the few examples of territory known from the molluscs were found in limpets which also grazed on algae on rocks. Among the marine crustaceans, the lobster's defended area is limited to its shelter whereas the amphipod *Erichthonicus* which feeds on small algae growing on the surface of seaweed defends a feeding area. This algal food is not concentrated temporally or spatially, nor does it shift from place to place. Such a pattern of food supply coupled with the habit of attachment to a site and easily conditionable two-dimensional life, made it possible, it seems, for algae-eaters of entirely different phylogenetic origin to develop type F territories.

(2) Attachment to a site (residentiality) of adult frogs has been reported by Martof (1953a, b) for *Rana clamitans*, by Kikuchi (1958) for *R. nigromaculata* and by Pearson (1955) for *Scaphiopus h. holbrooki* (Pelobatidae) but no defence has been observed in these species. There are several reports that the male between bouts of croaking chased approaching individuals (e.g. Emlen, 1968, on the bullfrog *Rana catesbeiana*) but it is not known if territoriality exists in such species. Wilson (1975) wrote that 'territoriality [in the Anura] is common in Dendrobatidae, Hylidae, Leptodactylidae, Pipidae and Ranidae', but most of the examples cited by him pertain to the family Dendrobatidae in which parental care is known for all species whose life history has been studied (Sexton, 1962; Duellman, 1966; Bunnell, 1973). Many of the frogs living around ponds have only a home range and no territory; territoriality has developed in a small portion of this fauna. On the other hand, as ecological studies of tropical species are advanced it has become known that many species from different phylogenetic groups have been emancipated from the pond water and become adapted to terrestrial life. It may be in this process that territoriality has evolved.

According to Jameson (1957), parallel evolution seen in the life history of anurans is characterized by an increased role of females in courtship, a transition from the aquatic to terrestrial habits in courtship and egg laying, and an increased protection of eggs. First, *Ascaphus truei*, a primitive frog with a vestigial tail, belonging to the Ascaphidae, has no vocalization; the male, instead of calling, wanders in search of a female. Secondly, most members of the Ranidae and others that depend on water for reproduction contain gregarious species, dispersed species and some territorial species. The males of this group call singly or in chorus to attract females. Finally, in the Denrobatidae females chase males for a long time prior to mating. In this group (e.g. *Dendrobates*, *Phyllobates*), females oviposit within the territory, and males carry larvae on their backs and take them to water (Silverstone, 1975, 1976).

An interesting point is that the colonization of land with the adoption of the low-fecundity/parental-protection strategy has occurred independently in many (or most) suborders and families. This is shown in Table 4.3 from Jameson (1957) and Vial (1973). Even in the most primitive family, Leiopelmatidae, *Leiopelma* lays large eggs on land. Its larvae complete metamorphosis inside the eggs and hatch out as fry. In the Discoglossidae, while *Bombina orientalis* breeds in water, the midwife toad *Alytes* carries eggs in the hind leg of males, from which the hatched larvae are released into water. In the Pelobatidae, to which a common American species, the spade foot toad *Scaphiopus holbrooki* belongs, there is *Megophrys longipes* which lays about ten large eggs in the moist ground. It has direct development and fry hatch out of eggs. In the Hylidae the Japanese tree frog *Hyla arborea japonica* and many others oviposit in water but even within the same genus *H. faber* lays in circular nests built with mud by the water. Among the members of different genera, *Fritziana goeldii*, *Cryptobatrachus* and *Hemiphractus* carry exposed eggs on the backs of males and *Flectonotus* and *Amphignathodon* hide eggs in the folds on the backs of females. The larvae of the pygmy marsupial frog *Flectonotus pygmaeus* obtain nutrients from blood vessels of the female and develop into frogs on the female. In the Rhinodermatidae, the male of the Darwin's toad *Rhinoderma darwini* picks up eggs laid on the ground and carries them inside the vocal sac. In the Rhacophoridae most of them lay their foam-covered eggs in trees; of the two species in Japan *Rhacophorus schlegelii* oviposits in the soil but *R. arboreus* oviposits in trees and lets its tadpoles fall from the foam into water below. The tadpoles of an Indonesian species, *R. reinwardtii*, swim inside the liquescent foam hung between leaves. In *Hoplophryne* (Microhylidae) of Africa the eggs are laid in foam on leaves but the larvae have acquired an extraordinary habit of eating larvae of other species, similarly living among leaves. A Cuban species, *Sminthillus limbatus*, said to be the smallest frog in the world (body length less than 12 mm), lays only one large egg once, in a nest constructed on the ground, and its development is direct. An extreme degree of parental care is seen in the Surinam toad *Pipa pipa*; the eggs laid in water are picked up on the back of the female by mid-water 'turnover' and the larvae complete metamorphosis in special depressions in

Table 4.3. *Parallel evolution in the Anura towards terrestrial life*

Aquatic development					Direct development	
Family name	Without nests	Aquatic 'nests'	Terrestrial nests	Tadpoles carried to water	Terrestrial nests	Embryo carried until birth
Ascaphidae	*Ascaphus*					
Discoglossidae	*Bombina*[a]			*Alytes*[b]		
Pipidae	*Xenopus*					*Pipa*[c]
Rhinophrinidae	*Rhinophrynus*					
Pelobatidae	*Scaphiopus*[d]				*Megophrys*[e]	
Hylidae	*Hyla*[f]	*Hyla*	*Phyllomedusa* *Hyla*	*Hemiphractus* *Flectonotus*[h]		*Fritziana*[g] *Flectonotus*[i]
Bufonidae	*Bufo*[j]				*Nectophryne*	*Nectophrynoides*
Rhinodermatidae						*Rhinoderma*[k]
Dendrobatidae				*Dendrobates* *Phyllobates*		
Leptodactylidae	*Helophryne*	*Leptodactylus* *Limnodynastes*	*Zachenus* *Heleioporus*		*Eleutherodactylus*	
Ranidae	*Rana*[l]	*Rana*			*Syrrhophus* *Rana*	
Rhacophoridae	*Polypedates*[m]	*Rhacophorus*[n]	*Rhacophorus*[o]		*Rhacophorus*[p]	
Microhylidae	*Microhyla*		*Breviops* *Hoplophryne*		*Oreophryne* *Asterophrys*	
Leiopelmatidae					*Leiopelma*	

Mainly based on Jameson (1957) and Vial (1973).
Example: [a]*B. orientalis.* [b]*A. obstetricans.* [c]*P. pipa.* [d]*S. holbrooki,* [e]*M. longipes.* [f]*H. arborea.* [g]*F. goeldii.* [h]*F. marsupiata.* [i]*F. pygmaeus.* [j]*B. bufo.* [k]*Rh. darwini.* [l]*R. catesbeiana.* [m]*P. buergeri.* [n]*Rh. aboreus.* [o]*Rh. reinwardtii.* [p]*Rh. microtympanum.*

the female's back. In an African species, *Hymenochirus boettgeri*, egg-laying 'turnover' occurs at the surface film. However, there are others among the Pipidae (e.g. *Xenopus*) which lay eggs on plants in water (see Rabb, 1973, and references therein).

The invasion of land, as seen in the above examples, meant adaptation to an environment in which the procurability of food by the young was even more reduced than in the running-water environment which was colonized by *Polypedates buergeri* and *Bufo torrenticola* in Japan (p. 13). I have already shown in Table 1.4 (p. 14) that in this process of adaptation the eggs enlarged, special parental care developed and the clutch size decreased.

Test (1954) was the first to report on territorial behaviour in amphibians. This was seen in *Colostethus trinitatis* of the Dendrobatidae, in which the male carries the young on its back. Sexton (1960) investigated this species in detail and found that the adult males, adult females and subadults all establish territories individually along the river bank and prevent invasion by other frogs of the same species. As shown in Fig. 4.10 the territory of a female is about 30 cm in diameter and is usually defended by the same individual for a long time. According to Sexton territorial behaviour is strongest in the female and this helps to make the distribution of females uniform.

Territoriality among the Urodela is little known (Porter, 1972). It has

Fig. 4.10. A map showing territorial defence by a female of the dendrobatid frog *Prostherapis trinitatis* (modified brom Sexton, 1960). White circles with arrows, directions of movements by the female; black circles with arrows, intruders; crosses, attacks. The intruder 3 left without fighting. The female was often found at A. The apex of a protruding stone, such as A, is often used as a sentry position.

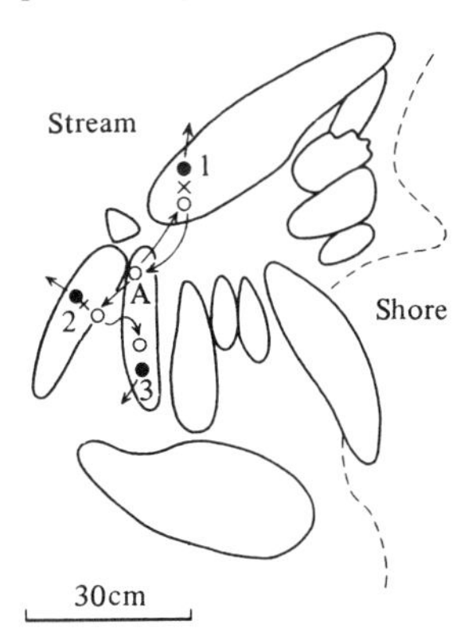

been reported for two species of lungless salamander *Desmognathus* and for newts *Triturus*. Development of territoriality in tailed amphibians will be found to be at a lower level than that of the anurans.

(3) In the reptiles there are many species of lizard which maintain territories. On the other hand no territory has been found in turtles. In the snakes the king cobra *Ophiophagus hannah*, which breeds in pairs, establishes a territory around the nest built by the female (Oliver, 1956). Territorial defence is directed at all intruders including man and domestic stock. That is why this snake is so dangerous in spite of its specialized feeding habit of preying on other snakes. Most snakes probably have no territories.

An example of territory in geckos (Bustard, 1970) has already been given (p. 219). Fitch (1940) found territoriality in a species of lizard, *Sceloporus occidentalis*, living in the mountains of California. The male of this species, when coming out of hibernation, selects a vantage-point, such as a stump, and looks out for females and rivals from there. If another male approaches fighting ensues. However, the boundary of the territory is not rigid; the centre of the territory is strongly defended but the periphery is only weakly defended. There is an effect of prior residence but the holder may evacuate his territory if a strong male appears in his neighbourhood. Since this species does not protect its eggs, the territory is used only for the centre of activity and courtship. Similar observations of territorial behaviour were made by Stebbins & Robinson (1946) on a related species, *S. g. gracilis*, whose territorial defence is milder than that of *S. occidentalis* and permits small individuals to live in the territory of an adult male.

Evans (1936a, b, c, d) studied the social behaviour of *Anolis carolinensis*. If several individuals of this species are put together in a container,

Fig. 4.11. *Anolis carolinensis*; males fighting with inflated thoracic sacs (drawn from photographs in Greenberg & Noble, 1944).

characteristic fighting develops and a hierarchy will be formed. This is not a social hierarchy as an integrating mechanism of the group but an extension of behaviour used in establishing territories in the field. The agonistic behaviour of this species is ritualized and strictly follows a particular sequence: (i) display of the thoracic sac (Fig. 4.11), (ii) raising of the dorsal skin, (iii) lateral approach to the intruding male, (iv) flattening of the body, (v) biting, (vi) withdrawal of the loser, (vii) chasing by the winner and (viii) return of the winner to the centre of the territory with flagging of the thoracic sac.

That such behaviour functions in the field has been confirmed by Evans (1938) himself for a Cuban species, *Anolis sagrei.* This species is common in gardens, living along a hedge, and it sits on guard on top of a post. In an area of the highest density Evans counted twenty-five fence posts with lizards distributed at almost equal distances along 240 m of barbed wire. Thus the territory size of this species is about 9 m in diameter round the fence post. The territory is mostly occupied by a male, sometimes accompanied by a female. Occasionally a female had a territory of her own. A large male occupied a large territory containing several females. The eggs are thought to be buried inside the territory but this has not been confirmed. Young lizards hatching out of eggs are tolerated witin the territory.

Parental care is more clearly shown by the iguanas of the Galapagos; in both *Amblyrhynchus cristatus* (Eibl-Eibesfeldt, 1966) and *Iguana iguana* (Rand, 1967) the females remain near the eggs and defend the surrounding area. In recent years many more examples of territories in tropical lizards have been discovered but most of them are used for mating and feeding (Milstead, 1967).

Crocodiles occupy a unique position among the reptiles. As it has become known recently all twenty-one living species build nests and defend an area around the nest (Greer, 1971; Pooley & Gans, 1976). The Indian ghavial *Gavialis gangeticus*, said to be the most primitive, and the crocodiles that have differentiated early in evolution (e.g. the Nile crocodile *Crocodylus niloticus*) dig a hole to lay and cover the eggs with mud. Other species of crocodiles and alligators build a mound with leaf litter, etc. and lay eggs in it. During the laying season a pair occupies a territory and the male drives other males out of the area. The female digs a hole or builds a mound after mating and defends such a site against other females. Young crocodiles call from inside the egg shells when nearing hatching. The female hearing the calls digs the eggs out and carries all hatched young in her mouth and takes them to water.

Territories of birds

In recent years field observations of birds in Japan have grown rapidly. The Japanese work on territories, courtship and parental care of birds may be found listed in the reference section of Toru Nakamura's (1976a) *The Social Life of Birds* [*Tori no Shakai*]. Also *Life of Wild Birds* [*Yacho no Seikatsu*] edited by Haneda (1975) provides crude summaries of the life histories of many species.

(1) We have already seen that ground birds, such as Struthioniformes and Galliformes, and Anseriformes have adopted the high-fecundity strategy whereas Passeriformes and other small birds, raptors and oceanic birds have adopted the low-fecundity/parental-care strategy. Territoriality is advanced in the latter group. The territory size, however, is closely related to the distributional pattern of food items and the territory of the low-fecundity group is not necessarily large.

First, the species with type A territories include the song sparrow *Melospiza melodia* (Nice, 1941), the robin *Erithacus rubecula* (Lack, 1946b), the wren *Troglodytes troglodytes* (Nice, 1941), the great tit *Parus major* (Krebs, 1971), the scrub jay *Aphelocoma coerulescens* (Brown, 1974) and many other species of passerines and the tawny owl *Strix aluco* (Southern, 1970), the golden eagle *Aquila chrysaëtos* and other birds of prey.

In the great tit the size of the territory, though it depends on the

Fig. 4.12. Dispersion and re-arrangement of type A territories held by the great tit *Parus major* (redrawn from J. R. Krebs, 1971). Unbroken line, boundary of woods; dotted line, territories. The six territories hatched in the left figure were vacated artificially and adjustment of remaining territories and establishment of four new territories (stippled) in the third day are shown in the right figure.

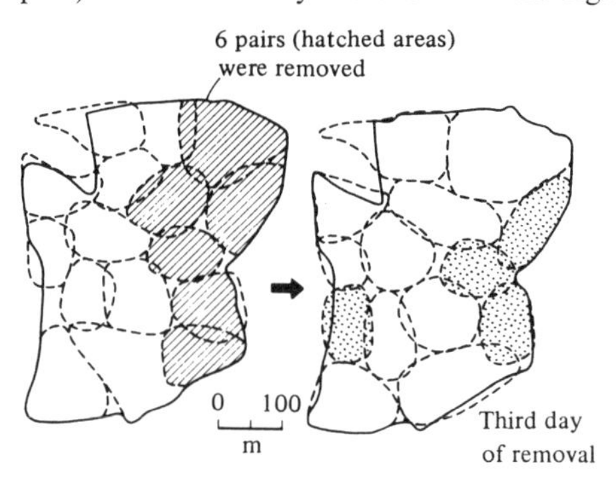

conditions of the wood, is about 0.5–1 ha under favourable conditions and when the density is high the wood is covered with a mosaic of territories (Fig. 4.12). The territory is defended by the male (the male has a characteristic territory song, as in many other passerine birds). The female is mated in the male's territory and reproductive activities are shared by both sexes. Males without a territory cannot take part in reproduction. Krebs (1971) eliminated six pairs of territory holders from such a mosaic of territories and found that within three days the remaining territories shifted their boundaries somewhat and four new pairs occupied territories in parts of the vacated area. As a result the wood was again covered with a mosaic of territories.

According to the investigation of Errickson (1938) the North American wren-tit *Chamaea fasciata* maintains the same territory for some years. Of the eighteen pairs studied, eight pairs maintained their territories for three years and ten other pairs for two years. Fig. 4.13 shows the activities and movements of this species within a territory. Most activities take place within a territory of 3200 m^2 (average size).

Type B territories are held by the tree sparrow *Passer montanus* (Sano, 1973), the grey starling *Sturnus cineraceus* (Kuroda, 1955, 1956a, b, 1957), the reed warbler *Acrocephalus scirpaceus* (Hinde, 1956), the blue jay *Cyanocitta cristata* (Brown, 1974), the magpie *Pica pica* (Kubo, 1975) and others among the passerines and the kestrel *Falco tinnunculus*

Fig. 4.13. A map showing activities of the wren-tit *Chamaea fasciata* within a territory (after Errickson, 1938). More than one nest was built as a result of destruction and desertion.

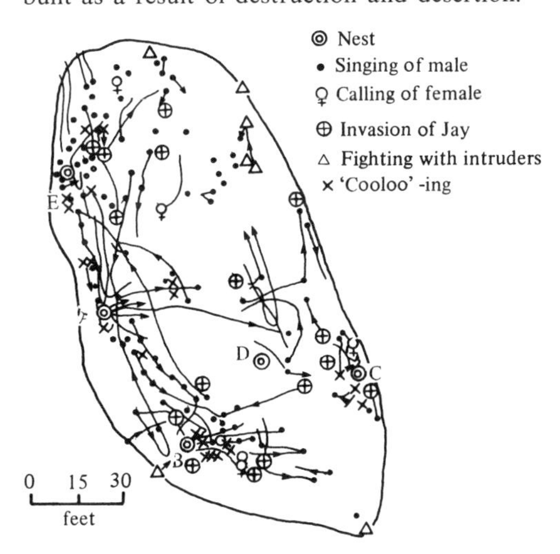

(Kagasaki, 1973) and others among the raptors. They defend their nests and narrow surrounding areas, in which mating, egg laying and rearing of young take place and from which most of their food is collected. Territories intermediate between type A and type B are found in the blackbird *Turdus merula* and the magpie. The blackbird is a resident with a territory of about 20 *ares* held by a pair. They feed partly inside the territory and partly outside. During the breeding season most food is collected in common feeding grounds such as flower gardens. The magpie occupies a large territory around the nest (several hectares) but often shares a common feeding area (Fig. 4.14). According to Kubo (1975) territorial defence in the magpie is limited to the higher parts of the area (e.g. tree tops) and does not concern the ground level (similar to grey starlings observed by Kuroda, 1956b). Birds holding type B territories often use such common feeding grounds. The differences between the type A and the type B may be related in that intermediate types are found within large genera, such as *Turdus* (Hinde, 1956), *Parus* (Gibb, 1956; Hinde, 1956) and *Emberiza* (Nakamura, 1976a), which contain species with territories ranging from typical type A to type B and even close to type C.

In effect there is no clear-cut distinction between type A and type B. Usually, the boundary of a territory is not so well defined as shown in many published diagrams of territorial arrangements. This applies not only to type B territories but also to those close to type A. Yamagishi

Fig. 4.14. Arrangement of territories (areas surrounded by bold lines) and home ranges (areas surrounded by broken lines) held by four pairs of the magpie *Pica pica* (after Kubo, 1975). Pairs visited common feeding grounds (stippled areas) within their home ranges. Five other pairs overlapped in their use of common feeding grounds. Hatched area, buildings and trees; black circles, nests of pairs A–D; white circles, nests of other pairs.

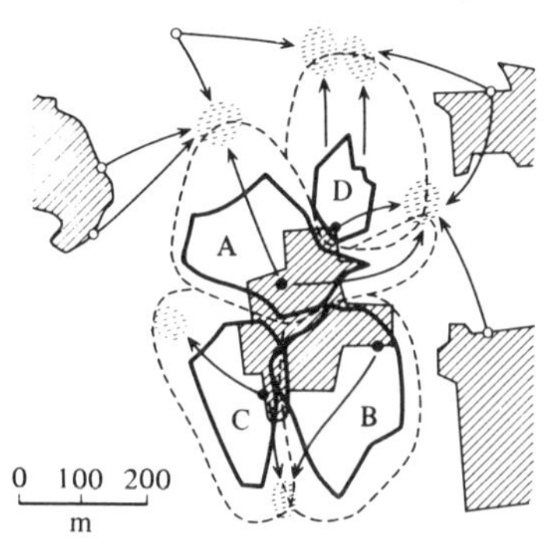

(1970, 1971) investigated the type A territory of the meadow bunting *Emberiza cioides*. He plotted the positions and outcome of territorial fights for a typical male holding a territory contiguous with other territories (Fig. 4.15). In this figure (left) the outermost polygon is the home range of this male; what Yamagishi called territory excludes that part of this home range in which the bird never won a territorial fight. The territory was thus confined to the area within the second outermost polygon in the figure. This area approximately corresponded with the area in which the song posts were located. In a low-density area there was a correspondence between the area of a territory and the song posts, but the home range was four times the size of the territory and overlapped widely with home ranges of other individuals. In the meadow bunting the territory seems to be of type A in high-density areas but in low-density areas it changes to type B as birds often go out of the territory to feed.

Type C territories are held by colonial breeders, such as the black-headed gull *Larus ridibundus* (Tinbergen, 1956), the black-tailed gull *Larus crassirostris* (Komatsu, 1935; Kawaguchi, 1937) and other gulls, the streaked shearwater *Calonectris leucomelas* (Yoshida, 1975), the barn swallow *Hirundo rustica*, herons and egrets. These birds defend a small

Fig. 4.15. Maps showing territorial behaviour of a male meadow bunting *Emberiza cioides* (left) and arrangement of male home ranges of five surrounding pairs (right) (drawn from Yamagishi, 1971, 1973). The points at which the bird was defeated (crosses), had combats (triangles) and had one-sided victories (circles) are shown on the map. The area surrounded by the second outermost line may be considered the territory of this bird; this polygon generally corresponds with the area of territorial singing.

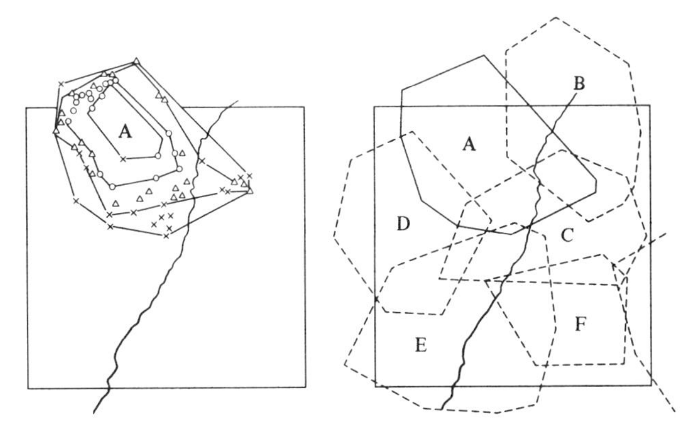

area round the nest within the colony only during the breeding season. Thus the territory has no connection with food supply and its function is considered to be the maintenance of monogamy and the defence of eggs (in gulls eggs are often eaten by other members of the same species). The distinction between type B and type C is difficult to make; for example, we may classify the territory of swallows as type B or conversely that of sparrows as type C (particularly in high-density areas).

Type D territories are found among the species that separate territorial functions of courtship and mating from other activities, e.g. lek-forming grouse, manakins (each male defending a small territory within a lek and mating in the territory), and some gulls mentioned earlier. According to Tinbergen (1956) a male black-headed gull coming to the breeding ground in early spring occupies an area of rocky ground several metres in diameter and invites a female there to mate. Apart from this territory they would have a type C territory for breeding in the colony. Among other gulls the herring gull *Larus argentatus* has only a weak territory for mating and the kittiwake *Rissa tridactyla* has no mating territory.

Type E territories are formed around the overwintering sites or roosting sites of many species. The starling *Sturnus vulgaris* defends a roosting territory (Nice 1941) and gulls defend resting sites in winter. The long-tailed tit *Aegithalos caudatus* is known to exhibit a combined type E and type F territory in winter (Nakamura, 1969, 1976a).

Famous among the type F territories is the one around flowering plants defended by hummingbirds. Wolf & Stiles (1970) found that the male of the Irazú hummingbird *Panterpe insignis* in Costa Rica defended a small area containing several flowering trees from which he obtained nectar. The female associated with a particular male takes nectar from specific flowering trees inside the male's territory but not from the ones that the male usually occupies. The female nests on the *periphery* of the male's territory. As the male does not necessarily defend the area against other females more than one female may enter his territory. He probably mates with the most frequent female visitor. The male does not take part in incubation or feeding of the young.

Birds without territory: Strictly speaking all birds show some form of territorial behaviour at least at the nest in which eggs are incubated. Cuckoos and other parasitic breeders often defend areas containing nests of host species. Even in the mound-builders (megapodes), the only group of birds that does not require incubation by birds for the development of its eggs, the male attends to the mound during the breeding season. Lack of incubation in this group is considered to be a secondary loss of the habit, not a carry-over of the reptilian character (Lack, 1968). The question in birds is whether only the nest is defended or to what extent the defended area extends into feeding areas. However, the emperor penguin *Aptenodytes forsteri* has neither a territory nor a nest in the true sense of the word. It incubates an egg on its feet between the belly and tail in the long nights of antarctic winter. Breeding occurs in a colony of several thousand birds and five to twelve birds share this task. When one gets hungry the egg is dropped and picked up by another bird. The unmated individuals take part in incubation, brooding and feeding of chicks (Allee *et al.*, 1949). Most other penguins breed in pairs and defend an area round the nest.

(2) How did type A and type B territoriality develop? There are two approaches to this question.

One approach is given by Schoener (1968) who investigated the feeding territory and its size according to the feeding habit of birds. Here the feeding territory includes all type A territories and some type B territories in which part of the feeding area is defended; the feeding territories not associated with breeding (type F) are not included. He examined 103 species belonging to twenty-three families of North American birds and calculated the percentages of species having feeding territories within each group. His results are summarized by Morisita (1976) as in Table 4.4. In the Accipitridae, Strigidae, Troglodytidae and Parulidae, in which the percentage of species with feeding territories is large, all species depend for more than 90% of their food on animal food. On the other hand, the Corvidae, Icteridae and Fringillidae, in which only a few species hold territories, depend little on animal food. Fig. 4.16 shows that the territory size increases with increased body weight and that the territory-size of carnivorous species is more than ten times that of herbivorous species (animal food 0–10%) of similar weights. Since the angle of the regression line differs between the two groups large carnivores would require even greater territories than would be expected from the above ratio. It may be argued that catching prey is a difficult task compared with herbivorous life and that territoriality has developed further in carnivores in order to secure food.

A somewhat different approach to the question of territoriality is given by Horn (1968). He considered the conditions which make territoriality advantageous to a species by studying the relations between spatial distribution of food items and energy efficiency of feeding with or without a territory. First, consider the four states shown in Fig. 4.17.

Table 4.4. *Distribution of feeding habits during the breeding season and percentages of species which have feeding territories in some North American land birds*

Family	No. of species	No. of species examined	Feeding categories[a]					Total	See below[b]	Percentage of species with feeding territories[c]
			A	AO	O	HO	H			
Columbidae	11	3					1–2	1–2		33
Accipitridae	22	9	5–2					5–2	2	56
Falconidae	7	1	1–0					1–0		
Tetraonidae	10	2					2–0	2–0		
Strigidae	18	6	5–1					5–1		83
Tyrannidae	31	5	5–0					5–0		100
Alaudidae	2	1			1–0			1–0		
Corvidae	15	8		1–2	1–3	0–1		2–6		25
Paridae	14	4	3–0	1–0				4–0		100
Sittidae	4	2						(2–0)		
Chamaeidae	1	1		1–0				1–0		
Troglodytidae	9	4	4–0					4–0		100
Mimidae	10	2		1–0	0–1			1–1		

Turdidae	13	3	2–0		1–0			3–0		100
Motacillidae	4	1	1–0					1–0		
Bombycillidae	2	1				0–1		0–1		
Laniidae	2	1	1–0					1–0		
Sturnidae	2	1			0–1			0–1		
Vireonidae	12	4	4–0					4–0		100
Parulidae	53	17	16–0					16–0	1	94
Ploceidae	2	1			0–1			0–1		
Icteridae	20	10	1–1	1–2	0–5			2–8		20
Fringillidae	78	16		2–3	0–4	2–2	0–2	4–11	1	25
Total	342	103	48–4	7–7	3–15	2–4	3–4	65–34	4	
% having feeding territories			92.3	50.0	16.7	33.3	42.9	65.7		

After Schoener (1968).

[a] A, 90–100% animal food; AO, 70–90% animal food; O, 30–70% animal food; HO, 10–30% animal food; H, 0–10% animal food. For each family and category, the number of bird species having feeding territories (first number) and those which do not (second number) for the species sampled are listed and summed.

[b] Number of species with both territorial and non-territorial feeding.

[c] For calculation of percentages families for which less than three species were examined were omitted.

There are four nests and sufficient food for the young of all nests is found at sixteen places in the square. Let k be the shortest distance between food sources and consider two situations: (1) stable food distribution, in which food is always found uniformly over sixteen places; and (2) moving food distribution, in which food is concentrated at one place at a given time, and the probability of finding it at any particular place is one-sixteenth. Though it shifts from place to place, after a long time it will

Fig. 4.16. Relation between body weight and territory size according to feeding habits of North American birds (from Schoener, 1968).

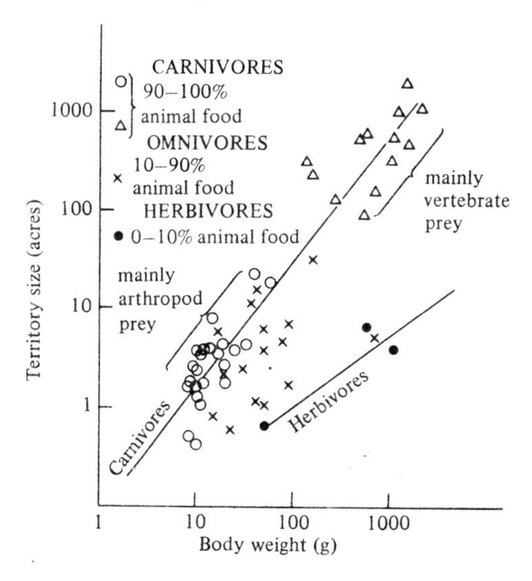

Fig. 4.17. Theoretical distribution of food and nests (modified from Horn, 1968). See text for explanation.

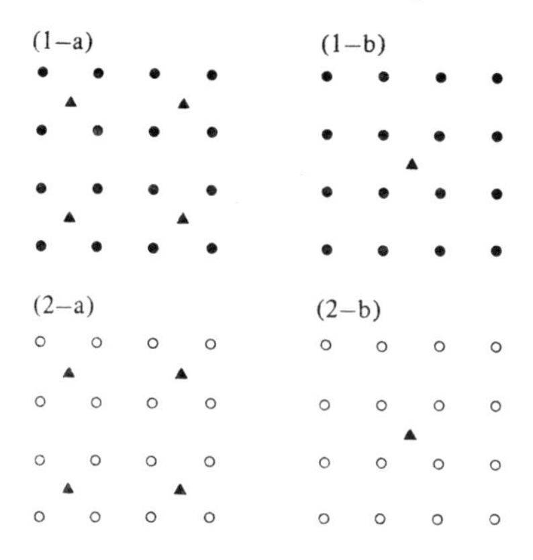

have been at each place for the same number of times. The total amount of food is assumed to be same between the two types of distribution.

In the case of (2), the amount of food found in concentration is sixteen times the amount found at any place in (1). For the distribution of the nests, two conditions are considered: (a) four nests are dispersed with equal distances (territorial breeding) and (B) four nests are concentrated in the centre of the square (colonial breeding). Then the mean distance that the parent travels to and from the nest for bringing food back to the young has been calculated. The time or energy spent during this period is assumed to change in proportion to the distance travelled.

Case (1–a); food distribution uniform, nest distribution territorial;

$$2\left[\left(\frac{1}{2}k\right)^2+\left(\frac{1}{2}k\right)^2\right]^{1/2}=1.42\ k$$

Case (1–b); food distribution uniform, nest distribution colonial:

$$\frac{2}{16}\left\{4\left[\left(\frac{1}{2}k\right)^2+\left(\frac{1}{2}k\right)^2\right]^{1/2}+8\left[\left(\frac{1}{2}k\right)^2+\left(\frac{3}{2}k\right)^2\right]^{1/2}+4\left[\left(\frac{3}{2}k\right)^2+\left(\frac{3}{2}k\right)^2\right]^{1/2}\right\}=3.00\ k$$

Comparing the above two, time or energy is much saved in the case of territorial nest distribution. Thus so long as the time and energy used for the defence of the territory does not exceed the amount saved in foraging, natural selection favours territoriality.

Case (2–a); moving food distribution, nest distribution territorial:

$$\frac{2}{16}\left\{4\left[\left(\frac{1}{2}k\right)^2+\left(\frac{1}{2}k\right)^2\right]^{1/2}+4\left[\left(\frac{3}{2}k\right)^2+\left(\frac{1}{2}k\right)^2\right]^{1/2}+4\left[\left(\frac{5}{2}k\right)^2+\left(\frac{1}{2}k\right)^2\right]^{1/2}+\left[\left(\frac{3}{2}k\right)^2+\left(\frac{3}{2}k\right)^2\right]^{1/2}+2\left[\left(\frac{5}{2}k\right)^2+\left(\frac{3}{2}k\right)^2\right]^{1/2}+\left[\left(\frac{5}{2}k\right)^2+\left(\frac{5}{2}k\right)^2\right]^{1/2}\right\}=3.86\ k$$

Case (2–b); moving food distribution, nest distribution colonial: This is the same as the case (1–b): 3.00 k.

Comparing the two under the moving food distribution colonial nesting is advantageous. Many insects have protective coloration and are dispersed in the forest. On the other hand, plant seeds or insects in clearings tend to be concentrated. If food was the selective factor of territoriality, territoriality should be developed more in forest birds than in grassland birds and more in insectivorous birds than in graminivorous birds. This in fact is supported by the facts presented. Predatory sea-birds breed in colonies. This is an adaptation to the distribution of food in the sea as well as to scarce nesting sites. The prey fish are not uniformly distributed in the sea, schools appearing on one side of an island one day and on another side the next day.

Territoriality and mating systems of birds

(1) Territoriality is not well developed in ground-living birds with the high-fecundity strategy. In this group there are many polygynous and promiscuous species – the latter being combinations of a few males and many females, which I once called polygyn-oligandrous.

Fig. 4.18. Maps showing home ranges of the partridge *Perdix perdix* (after Blank & Ash, 1956). Top, home ranges of a winter covey 1952–53; bottom, home ranges of pairs in spring 1953.

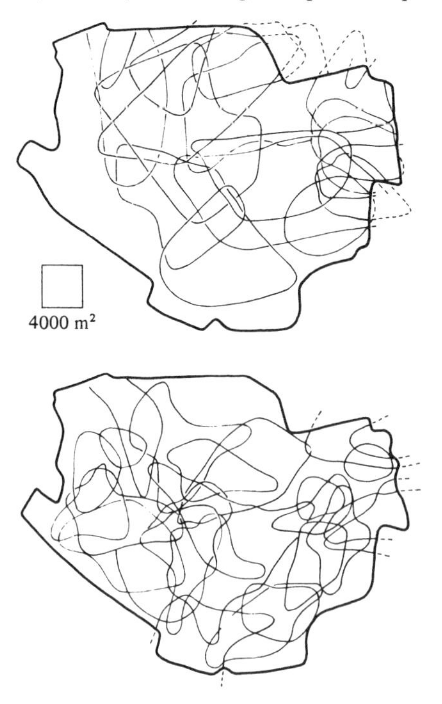

In the ostrich *Struthio camelus* and the rhea *Rhea americana*, more than one female may lay in the same nest while the male incubates the eggs and defends the nesting area. In the case of tinamous the male also incubates, but the defence of a territory may be a role of the male or female depending on the species (Kuroda, 1962; Lack, 1968).

In the Phasianidae most species do not defend feeding areas. Howard & Emlen (1942) found that the California quail *Lophortyx californica vallicola* is sedentary and lives in groups but does not expel intruders. Blank & Ash (1956) studied the behaviour of the partridge *Perdix perdix* and the red-legged partridge *Alectoris r. rufa* in England. Both species breed in pairs and live in a covey outside the breeding season. The covey consists of one or two adults and their young. The home range of a family is limited to 2–3 ha and, since it is not defended, can overlap with home ranges of other families. In December the covey breaks up and new pairs are formed. Although an area around the pair is defended this area is not fixed. Thus the aggressive behaviour observed should probably be regarded as a repugnant response to other individuals. In effect home ranges of different pairs overlap and sometimes the nest is built outside the normal range (Fig. 4.18).

According to Taber (1949) and Collias & Taber (1951) the pheasants *Phasianus colchicus* and *P. torquatus* form a 'harem' and each male has a more or less fixed territory in which he holds five to eight females. The territory may be as large as several hectares in area and feeding occurs within it. However, the area of the territory is not strictly fixed as in territorial forest birds.

The most interesting group in the Galliformes is the Tetraonidae which exhibits evolutionary trends of territoriality and mating behaviour. This group is considered to have originated in northeastern Asia or North America and some have adapted to taiga and associated habitats while others have been established in the alpine region and tundra. Different modes of their life are represented by the sage grouse *Centrocercus urophasianus* and the greater prairie chicken *Tympanuchus cupido*, which form large mating assemblages and perform characteristic courtship displays, and the blue grouse *Dendragapus obscurus* which breeds in pairs and defends a large territory. Since the 1950s many investigations have been conducted on the social behaviour of these species, which have been summarized by Wiley (1974) in his treatment of social structure and evolution in the grouse (Table 4.5).

According to Wiley the grouse may be divided into three groups based on their sex-relations. Type I is a group formed at a place of group

Table 4.5. *Habitat, sex-relation and territoriality among grouse and ptarmigans*

Sex-relation	Species	Habitat	Territoriality	No. of males attending a lek
I. Species that form leks	Sage grouse/*Centrocercus urophasianus*	Semi desert[a]	Small territory of males within lek	several tens to several hundreds
	Black grouse/*Lyrurus tetrix*	Periphery of forest	Small territories of males within lek	1–30
	Caucasian black grouse/*L. mlokosiewiczi*	Periphery of forest	?	
	Greater prairie chicken/*Tympanuchus cupido*	Grassland[a]	Small territories of males within lek	
	Lesser prairie chicken/*T. pallidicinctus*	Grassland[a]	?	
	Sharp-tailed grouse/*Pedioecetes phasianellus*	Grassland[a]	Small territories of males within lek	1–40
	Capercaillie/*Tetrao urogallus*	Coniferous forest	Large territories of males within lek	1–12
	Small-billed capercaillie/*T. parvirostris*	Coniferous forest	?	

II. Promiscuous species with dispersed males	Blue grouse/*Dendragapus oscurus*	Coniferous forest[a]	Mating territories of males	1–4
	Spruce grouse/*Canachites canadensis*	Coniferous forest[a]	Mating territories of males	
	Sharp-winged grouse/*Falcipennis falcipennis*	Coniferous forest	?	
	Ruffed grouse/*Bonasa umbellus*	Mixed forest[a]	Mating territories of males	
III. Species that form pair-bonds	Hazel grouse/*Tetrastes bonasia*	Coniferous forest	Mating and nesting territory	
	Amur grouse/*T. sewerzowi*	Mixed forest	?	
	White-tailed ptarmigan/*Lagopus leucurus*	High mountain[a]	Large territory; males sometimes care for young	
	Rock ptarmigan/*L. mutus*	Tundra; high mountain	Large territory; males sometimes care for young	
	Willow grouse/*L. lagopus*	Tundra; high mountain	Large territory; probably the care for young same with the next species	
	Red grouse/*L. lagopus scoticus*	Heath	Large territory; males care for young with females	

From Wiley (1974) and Nakamura (1976b).
[a]North American species.

courtship called the arena or the lek. A typical example is seen in the sage grouse, which lives in sage fields of semi-desert. They are sexually dimorphic and dozens or sometimes as many as 400 huge males gather at a lek in the mating season. In the lek each male defends a territory of about 10–20 m^2 in the centre (or larger at the periphery), large old males being positioned in the centre and young males at the periphery (Fig. 4.19). Peripheral territories are large and sometimes the boundaries are obscure. Or rather, weakly territorial males and males without territories (peripheral males) surround the territorial males (central males). The boundary of territories is determined by 'facing-past' (pushing each other in the displaying position with expanded breast feathers) or by beating with wings. The females visit the lek for several weeks and take part in the characteristic courtship dance. The mating centre is always around the territory of the most dominant male (Fig. 4.20). Since mating occurs almost exclusively in the mating centre, only the dominant male and one or a few other males can mate. However, the dominant male, once satisfied, may leave the mating centre and permit subdominant males to enter the area. Thus several males can mate in the end. This is somewhat different from the account of Scott (1942, 1944), which I cited in the first edition of this book. The area of a lek and the mating centre are fixed and for many years the same place is visited by several hundred sage grouse. In a large lek there are several mating centres (Fig. 4.19). According to Wiley (1974), 10–75% of the male sage grouse coming to a lek mated successfully. Young males cannot matc as thcy cannot rcach

Fig. 4.19. A mating centre of the sage grouse *Centrocercus urophasianus* (drawn from a photograph in Wiley, 1973). Females gather round the central males. Two central males are displaying in competition; one with a frontal view shows orange patches of exposed skin. Another mating centre with males in the same lek is seen in the background.

the centre. In an experiment with the sharp-tailed grouse *Pedioecetes phasianellus*, which also has a type I lek, the first-year male was shown to be able to mate (physiologically capable) when the old males were removed from the lek. Similar lekking behaviour is known for the greater prairie chicken, which has orange markings on both sides of the neck used for display.

In type II the males are dispersed. In the blue grouse *Dendragapus obscurus* each male establishes a large territory in the forest while females have overlapping home ranges larger than the males' territories. Thus there are several territories of males within home ranges of several females and the female visits different males in turn. To attract females males use 'drumming' in addition to the visual display of fluffing feathers, typical of type I birds. Particularly well known is the display of the ruffed grouse *Bonasa umbellus* which selects a suitable tree in the forest to perform drumming. Drumming is stronger in the males near the centre of the lek where the rate of mating is thought to be high. In the blue grouse if these males are eliminated the first-year birds may occupy central positions and perform drumming. They too were shown to be able to mate.

Type III is a pair relation, of which the red grouse *Lagopus lagopus scoticus* has been studied in most detail. This species lives on the moors of Scotland. In this tundra-like vegetation the male red grouse forms a territory of several hectares in autumn. During the winter females appear and pairs are formed in spring. There is an effect of prior residence in a territory. When the female lays the male takes the role of territorial defence and both parents accompany chicks. They feign injury to distract

Fig. 4.20. Territorial arrangement of males around the mating centre in a lek of the sage grouse *Centrocercus urophasianus* (from Wiley, 1973). At first the males A and C could mate, then A left and N was promoted. The position of the mating centre changed somewhat and 2W could now also mate. Circles indicate defended positions.

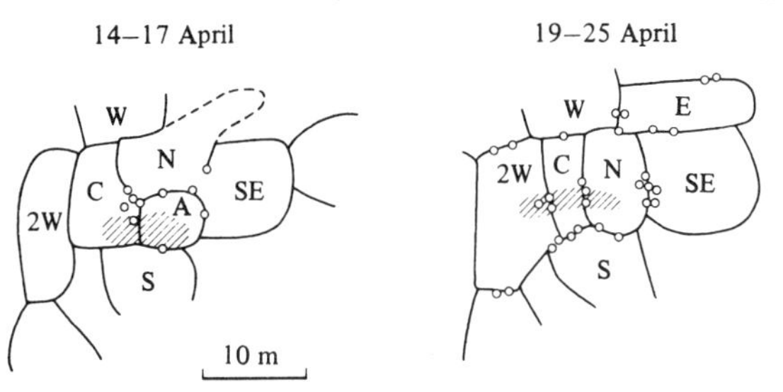

the attention of predators from the chicks. Thus the female and young spend a few months in the male's territory and leave the male in summer. In this case also, the older male establishes a territory earlier than younger males and young males sometimes fail to establish a territory. The individuals with territories have a high rate of survival (see p. 176).

In the rock ptarmigan *Lagopus mutus* (Hirabayashi, 1976) and the white-tailed ptarmigan *L. leucurus* a pair establishes a large territory which persists for at least one breeding season. The male defends the territory while the female is incubating but by the time the young hatch most males will have left the family. Parental care by the male is rare in these species.

Thus the mating system of grouse varies a great deal from pairing to promiscuous lekking. The former is found in the species of tundra and alpine regions and the latter among the plain-dwellers (grassland to semi-desert). The black grouse *Lyrurus tetrix* and the Caucasian black grouse *L. mlokosiewiczi* live not in large grassland but along the forest edges in grassy fields or old fields. The lek is always formed in open places. As discussed in Chapter 1, grassland favours the high-fecundity strategy while high mountains and polar regions favour the low-fecundity/parental-protection strategy by being a harsh environment in which food is not easily procurable by the young. It is in this latter habitat that territoriality with monogamous pairing has developed. Wiley contends that the lek probably developed in a region where food was easily obtainable and the habitat was uniform and extensive. That the individuals of forest species are dispersed, though not as much as in the alpine species, is also likely to be related to the fact that in the forest habitat it is more difficult for the predators to discover scattered prey than in grassland.

Here, the problem is the capercaillie *Tetrao urogallus* which is known to form a lek in the coniferous forest. The sex-relation of this species may be intermediate between type I and type II, for the male defends a very large territory of 100 m in diameter (larger than other type I species) within the lek and displays in trees.

Wiley, in addition to the above consideration of food distribution, discussed the reasons for the sexual dimorphism in size of promiscuous species and for the great individual variation found in the maturing age of males. This aspect will be treated later.

(2) Next, I will mention the ecology of primitive cuckoos. According to Davis (1940a, b), the smooth-billed ani *Crotophaga ani* and the guira

cuckoo *Guira guira* found in the pampas build communal nests in large trees. In the first species the colony consists of 10–20 birds and several females lay 4–7 eggs in one nest. Feeding of young is also communal and the colony's territory is defended against individuals of other colonies (Fig. 4.21). However, some individuals are allowed to join the colony and membership often changes between colonies. Young birds remain in the colony for several months and some of them help feed young that hatch later (like worker bees!). This phenomenon, akin to the worker caste of insects, seen in birds will be discussed later.

The territory of *Guira guira* is larger and some members form a pair and defend their own territory within the colony's territory. There are also communal nests used by several females.

According to Davis (1942) the stages of development of communality in cuckoos are as follows:

> *The first stage – Guira guira.* Some communal nesting, others pair nesting; colony territory with internal pair territories, defence of colony territory mild.
>
> *The second stage – Crotophaga major.* Mostly communal nesting but some pair nesting, defence of colony territory mild.

Fig. 4.21. A map showing communal territories of the smooth-billed ani *Crotophaga ani* (modified from Davis, 1940 a). The group size fluctuates; the numerals indicate the numbers of birds in groups on 5 June.

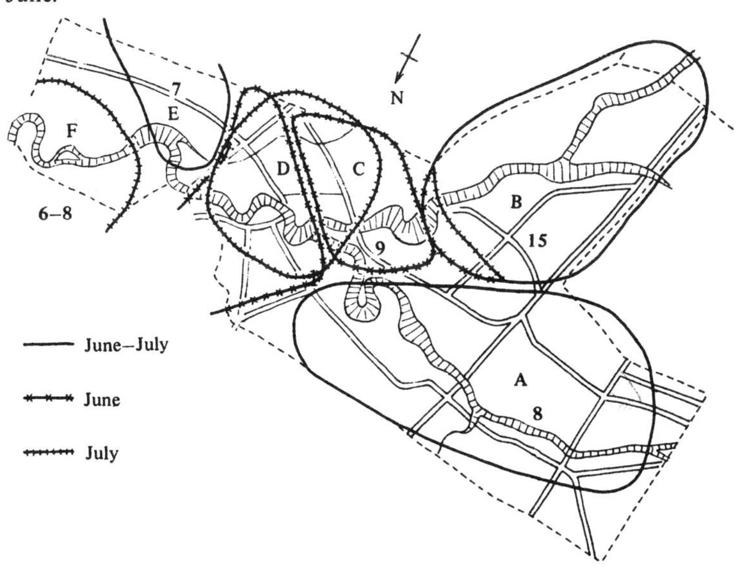

The third stage – *C. ani* and *C. sulcirostris.* All communal nesting, promiscuity common, defence of communal territory strong.

The above habit seems to have developed because of the scarcity of large trees in pampas. Davis states that the nesting habit and the pair bond are not well developed, permitting laying of eggs in the nest of others, which probably developed later into parasitism on other species.

Tropical grassland and rain forest harbour many other species, which lack feeding territories or which perform group courtship, though perhaps not so pronounced as in communal cuckoos. Of the colonial species in tropical grassland, there are, apart from the anis, cowbirds (*Molothrus*; some species parasitic), the budgerigar *Melopsittacus undulatus* and weaver-birds (Ploceinae). Of the tropical rain forest species there are manakins (e.g. *Manacus*, *Pipra*), birds of paradise (e.g. *Paradisaea*) and cotingas (Cotingidae). Manakins and some birds of paradise form a lek and perform group courtship. In manakins there is no defence of territory and there is no fighting over females (Chapman, 1935). In tropical savannah nesting sites are limited for tree-nesting birds and the fruit of trees like the baobab is concentrated in one tree at a time in every so many thousand metres (like Horn's moving food distribution). In the tropical rain forest, although the canopy is continuous any particular species of tree is distributed sparsely and fruits of one species mature in concentration at one time in one place. The above mentioned groups of various forest birds are all fruit-eaters and their feeding habits may have promoted the evolution of colonial life under such environmental conditions (see Chapter 1, p. 41).

(3) The status of insectivorous birds and seed-eating birds in the tropics is different. The weavers are grassland birds of Africa and southern Asia and are well known for the intricate nests that they weave. Of these the sociable weaver *Philetairus socius* weaves a large communal nest but the chamber inside is used exclusively by a pair. On the other hand, the savannah weaver *Ploceus jacksoni* builds a cluster of dozens of nests in one tree and is polygynous. Some weavers live in forest but one of these, the forest weaver *Melimbus nitens* builds hidden nests separately by each pair. Some of them build such independent nests dispersed within a colony territory. Crook (1964) discussed the evolution of such behaviour of weavers (see Uramoto, 1967, for Japanese summary). Later, Lack (1968) examined the same problem and disputed some of Crook's points. Lack summarized the comparative ecology of ploceinae as in Table 4.6.

Table 4.6. *Relationship between food, habitat and modes of life in ploceid birds*[a]

		Number of species in each category							
		Dispersion of nest				Pair-bond			Mean clutch size on tropical mainland (no. of species)
Main food	Main habitat	Solitary (or 2–4 pairs together)	Grouped territories	Colonial	Solitary or colonial	Monogamous	Polygynous[b]	Probably both	
Ploceinae									
Insects	forest	15(+2)	0	1	0	17	0	0	2.1(8)
Insects	savannah	4	0	2	0	5	0	1	2.4(5)
Insects and seeds	forest	2	0	0	1	3	0	0	—
Insects and seeds	savannah	1	0	4	3	1	4	3	2.2(7)
Insects and seeds	grassland	1	0	0	1	1	(1?)	0	2.9(2)
Seeds	savannah	0	1	16	0	2	10	1	2.6(14)
Seeds	grassland	0	13	2(+1)	0	0	14 (+1Pr.)	0	2.7(17)
Bubalornithinae and Passerinae									
Insects and seeds	savannah	3	0	17	4	14	1	0	3.6(17)
Insects and seeds	alpine	1	0	0	0	1	0	0	—
Seeds	savannah	0	0	2	0	1	0	0	3.0(2)

From Lack (1968).

[a]Sum of categories are not necessarily the same because of insufficient data on modes of life.

[b]Includes promiscuity (Crook, who presented original data, included promiscuity in the category of polygamy).

According to this table, the forest species are monogamous and mostly build nests singly while the grassland species, particularly seed-eating species, nest colonially and are polygynous or promiscuous. Crook (1964) explained this fact as follows:

In the forest insects are generally dispersed or hidden. The density of any particular prey is not large. Under such conditions it is advantageous for the pairs to be dispersed and to work together in rearing young by collecting low-density food items secured in the territory. On the other hand, in the tropical grassland the grass seeds are abundant locally in a particular season. This is quite different from the seeding of plants in the mountain grassland of Japan. In tropical grassland there is always a dry season in which most plants die and herbivorous animals go into diapause (or aestivation) or migrate. The plants with the arrival of rain grow, flower and seed rapidly. Insects often outbreak. Thus the low-density populations that have survived the dry season must maximize their reproductive effort in the face of an abundant food supply. Under such conditions there is no need for the male to secure food but it is advantageous for him to build as many nests and mate with as many females as possible. In the weaver-birds, because of the shortage of trees, only a few males can secure a tree for nesting. Females congregate in such trees but this aggregation promotes synchronization of breeding by mutual stimulation. Thus the polygynous or promiscuous mating system has evolved. Lack questioned this interpretation by pointing out that in the Passerinae many seed/insect-eating grassland birds are monogamous. The danger of predation in grassland probably made them choose colonial or communal nesting. In the forest it would be advantageous to disperse and hide but in the grassland where one is conspicuous anyway it would be advantageous to defend as a group.

If there are different mating systems among related species, ranging from monogamy to polygyny or communal life, the latter type is often found in swamp vegetation. According to Verner & Wilson (1966), of the fifteen polygynous passerine species known from North America eight are found in swamps. Since only 5% of the passerines are found in the swamp habitat it follows that in this habitat the proportion of polygynous species among the passerines is twenty times as great as in other habitats. In the case of American blackbirds studied by Orians (1961), the red-winged blackbird *Agelaius phoeniceus*, which breeds in pairs and has a type A territory, inhabits grassland whereas the tricoloured blackbird *A. tricolor*, which is gregarious and often polygynous and has very small territories of 4 m^2 in size, lives in swamps. Verner & Wilson (1966) and Orians (1969) attributed this fact to the environment of swamps, which have much greater productivity than dry grassland but which often undergo drastic changes due to floods, so that the opportunist strategy of finding vacant places in which to breed would be advantageous in the swamp habitat. Holm (1973) found the highest survival-rate of chicks in the larger 'families' of these birds; the greater the number of females in a 'harem' of polygynous blackbirds the greater the number of young fledging per female as well as per male.

The above discussion leads us to the conclusion that communal nesting and the polygynous or promiscuous mating system in birds are a manifestation of the high-fecundity strategy to cope with unstable but temporarily highly productive environments, in which food is abundantly available but the risk of predation is very high. As can be seen in Table 4.6 the clutch size of weavers increases from forest to grassland habitats and from insectivorous to granivorous species. This is a tendency towards the *r*-strategy to use Pianka's expression. However, in polygynous or promiscuous species this is often accompanied by the retardation of reproduction in males. In colonial weavers most males do not breed in their first year while females do, so that the sex-ratio of breeding adults is a favour of females. In the grouse that form large leks the males cannot breed till they are about three years old. Though *r* increases logarithmically with increase in clutch size it is inversely proportional to the mean generation time. Thus to retard the reproductive age of males is inconsistent with the *r*-strategy. However, the above examination shows that the colonial breeders have clearly succeeded in the '*r*-type' environment. For this reason alone the proposition of the high-fecundity strategy has a greater generality than the interpretation made in terms of *r*-strategy'.

Territories of mammals

(1) In mammals, as the name suggests, there is mother–infant contact through suckling in all species including egg-laying monotremes and early-parturient edentates. Because of this bond, however, the existence of nests is not universal as in birds. Among the mammals those that build nests are phylogenetically older, not younger, groups such as the Monotremata, Insectivora, Rodentia and lower primates. Those that have evolved since the mid-Tertiary, such as large ungulates, Chiroptera, Cetacea and higher primates, do not build nests. Of course the nest-building species would at least defend their nests but the defence might not be strong.

The Prototheria (Monotremata: echidnas and platypus) make nests and possibly defend an area around the nest. Wilson (1975) states that the echidna probably defends territories but this is perhaps restricted to a small area around the nest.

Marsupials, because of their peculiar reproductive method, do not make nests for parturition and territoriality is little developed. Wilson (1975) mentions the ringtail possum *Pseudocheirus peregrinus* as a nest-building, territorial species. In this species nests are often made jointly by both sexes and the sexes remain together throughout the year but, since they also build nests unisexually and the membership of a nest group often changes, it is still doubtful if territoriality exists (Thomson & Owen, 1964).

Dunnet (1964) suggested the presence of territories in the brush-tailed possum *Trichosurus vulpecula*, in which home ranges of males did not overlap in his study. In New Zealand where this species was introduced, Crawley (1973) found overlapping home ranges in both sexes suggesting the absence of territories. Social organization of macropods has been studied by Caughley (1964) for the red kangaroo *Megaleia rufa* and the eastern grey kangaroo *Macropus giganteus* and by Kaufmann (1974) for the whip-tailed wallaby *Macropus parryi*. The latter work is particularly detailed but in neither case was there any sign of territory found among free-ranging groups.

In the Lagomorpha, the existence of territories was suggested for the Japanese pika *Ochotona hyperborea* from its specific vocal communication in the study of Sakagami *et al.* (1956) at Mt. Daisetsu, Hokkaidо and territoriality was confirmed for the American species, *O. princeps*, by Kilham (1958). The pika aggregates in a suitable place for nesting. In *O. priceps* each animal has a territory around a special rock and guards it. They utter a sound 'coack' in display. The territory includes food stores but feeding takes place outside (type B).

The territories of rabbits differ according to the species. In the brush rabbit *Sylvilagus bachmani* of the chaparral in California, studied by Connell (1954), movements during the breeding season (Fig. 4.22) revealed that the home range of the male was large and overlapped with others of the same sex while that of the female was small and did not overlap. Connell did not observe defence of territories but suggested that in this species the female might be 'semi-territorial'. According to Kawai (1955, and personal communication, 1978) the domestic rabbit *Oryctolagus cuniculus* forms a polygynous family group and the male defends an area containing the nests of several females. The female defends only around the nesting hole and overlaps in her range with other females. This study was made in an enclosure but studies made on the feral domestic rabbit in Australia and the wild form revealed that polygynous groups also existed in field populations and the defence of an area at least around the nest occurred.

Among the rodents most of the mice (Cricetidae and Muridae) have nests built by females. No males are known to have territories but females defend territories in some species. Burt (1940) described territorial behaviour of the white-footed mouse *Peromyscus leucopus noveboracensis.* This species was solitary most of the year but became somewhat gregarious in autumn and winter. In this period home ranges of males were 1300–1800 m^2 in size and those of females were 800–1500 m^2. The female had a territory of about 50×20 m in size within her home range and defended it keenly against other individuals of the same sex. Tanaka (1953) deduced from his study of home ranges in the vole *Clethrionomys rufocanus bedfordiae* that only the females had territories (their home ranges did not overlap). According to Izumi (1976) the male of the Norway rat is often absent from the nest but the female defends a relatively large area around the nest. Since this territory sometimes contains sources of food and the female does not often wander out of it, it may be said to be of type B. However, the defence of territories often breaks down as a result of artificial influences such as the creation of a

Fig. 4.22. A map showing spring home ranges of the brush rabbit *Sylvilagus bachmani* (after Connell, 1954).

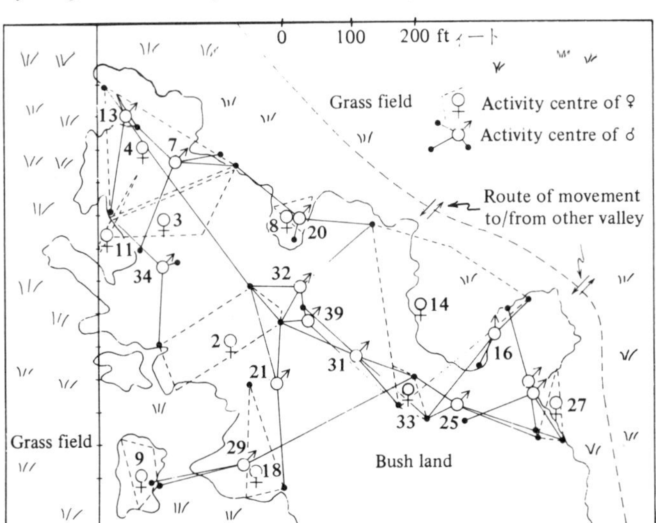

highly concentrated supply of food at a rubbish tip, etc. where many adults congregate and build nests.

Among squirrels there are many that defend territories. Arboreal species are solitary; in the two North American species studied, *Tamiasciurus douglasii* and *T. hudsonicus*, the sexes establish territories independently and store food within the territory. The function is of type A (Fig. 4.23). Territorial defence is achieved mainly by vocalization. During the mating season the males wander outside the territory and are tolerated in females' territories while the females after mating rear young alone (Smith, 1968). Carl (1971) reported on the territory, close to type A, of the arctic ground squirrel *Spermophilus undulatus* from the Alaskan tundra. Unlike the lemmings this species maintains a very stable population through territoriality (see p. 183).

Thus the sexual bond of rodents is very weak even during the breeding season and in many species males do not take part in the defence of the nest and its vicinity. In some species, however, pairs are formed during the breeding season or permanent groups are maintained. In these species territorial defence by the male is known. For example, in the yellow-bellied marmot *Marmota flaviventris* the male defends a territory of a 'harem' (sometimes one female) (Armitage, 1962). Sometimes these harems occur crowded in a small area, but each harem is defended independently. Similar territories are known for the guinea-pig *Cavia porcellus*. As for the large rodents, the American beaver *Castor canadensis* and the prairie-dog *Cynomys ludovicianus* (Sciuridae) defend the ground of a permanent group. The prairie-dog will be discussed again in Chapter 6.

Fig. 4.23. A map showing territories of two species of squirrel *Tamiasciurus hudsonicus* (*T.h.*) and *T. douglasii* (*T.d.*) (after Smith, 1968). Territories are interspecific as well as intraspecific. The sex is unknown if not shown.

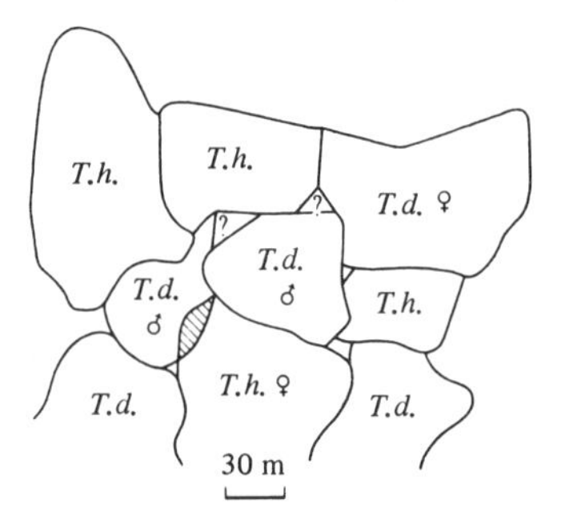

(2) The ungulates produce a kind of pheromone from the suborbital gland and rub it into a tree trunk to declare their territory (see Stoddart's (1976) review for mammalian pheromones used in territorial defence and individual recognition). Thus they are basically territorial. The problem, restricted to the breeding season, is probably to ascertain whether there is a territory for a pair as a unit or whether territorial defence is carried out by a polygynous group or a herd as a unit and, in the species that remain in herds during the breeding season, whether there is a territory defended individually within a herd. The home range of ungulates is extensive and it is not easy to observe territorial defence. For example, the Japanese serow *Capricornis crispus* has not been observed to defend territories but marking of the home range with pheromones and characteristic dunghills suggests at least some mutual avoidance (Miyazaki, 1974). As in the case of the lion to be discussed below it is possible that while the territory holder is at one end of the territory another individual may violate it at the other end.

Among the ungulates, those living in the forest interior, such as mousedeer and duikers, are solitary and they probably form a monogamous pair during the breeding season. Estes (1974) reported that in small forest-dwelling antelopes territories were maintained by the incessant application of pheromones. He contends that territoriality is held by the duikers and the dik-dik (a pigmy antelope). From observations in

Fig. 4.24. A map of Nara park showing male territories of the sika deer *Cervus n. nippon* (from Kawamura, 1950, 1952). A, boundary between two large male groups in the non-breeding season; I-VI enclosed by crosses, mating territories; B (hatched area), a pseudo-territory occupied by a low-ranking male which failed to establish a mating territory; B.G., Botanic Garden: D.G., Deer Garden.

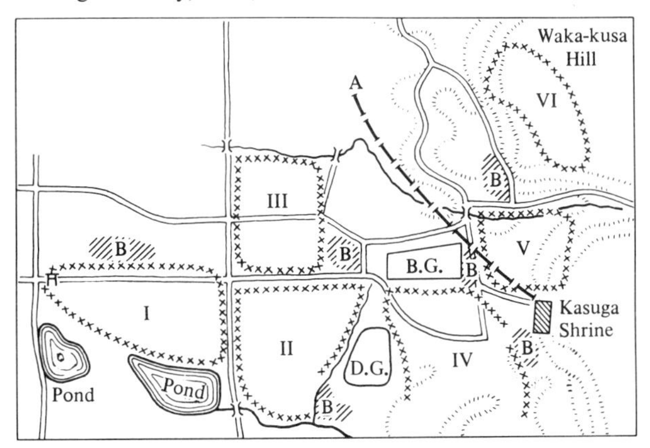

captivity, Wilson (1975) considers that the male of the mousedeer *Tragulus napu* has a territory or at least it defends an area around the female.

No pair relation has been confirmed for the Cervidae. During the non-breeding season the sexes segregate into different herds (the yearlings stay with the females) and sometimes territorial defence is seen within the female herd; e.g. the black-tailed deer *Odocoileus hemionus* (Dasmann & Taber, 1956), the wapiti *Cervus canadensis* (Graf, 1956). Kawamura (1950) observed mating territories for the sika deer *Cervus n. nippon* in Nara Park (Fig. 4.24). During oestrus in the female, large males establish mating territories and attack intruding males vigorously.

The best developed territory among the ungulates is seen in the vicuña *Vicugna vicugna* (Camelidae). This species lives in the semi-desert of the Andes and forms a polygynous group (harem) led by the male. Each harem occupies the same territory throughout the year (Fig. 4.25). Defence is carried out by the male. Since one male owns four to five females many surplus males appear and they form male groups without a leader. They graze in loose herds and do not defend a territory. The area of a territory occupied by a harem is about 4 ha.

While the ungulates of the forest and mountains are dispersed, the antelopes of the extensive grasslands of Africa and the caribou of tundra live in large herds of some permanency. The wildebeest (gnu)

Fig. 4.25. A map of Huaylarco, Peru, showing territories of *Vicugna vicugna* (after Koford, 1957). Each territory (enclosed by broken line) is occupied by a polygynous group, defended by the male.

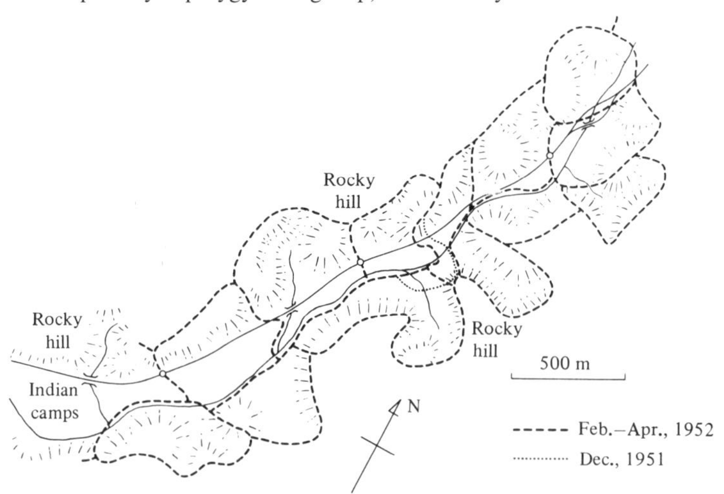

Connochaetes taurinus lives in grassland in a herd of thousands but in the breeding season the males establish mating territories which are defended against one another. Groups of females and young are tolerated. As a rule only the males with established territories can mate with a number of females (Estes, 1969). Thus the territory of males has a radius of 100–150 m and shifts gradually with the herd as it migrates over the plain.

Among other antelopes, the Uganda kob *Kobus kob* and the waterbuck *Kobus ellipsiprymnus* do not form a large herd but congregate during the mating season like a lek of grouse. The male establishes a territory for his polygynous group within the mating assemblage (Buechner, 1961; and others).

Geist (1971) investigated in detail the ecology of the mountain sheep (the Dall mountain sheep *Ovis dalli dalli*, the Stone bighorn *O. dalli stonei* and the American bighorn *O. canadensis canadensis*) but there is no description of territorial defence by the flock. They also secrete from suborbital glands and frequently perform rubbing. Would the functions of this behaviour be other than the territorial display (attraction, route marking, etc.)? I have no solution to this problem, but even if there were territorial defence by the group the group would not be very closed as there were individuals joining another group across the valley.

In the Carnivora the territory is known for the American black bear *Ursus americanus*. This species is solitary but the female has a territory in which she produces young. The young bear is said to be able to hold a territory of its own within the mother's territory. When the mother is killed in such a territory one of her female offspring succeeds her territory. This is akin to the mother–daughter relation of social hymenopterans (see p. 294) (Wilson, 1975).

In the Canidae, apart from the foxes (*Alopex*, *Urocyon*, *Vulpes*, etc.), they are mostly group hunters. I shall comment on group hunters in Chapter 6. As for the solitary hunters, the red fox *Vulpes fulva* is usually monogamous and both sexes take part in parental care, suggesting territorial defence of at least some area. [Takedatsu (1973) observed some polygynous cases in the northern Japanese fox.]

In contrast to the Canidae, the Felidae are mostly solitary (except the lion) and there is no parental care or nest defence by the male. Only in the cheetah *Acinonyx jubatus* is there a report of the male staying with the female during the gestation period, but even this is not found in other cats. They always mark their home ranges with urine, suggesting individual type F territories for both sexes, but there is no proof of it. *Mayailurus iriomotensis*, a wild cat recently described from Iriomote

Island, Okinawa, has overlapping home ranges, judging from the distribution of individually recognized dung (Imaizumi, 1976), but they may have territories overlapping temporally.

The lion *Panthera leo* is the only cat that leads a group life. If we take the Feloidea, instead of the Felidae, then the spotted hyaena *Crocuta crocuta* has a group territory (see Chapter 6) though a closely related species, the striped hyaena *Hyaena hyaena* is not social. The lion has a pride consisting of mothers and daughters or sisters. The pride (4–37 lions) defends the same territory over many generations (Fig. 4.26). In the society of lions the male is only a parasite, eating the prey caught by females but only specific males (usually two) are allowed to depend on the pride. I shall discuss social relations within a pride later, but the fact which concerns us at present is that the pride has a territory of several

Fig. 4.26. A map showing the territories occupied by two prides of lions *Panthera leo* and typical positions taken up by individual members on one day (redrawn from Bertram, 1975). Female signs show positions of adult females (black circles, peripheral females), figures of lions indicate adult males (black figures, intruders and wanderers), and white circles indicate juveniles and cubs.

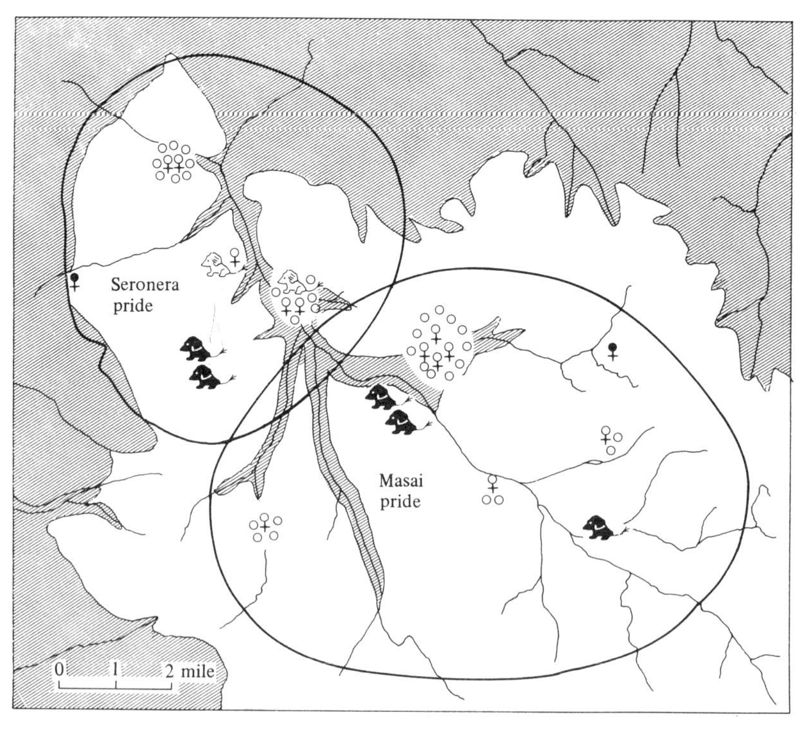

kilometres or more in diameter and excludes other lions, particularly females not belonging to the pride. Strange males gradually get acquainted with the members of the pride (particularly if the pride has lost its dependant male) and can join the pride (see intruding males in a Masai pride in Fig. 4.26). However, females cannot join the pride though they may be allowed to stay near the periphery of the territory. Females born in the pride will be driven out when they are about three years of age. Since females outside the pride have little chance of producing cubs and since this large carnivore has almost no natural enemies, the territories are possibly serving the function of population regulation.

Functions of territoriality

As the functions of territoriality in individual cases have already been discussed, here a brief summary should suffice.

The eight items listed by Allee *et al.* (1949) as being significant in territoriality are as follows:

1. The more or less automatic organization of a local population into a sort of well-spaced aggregation.
2. The promotion of monogamy, which is often important in rearing helpless young.
3. Limitation of the breeding population and hence partial control of an increase in numbers beyond the carrying capacity of the habitat.
4. The provision of a reserve of unmated males and females, making possible the prompt replacement of a lost mate in a breeding territory.
5. With larger territories, the insurance of an adequate supply of easily accessible food.
6. A reduction in the rate of spread of parasites or diseases.
7. Close acquaintance with the locality, giving an advantage to the territory holder in hiding from predators.
8. For psychological reasons that are not wholly known, there is an increased vigour of defence by the occupant and decreased aggressiveness by the invader that make for social stability.

Hinde (1956) listed the following ten:

1. Familiarity with food sources, refuges from predators, etc. and possible increase of fighting potentiality (circumstantial evidence).
2. Territorial behaviour produces over-dispersion and can regulate density in favoured habitats (strong evidence).

3. The formation and maintenance of the pair-bond (evidence for many species).
4. Reduction in interference with various reproductive activities (including courtship and mating) by other members of the species (much evidence).
5. Defence of the nest-site, an important consequence of territorial behaviour in many species, with aggressive behaviour often clearly specialized to this end.
6. In a few species the territory is primarily concerned with food but in most the food value of the territory is not significant.
7. In some species the over-dispersion produced by territorial behaviour may reduce predation (little evidence).
8. Reduction in the despotism of other males.
9. Reduction in disease, unlikely to be significant except in some colonial species.
10. The prevention of inbreeding and the promotion of range extension, unlikely to be significant consequences of territorial behaviour.

Both authorities gave many similar factors but one significant difference is that Hinde questions the function of food acquisition.

Lack (1954b) criticized the theory that territoriality promotes dispersal [Hinde (1956) and Tinbergen (1957) counter-attacked this idea]. As already mentioned, Lack held the view that food was the basic regulatory factor of animal numbers. The reason for denying the territorial role in dispersal was that while in many species more than enough food was held in the territory, others fed outside the territory. On this point Lack and Hinde shared the same view in spite of their differences on other points.

However, as we have seen type A territories are developed in insectivorous forest birds, predatory birds and alpine mammals, and their food organisms are characterized by low density and unclumped spatial distribution – in other words, food is sparse and uniform (see p. 246). These food organisms are not abundantly spread like the grasses consumed by the ungulates of the plains. They are not concentrated in one place at one time, nor do they fluctuate drastically from year to year in abundance. Rather, the food is tending to be short but stable in supply, which selects for the low-fecundity/parental-protection strategy. [Geist (1974) who compared territoriality among the ungulates also claimed that the temporally and spatially stable habitat caused the evolution of territoriality.] From this point of view the role of territoriality in securing food is undeniable for some animals. As already mentioned, there is more food than necessary in the territory but this is because territorial behaviour has been selected for in such a way that animals can survive the year of least abundant food supply in its yearly fluctuations. Many elimination experiments, as exemplified by the territorial great tits cited earlier (Fig. 4.12), have clearly demonstrated the role of territoriality in the maintenance of population density. This seems to be ultimately connected with the securing of food.

Noble (1938, 1939b) distinguished two types of territory; one for sexual relations and the other for nest-building, which he thought would be based on different motivations. He stated that in lower animals the territory, if it existed, would only be for sexual relations. However, this view is contrary to the fact. Territories of the ayu were concerned with food only and so was the territory of ants on the whole. Even in the dragonflies in which the sexual tendency is strong in territorial behaviour it promotes dispersal of the population (leading to the avoidance of the eating-out of food supply).

Parental care is another important function of the territory for many species (most territories of fish are related to this function). There are also cases where the maintenance of the pair relation is considered to be the main function.

In effect I think that the maintenance of monogamous or polygynous relations, parental care and the securing of food are the three major factors. From the knowledge of territories in lower animals it seems possible that territoriality could evolve independently in relation to any one of these factors. Also territoriality that has evolved in relation to one factor may change to serve other functions. In the case of the benthic darters, we have seen that as territoriality developed the defended area increased and approached the entire home range used for feeding. Yet, further evolution took the form of nesting territories, decreasing the size of the defended area, but with intensified defence. Even so, it is clear from the reduction in egg number and other tendencies that this series of territoriality shows a direction of increasing ecological homeostasis.

In Chapters 1 and 2, I said that the high-fecundity strategy has developed in 'continuous' habitats in which food is easily procurable by the young but the risk of predation is high and that the low-fecundity/parental-protection strategy has developed in 'discontinuous' habitats in which it is difficult for the young to obtain food. Although three main factors are involved in the development of territoriality, territoriality may be considered to have evolved along the series of the low-fecundity/parental-protection strategy.

5

Social insects

What is a social insect?

Let us digress a little here; this side-track has a purpose. I have stated repeatedly that species have always faced the choice between two strategies: that of high fecundity, or low fecundity with parental protection. Of the two, only the low-fecundity/parental-protection strategy strengthens the mother–offspring bond, producing a family structure with the participation of the father and a persistent aggregation of parents and offspring. The human society, via a roundabout route, was established in the lineage of this familial evolution (Chapter 6). Here, the society is a group of individuals with approximately equal abilities at least between the sexes.

The social evolution leading to the so-called 'social insects' is different. Apart from the termites the social evolution of ants, bees and hunting wasps also started from the protection of young by the parents and then advanced to the co-existence of adults from two generations. At the next step they lost their independence as individuals. Within the colony which had originally been a form of family, individuals differentiated into those that could not survive alone (the queen and the males) and the degenerate adults that could live but could not breed or in which such ability was suppressed (workers). Thus in contrast to the vertebrate societies in which individuals of equal abilities have acquired complex adjustment functions and if we call such a population organization of vertebrates a society, the social insects have denied themselves an opportunity of establishing a 'society'.

Therefore, the individuals of social insects are fundamentally different from the individuals of a human society. At the climax of their advancement the individuals of social insects are nothing but the self-powered 'parts' of a supra-organism. Considering this point I used the expressions 'familial insects' and 'familial evolution' in the first edition of this book. Following Sakagami (1975) I would now use more customary expressions of 'social insects' and 'social evolution' (equivalent Japanese words). The reason for this change is partly a convenience that the expression 'social insects' has now been used widely but also a result of Michener's (1958) finding, published during the preparation of the first edition, that in some species females of identical morphology and equal ability emerging from different nests form a group to nest and co-operate in caring for the young, suggesting that the evolution of sociality in some bees started once division of labour had occurred in this communal life.

Here the polymorphism supporting the 'division of labour' in the social insects is called the caste system and each 'class' is called a caste. The typical castes of the hymenopterans are the queen and the worker. Among the ants there is sometimes a sub-caste of the worker known as the soldier. In the Isoptera the castes consist of the queen, the king, the worker and the soldier (in the termites the soldier is not a derivative of the worker). Of course such a caste is different from the class or the caste of human societies. By social insects I mean the insect species in which the mother and daughters cohabit in the nest to care for the young, in particular the species with the caste system.

Wheeler (1928) considered the evolution of the family (sociality) in insects to have occurred in the following order:

1. The insect mother merely scatters her eggs in the general environment in which the individuals of her species normally live (*atrophaptic* insects). In some cases the eggs are placed near the larval food (*dystrophaptic* insects).
2. She places her eggs on some portion of the environment (leaves, etc.) which will serve as food for the hatching larvae (*eutrophaptic* insects).
3. She supplies her eggs with a protective covering. This stage may be combined with (1) or (2); in some species this is done by the male.
4. She remains with her eggs and young larvae, and protects them.
5. She deposits her eggs in a safe or specially prepared situation (nest) with a supply of food easily accessible to the hatching young (mass provisioning).
6. She remains with the eggs and young, and protects and repeatedly feeds the young with prepared food (progressive provisioning).

7. The progeny are not only protected and fed by the mother, but eventually co-operate with her in rearing additional broods of young, so that the parent and offspring live together in an annual or perennial society.

Wheeler called the first five categories the infrasocial, the sixth the subsocial and only the seventh the social, and listed eight orders of insects containing the subsocial or social species. They are Coleoptera and Hymenoptera, both showing various stages of sociality; Isoptera, containing only the 'social' group; Blattodea (wood-eating cockroaches); Dermaptera; Orthoptera (part of Gryllidae); Embioptera (Embiidae); and Zoraptera.

In considering the origin of social insects, the important division would be whether or not the parents co-exist with the hatched larvae to provide protection (Sakagami, 1970). Some members of the Heteroptera and Homoptera (Membracidae) in addition to the above orders, provide such parental protection (some spiders also do this), while other dominant groups of insects with an enormous number of species, such as Diptera and Lepidoptera have not achieved this primordial stage of social development.

Of the above, only in the Hymenoptera and the Isoptera can we trace a consistent trend of social evolution culminating in the most advanced stage. Social evolution within the Hemiptera has gone no further than simple protection of the young by covering the larvae with the adult's body or carrying eggs on the back. In the Coleoptera we can see a series of early stages of social evolution but examples are scattered among phylogenetically different groups within the order (e.g. a chrysomelid species, *Omaspides specularis*, protects offspring from the egg stage to mature larvae; a scarabeid genus, *Copris*, remains near the rolled dung and eliminates fungi; female burying beetles (*Necrophorus*) feed larvae by regurgitation). The peak of social evolution in beetles is seen in *Xyleborus* (Scolytidae) and *Platypus* (Platypodidae), which carve galleries in tree trunks to culture fungi to provide food for the larvae as well as for the adults themselves. Two or more generations of adults co-exist. This mode of life is reminiscent of the origin of termites (see p. 289) but it is not known whether or not there are the beginnings of caste differentiation among the individuals.

Stages of evolution in social insects

The discussion in this section should really be placed at the end of the chapter but is given here for reasons of convenience. Since the social insects are characterized by a persistent aggregation of individuals that have their lost independence as a result of caste differentiation, what is important in their evolution is how individuals with different abilities have come to co-exist and what factors have operated in the course of it.

Let us first consider the process according to the theory of Wheeler (1928) who was the first to discuss the evolution of social insects. Here I examine the hymenopterans (ants, hunting wasps and bees), in which only the females have been differentiated into different castes. Here, the wasps mean those hymenopterans, excluding Formicoidea (ants), that hunt other animals; namely, Scolioidea, Bethyloidea, Pompiloidea, Sphecoidea, Vespoidea (Masaridae which belongs to Vespoidea and collects pollen and nectar, is an exception). The bees belong to the Apoidea which collect pollen and nectar. The Vespoidea is the only group of hunting wasps that has produced social insects. Thus the social hymenopterans consist of Formicoidea, Vespoidea and Apoidea. The phylogentic relationships suggested by Wilson (1971) for these groups are given in Figs. 5.1 and 5.9. They are considered to represent the most probable phylogeny based on present knowledge.

Wheeler (1928) considered the following stages of evolution.

1. Appearance of feeding of larvae by the mother only (stage 6 of social evolution mentioned above).

Fig. 5.1. Phylogeny of the aculeate Hymenoptera proposed by Wilson (1971). The taxa inside boxes contain 'social' species.

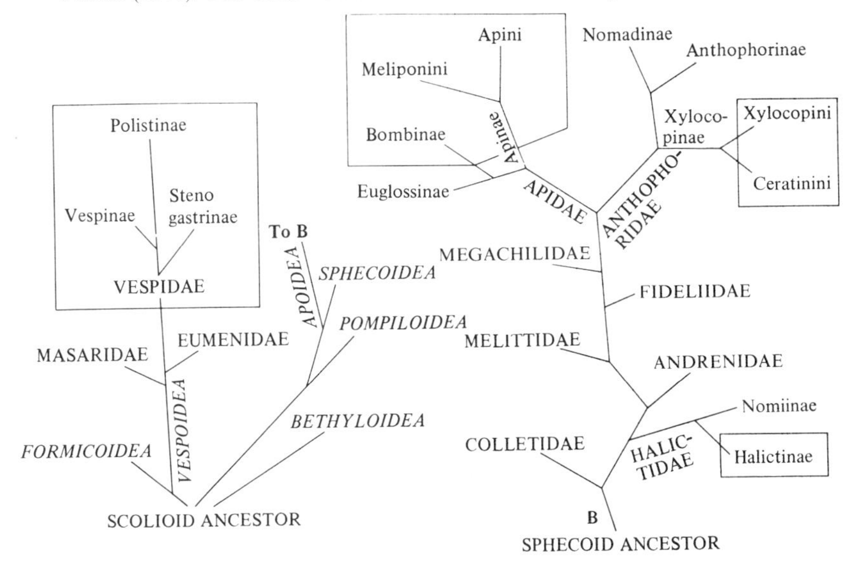

2. Lengthening of mother's longevity permitting her meeting with adult daughters.
3. Adult daughters remaining with the mother to assist her.
4. Caste differentiation, namely, the mother (the queen) and the daughters (the workers).

This scheme excludes 'trophallaxis', an important concept of Wheeler's, which will be discussed later.

For a long time Wheeler's evolutionary scheme has been accepted as the only possible path of social evolution in social insects. However, new facts discovered after the war made other interpretations possible. For example, it was discovered that in some species of Halictinae (primitive bees) adult females belonging to the same generation (sisters) use the nest left by their mother communally and some of them co-operate in rearing young. Marking of individuals revealed that in some cases daughters from different mothers were mixed in the same colony. Michener (1969) claimed that the Wheeler-type evolution is possible in ants, *Vespa* and *Xylocopa* but is unlikely to occur in bees and proposed the following as an alternative route of evolution:

1. solitary – showing neither co-operative caring for the young, caste differentiation, nor an overlap of generations;
2. subsocial – the adults care for their own nymphs or larvae for some period of time;
3. communal – members of the same generation use the same composite nest without co-operating in brood care;
4. quasisocial – members of the same generation use the same composite nest and also co-operate in brood care;
5. semisocial – as in quasisocial, but there is also reproductive division of labour, that is, a worker caste cares for the young of the reproductive caste;
6. eusocial – as in semisocial, but there is also an overlap in generations so that offspring assist parents.

Sakagami (1970) used a diagram similar to Fig. 5.2 to explain succinctly the two possible routes of evolution. The upper series is the subsocial route of Wheeler's and the lower series is the semisocial route of Michener's.

At first, in stage A each female nests independently (m indicates that the mother is dead). In stage B the mother now lives longer and co-exists with daughters but each daughter still nests independently. In stage C the daughters remain in the mother's nest and live with the mother. They build cells and oviposit in them. Only in stage D does division of labour occur between the mother and daughters in the same nest – the mother concentrates on ovipositing while daughters work.

In A′ the daughters form a nest aggregation instead of nesting in different places. In this aggregation there may be nests built by daughters from another mother (F with an arrow from outside the circle in the figure). In B′ different nests are combined within the mother's nest and communal nesting begins (each bee entering the ground of the mother's nest builds her cells independently). In C′ the independence of nest chambers (cells) is lost (small circles are removed from the F's in the figure) and communal caring for the young begins. Finally in D′ caste differentiation occurs among the sisters and their 'girl friends' of the same generation as the original individual differences are enlarged physiologically and behaviourally.

In Michener's terminology D′ is the 'semisocial' and D is the 'eusocial' (the matrifilial – mother–daughter – caste system of Sakagami). [In the following, 'social' refers to the cases where adults of two successive generations co-exist. In the Hymenoptera they pertain to mother–daughter co-existence, including the stages of C and D. In reality the stage C seldom exists.] As the terminal of either series is D there should

Fig. 5.2. Two hypothetical routes of evolution from solitary life to caste differentiation (modified from Sakagami, 1970). Above, the subsocial route; below, the semisocial route. M, mother (m, after death); F, adult daughter; Q, queen; W, workers. The large circle indicates the nest built by the mother. Black arrows from outside large circles indicate individuals joining the nest.

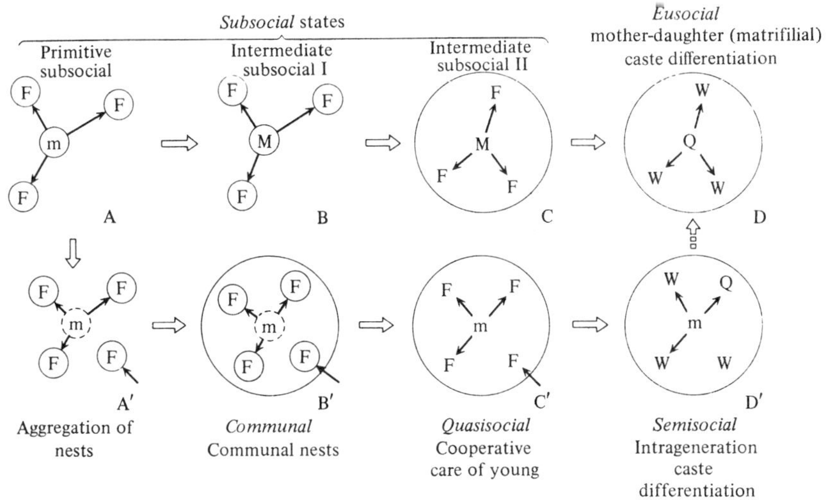

be an arrow from D′ to D. The reason for the broken arrow in the figure is that I find it difficult to believe that this route was a direct one (see below).

Recent reviews on the social evolution of the Hymenoptera are found in Iwata (1976), Sakagami (1970, 1975), Wilson (1971, 1975), Michener (1974 for bees). The following account is based on these reviews. As comprehensive references are found in these reviews (except in Sakagami), in many cases the original work has not been cited. Also, the following discussion has three limitations: first, future studies in the tropics may greatly alter the line of argument presented here (the study of social insects in the tropics is lagging behind in spite of the fact that the tropics is the home of the social insects); secondly, the side-chain of 'labour parasitism' that appears in various stages of social evolution has been *completely ignored* (it is of great interest on its own accord and the reviews of Iwata and Sakagami should be consulted for it); and thirdly, the special bypasses have been omitted from the discussion (only the crude main routes have been followed). These limitations were unavoidable in the present volume.

Social evolution of hunting wasps

Wheeler (1928) dealt with wasps that were relatively advanced in social evolution. It was Kunio Iwata (1942, 1955a, 1971) who admirably traced the evolution of parental care for the young from the Parasitica through the solitary hunting wasps to the social *Vespa*.

Iwata (1942) distinguished several basic habit-types in the parental behaviour of solitary hunting wasps and searched for the possible routes of evolution in their combinations. The basis of his habit-types is given by:

O: oviposition (Ovum parere),
P: paralysing of prey (Pungere),
T: transportation of prey (Transfere),
V: hunting (P plus T) (Venari),
I: nest construction (Instruere), and
C: closure of nest hole (Clandere).

Iwata called each unit of behaviour (represented by a letter) the *partial habit* and the total innate behaviour of any series of partial habits possessed by a species the *complete habit*.

The sawflies belonging to the Symphyta, considered to be the most primitive group in the Hymenoptera, lay eggs on host plants and leave them. Their larvae eat on their own (there is no species of Hymenoptera that scatters eggs). The parasitic wasps and gall-wasps among the Parasitica do the same. Their habit-type is therefore O.

In the Aculeata there are parasitic wasps (excepting social parastism) that have the habit-type of O (e.g. the dryinid wasps of Bethyloidea, which parasitize planthoppers, and the chrysidid wasps, which parasitize other wasps and bees). Most species of the Scolioidea, the most primitive group of the Aculeata, oviposit underground in the larvae of scarabeid beetles. As they pierce the thorax of the host to paralyse it their habit-type becomes PO. However, the paralysis caused by the tiphiid wasps is temporary and the host can move about soon. In the Scolioidea, the Tiphiidae has a more advanced habit-type. For example, in the case of *Methocha yasumatsui* the female enters the gallery made by a tiger beetle larva and, after paralysing the larva and ovipositing in it, plugs the gallery (POC).

In the Pompiloidea, which all hunt spiders, some (e.g. *Pompilus aculeatus*) paralyse their prey in the spider's underground burrow and oviposit in it and leave, but most dig holes, build nests and bring in the paralysed spiders (the hole is dug first, IPTOC; or later, PTIOC). Evans (1953) used Iwata's system to compare American spider wasps and recognized the following evolutionary routes (Evans used an additional partial habit of V, denoting 'searching for host', but this is omitted here; Iwata's O would be Evans' VO). Examples of spider wasps in the following diagram include Japanese species.

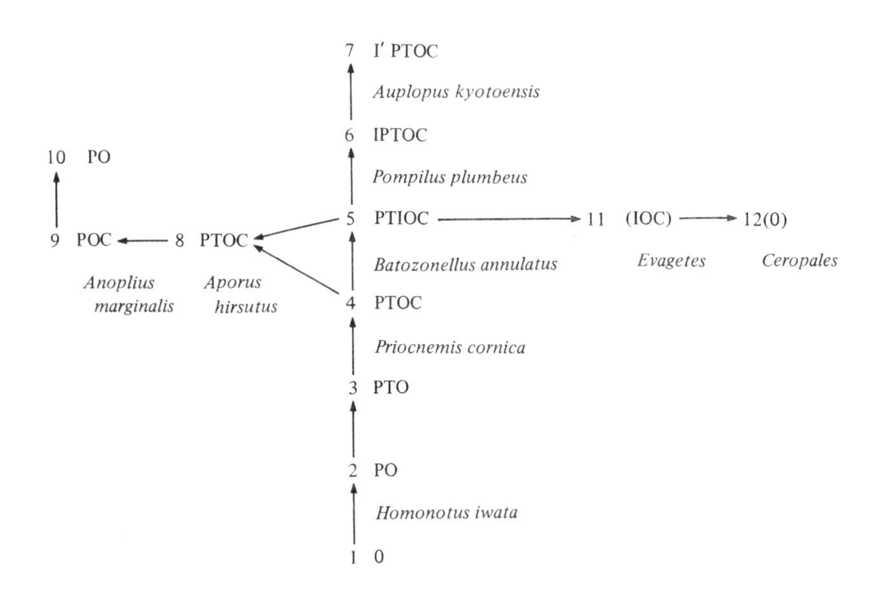

The species with the lowest grade of habit-type in the Pompiloidea do not build nests themselves. They include the habit-type PO (e.g. *Homonotus iwata*, which parasitizes primitive subterranean spiders, and *Notocyphus*, which parasitizes and lays in aerial spiders which regain mobility later), POC (e.g. *Pompilus apicalis*, which plugs the burrow) and PTOC (e.g. *Priocnemis cornica*, which carries the spider into a burrow of an insect or a snake and plugs it). Most other species of Pompiloidea (e.g. *Batozonellus annulatus*) have the habit-type PTIOC (shown as 5 in the above diagram).

There is another group which contains the same number of partial habits but with a different order. *Pompilus plumbeus* in Europe and *Priocnemis exaltatus* in America, for example, go hunting after digging a nest (IPTOC). As a bypass to this, *Tachypompilus analis* hunts a spider and then digs a nest hole but lays in the empty nest before transporting the spider. Strictly speaking this represents PITOC but was put under 6 in the diagram with IPTOC. Nest building prior to prey capture is an important change towards social evolution and has occurred in many groups (see Fig. 5.3), which will be discussed below.

In both the hunting-preceding type and the nest-building-preceding type, there are species constructing more than one cell in the nest for more than one larva. In most cases the nesting hole is dug first, then they hunt for the prey before building cells. The repeated operation of hunting, transporting and ovipositing may be expressed as I(PTO)C (not distinguished from IPTOC in the above diagram). Many examples of this are found in the genus *Anoplius* (e.g. *A. fuscus*).

Most members of the Pompiloidea dig a hole in the ground but the tribe Auplopodini (e.g. *Auplopus kyotoensis*) constructs nests above the ground, being distinguished from others in the diagram as I′PTOC (7). The specialized groups are shown on either side of the main route: 11 and 12 show labour parasitism; 8, 9 and 10 show degeneration of habit-types. Evans placed *Homonotus iwata* in 10 on morphological grounds but Iwata (1942) considered this species to be of the most primitive type among the Pompiloidea.

Iwata's habit-types and their application by Evans have achieved a high degree of success but the scheme contains parts which are not satisfactory for accurate description. For example, the habit of using other animals' nests is not apparent in this scheme. Later, Evans (1958) proposed a different scheme enhancing the habits of the Sphecoidea and Vespoidea. Fig. 5.3 shows *only the presumed main streams* of social evolution in hunting wasps, based largely on the stages of Evans. Of

Fig. 5.3. A schematic presentation of social evolution in hunting wasps (from Evans, 1958; Wilson, 1971).

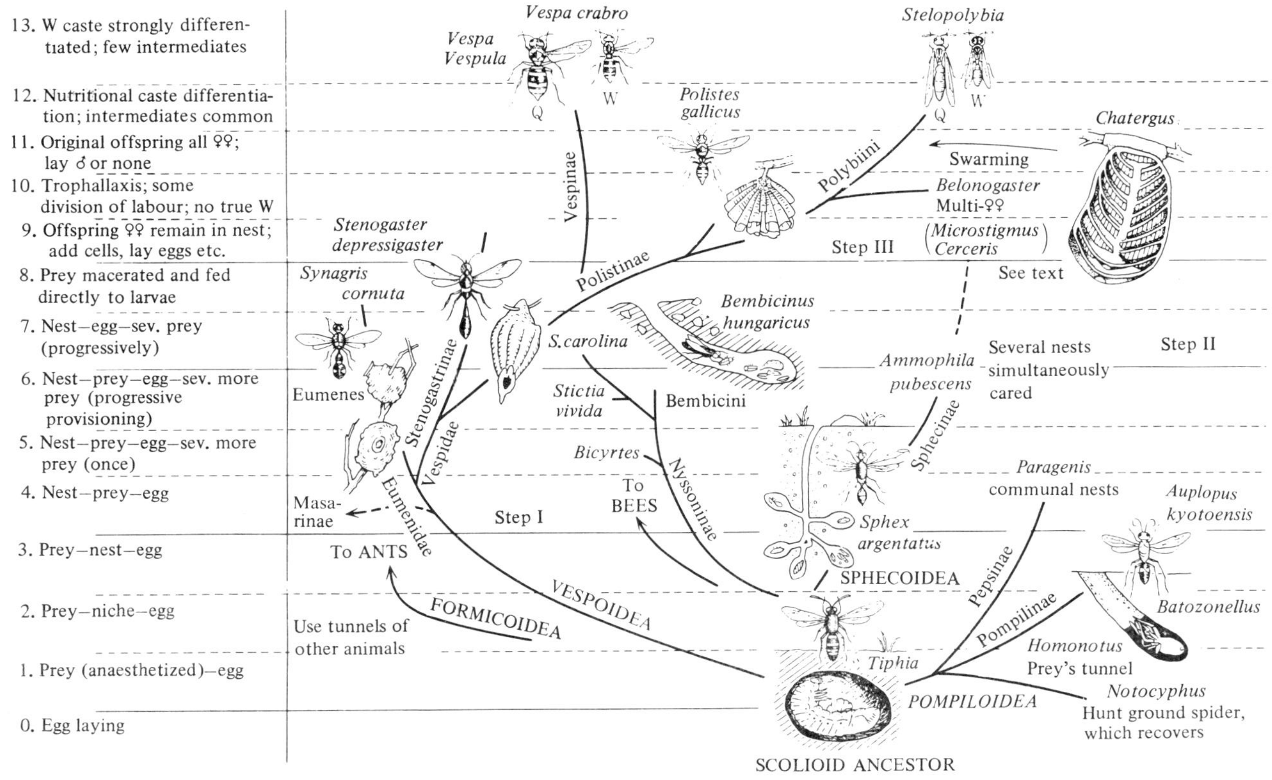

13. W caste strongly differentiated; few intermediates
12. Nutritional caste differentiation; intermediates common
11. Original offspring all ♀♀; lay ♂ or none
10. Trophallaxis; some division of labour; no true W
9. Offspring ♀♀ remain in nest; add cells, lay eggs etc.
8. Prey macerated and fed directly to larvae
7. Nest–egg–sev. prey (progressively)
6. Nest–prey–egg–sev. more prey (progressive provisioning)
5. Nest–prey–egg–sev. more prey (once)
4. Nest–prey–egg
3. Prey–nest–egg
2. Prey–niche–egg
1. Prey (anaesthetized)–egg
0. Egg laying

Vespa crabro
Vespa
Vespula
Q
W
Stelopolybia
Q
W
Polistes gallicus
Chatergus
Swarming
Polybiini
Belonogaster
Multi-♀♀
Vespinae
Stenogaster depressigaster
(Microstigmus
Cerceris)
Step III
See text
Polistinae
Synagris cornuta
Bembicinus hungaricus
S. carolina
Ammophila pubescens
Several nests simultaneously cared
Step II
Eumenes
Stenogastrinae
Stictia vivida
Bembicini
Sphecinae
Vespidae
Bicyrtes
Nyssoninae
Paragenis
communal nests
Auplopus kyotoensis
To BEES
Masarinae
Eumenidae
Step I
Sphex argentatus
To ANTS
SPHECOIDEA
Pepsinae
VESPOIDEA
Batozonellus
FORMICOIDEA
Pompilinae
Use tunnels of other animals
Homonotus
Prey's tunnel
Tiphia
Notocyphus
POMPILOIDEA
Hunt ground spider, which recovers
SCOLIOID ANCESTOR

course this crude scheme does not completely represent the evolutionary series but it seems certain that the social evolution of hunting wasps, apart from a few species in the bypasses and in degenerate forms, has occurred in several groups following approximately the stages shown here.

There were two important incidents in the development of social hunting wasps. One was the precedence of nest building. At this stage, however, feeding was by mass provisioning. In the Sphecoidea, an example may be found in *Ammophila*, which provides one large lepidopteran larva as the provision. In *Ectemnius*, which hunts small prey, more than one prey individual was provided even at this stage of development.

The next significant event was the conversion from mass provisioning to progressive provisioning. Apart from some sphecid wasps to be mentioned later, this conversion was indispensable for the social evolution of hunting wasps. Connecting the two methods of provisioning is the situation where the female remains with the hatched young before the nest is sealed (perhaps for defence) as seen in *Ammophila aberti* in Sphecoidea. The pathway to progressive provisioning might be found in the habit of ovipositing on the way to prey transportation (many examples in *Sphex*) or in mass provisioning after ovipositing. An example of the latter is seen in *Synagris spiniventris* which, when food is abundant, provides mass provisioning before sealing the nest but which, when food is scarce, may also provide additional food after the young hatch (Evans, 1958).

If oviposition is completed first and then food is brought in continually after the hatching of eggs, the co-existence of the mother and daughters is just around the corner. This stage is achieved by many members of Bembicini in the Sphecoidea and in *Cerceris rubida* in Italy the co-existence of two generations has begun (Wilson, 1971).

Noteworthy among the Sphecoidea is *Microstigmus comes* (Sphecidae,

Fig. 5.4. *Microstigmus comes*, a sphecid wasp close to eusocial (redrawn from Matthews, 1968).

Fig. 5.4) in Costa Rica, reported by Matthews (1968). Females of this species build suspended nests with very special material made of kneaded waxy powder from fern leaves. More than half the nests were occupied by two or more females (maximum eight) and one of the females had better developed ovaries than others in these cases (rudimentary functional caste?). Two females were seen bringing food (Collembola) into the same cell. Although the cells were sealed after mass provisioning the adult remained above the nest. There are two possibilities open to this wasp. The first is to develop functional castes between the mother and daughters while maintaining mass provisioning (see *Lasioglossum duplex* in the next section). The second is when the females belong to the same generation (sisters or 'girl friends'), in which case the evolution may take the semisocial route as has frequently occurred in bees. According to Evans (1964), since there are at least four species, including *Moniaecera asperata*, in Sphecidae, in which two or three females of the same generation use the same nest and rear young independently, the possibility of the semisocial route arising in hunting wasps cannot be dismissed. However, Wilson considers that the co-existence of many females in *Microstigmus comes* is the co-existence of two generations (mother and daughters) and that this is the only eusocial hunting wasp outside the Vespoidea and Apoidea.

In the Eumenidae of Vespoidea, oviposition always takes place before hunting. The only exception is the genus *Pseumenes* (one species in Japan on Iriomote Island), in which hunting precedes oviposition. In most members of the Eumenidae small lepidopteran larvae are brought into the nest over many days and by the time the last provision is made the larvae have grown considerably. Of these the females of *Orancistrocerus drewseni* feed young after crushing the prey with their mandibles (Iwata, 1971) and in *Synagris cornuta* they feed young by regurgitation as in *Polistes*. This is the only example known among the hunting wasps other than the Vespidae of feeding by regurgitation.

In the Vespidae, including the most primitive subfamily of Stenogastrinae, food is supplied to the larvae by means of progressive provisioning. The important point is that for the first time in this family we find a widespread tendency for feeding many larvae simultaneously in one nest. In the more primitive species of Stenogastrinae (e.g. *Stenogaster* (*Eustenogaster*) *micans*), only one female is found on the nest, suggesting an absence of mother–daughter co-existence, but in other species, such as *Stenogaster* (*Liostenogaster*) *nitidipennis* and most of the known species of *Parischnogaster*, several females are found on the nest (Iwata, 1971). For example, Yoshikawa *et al.* (1969) found two or more females at twenty-four nests of *Parischnogaster striatula* and *P. alternata* out of thirty-four collected in the Malay Peninsula. There was no difference in size or other characteristics among the attendant females, which mostly had developed ovaries and were fertilized (except in newly emerged females). Since wings or mandibles showed no functional differences among the females, Yoshikawa *et al.* concluded that they were mostly from one generation (sisters) though there might have been some cases of overlapping generations (mother and daughters). Wilson (1971), on the other hand, considered *Parischnogaster depressigaster* of Luzon to have overlapping generations.

E. Roubaud (1916; cited by Wheeler, 1928) reported that there was no caste differentiation in *Belonogaster dubius* (Polistinae) or *B. griseus* in Africa. Piccioli & Pardi (1970) re-examined *B. griseus* in the laboratory in Italy and Africa and found that there was a linear hierarchy among the females and that only the dominant female could oviposit. Yoshikawa *et al.* (1969) had earlier reported on hierarchical behaviour among the females of the same nest (probably founded by multiple females of one generation – i.e. pleometrotic) in another species of *Parischnogaster*, the dominant female with developed ovaries providing little labour. There is no doubt that functional castes exist in *Stenogaster* and some species of *Belonogaster*. In *Brachygastra scutellaris*, related to *Belonogaster*, a nest contains several dozen females, several hundred infertile females and workers, and several thousand cells. The colony is perennial and founded by swarming. Many polybiine wasps, of which this species is a member, seem to construct new permanent nests by swarming. *Stelopolybia flavipennis* of Brazil, the last member of this tribe, has morphologically differentiated castes.

The life history of the paper wasp *Polistes* is well known in Japan. In most members of this genus known from the temperate region, the females fertilized in autumn found their new nests singly (haplometrotic) in the following spring and become the queens. All the young out of the first batch of eggs (spring generation) function as workers but their morphology is little different from the queen's. They can be fertilized; the reason for not breeding is that there are no males at this time of the year. Might it be possible for them to mate and lay female eggs if males were made to emerge by controlling climatic conditions at the time when females of the spring generation emerge?

Caste differentiation becomes almost complete in the Vespinae (*Vespa, Vespula, Provespa*). That is, the workers are considerably smaller than the queen. Although the number of cells amounts to several hundreds or several thousands (in the case of *Vespula rufa*, more than 10 000), swarming is known only for a few species in the tropics. It is likely for all species of *Vespa* that the queen maintains an ability to found a colony singly by building a nest and collecting food.

Social evolution of bees

Apart from social parasites there is no species of Apoidea that collects food before nest building. If we follow Iwata (1971) and denote L as collecting of pollen (or nectar), the basic habit-type of bees is ILOC.

In the phylogenetic tree of bees, halictine bees are thought to be relatively old (Fig. 5.1). This group will be discussed later for reasons which will become clear below.

Not much is known about the Colletidae but their habit-type would generally be ILOC or I(LO)C. The female of *Colletes pateratus* in Japan digs a burrow and stores pollen in it before ovipositing. If only one cell is built, she seals the burrow after ovipositing and flies off to another site for a repeated operation. However, instead of building one cell she sometimes digs a long tunnel and works from the end, storing food, ovipositing and building a wall, then again storing food and so on. The tunnel is sealed after a few cycles. The maximum number of cells (chambers) found in such a tunnel is seven (Iwata, 1971).

In the Andrenidae the life cycle of *Andrena* is known (the nest is multi-celled and mass-provisioned). It is noteworthy that there is also a nest aggregation, in which many females aggregate in a small area to dig nest holes independently. This phenomenon, often seen in bees, is considered to develop into communal nesting.

In the Melittidae the habit-type is ILOC and females build single-celled (e.g. *Melitta leporina*) or multi-celled nests. Many species form nest aggregations and in the case of *Hesperaspis fulvipes* the same nest entrance is shared by two or more females building independent cells in different branches of the burrow.

In the Megachilidae the habit-type is usually I(LO)C and the nest is mass-provisioned (Fig. 5.5 bottom right). Typically the cells are separated by cut leaves but in *Lithurge collaris* (Lithurginae) the main burrow is not sealed.

Among the Anthophoridae, multi-celled underground nests are built by the bees of Anthophorinae, in which nest aggregation and the sharing of a nest entrance occur. In *Exomalopsis aureopilosa* of South America one huge underground nest contained 884 adult females, 46 adult males and 653 cells for the young (Sakagami, 1975). Most females were fertilized and took part in cooperative caring for the young. As there was no individual with tattered wings (the queen) this multi-female aggregation was considered to be of one generation (sisters). The large number of individuals involved and other evidence suggested that there were immigrant females from other nests.

In the Xylocopinae only few build nests underground. In most species the female builds a gallery in the woody tissue of an aerial stem and provisions the nest with pollen before oviposition and sealing the cell.

Fig. 5.5. Various types of nest in bees. Top left, *Allodape marginata* (progressive provisioning) (from Iwata, 1940); top centre, *Lasioglossum ohei* (common entrance to nest, mostly one female (a dot) per gallery for cells but the gallery F with two females suggesting the beginning of cooperative caring of young) (from Sakagami *et al.*, 1966); top right, *Melipona pseudocentris* (from Wilson, 1971); bottom left, *Bombus lapidarius* (inside a mouse nest; inside of communal larval cells are shown in the bottom left corner) (from Michener, 1974); bottom right, a megachilid bee (mass provisioning).

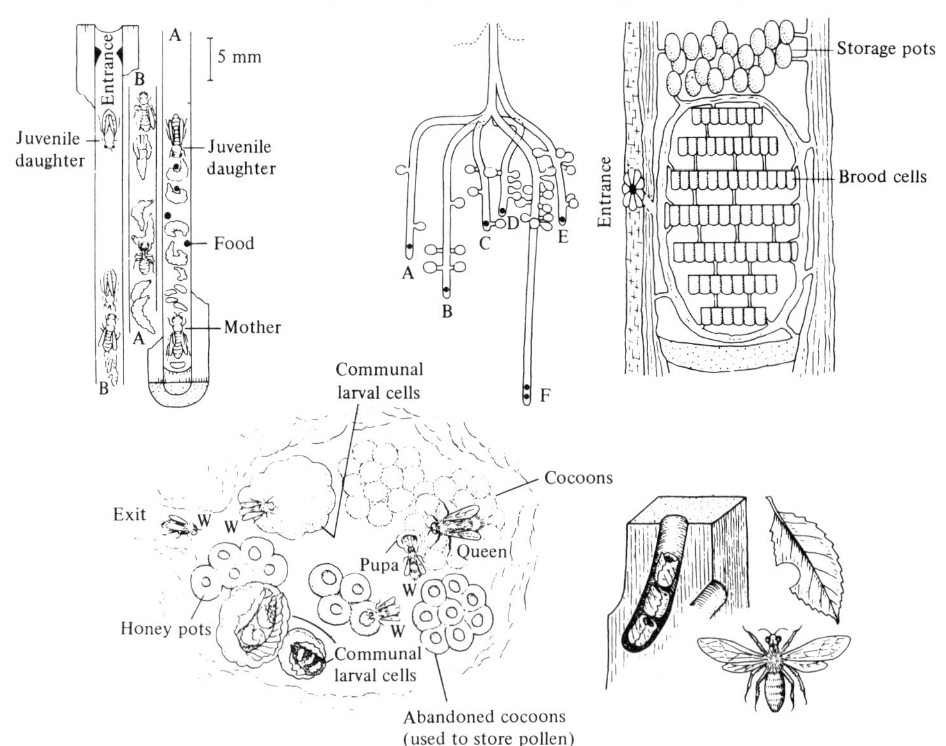

She repeats this procedure a few times. In the Japanese carpenter wasp *Xylocopa appendiculata* the female stays by the nest after completing half a dozen cells. In the meantime the young (both sexes) from the cells built at the beginning emerge and co-exist with the mother for a time. As the mother dies in autumn her offspring overwinter together and disperse in spring. In *X. frontalis* of South America Sakagami (1970) discovered one of the young of another species, *X. grisescens*, joining such a mother–offspring group! It shows a rudimentary form of 'joining' in that the strangers are not repulsed. In the Formosan carpenter wasp *X. pictifrons* they use the nest left by the mother and in the Sauter's carpenter wasp *X. sauteri* several females share the common entrance to their respective cells.

An interesting development within the Xylocopinae is seen in the tribe Ceratinini. In one species, *Ceratina flavipes*, the female nests in an aerial stem and builds cells, repeating the cycle of provisioning, oviposition and wall-making in the stem. In many cases the female removes the walls when the larvae grow to a certain size (Sakagami, 1970). In *C. japonica* the female mass-provisions the cell and builds a wall, but breaks it down later to remove the faeces of the young and then builds it again (Sakagami, personal communication, 1977). This may be the beginning of the abolition of the single-celled system. An extreme example of this is found in *Allodape marginata* (= *Braunsapis sauteriella*) discovered by Iwata (1940) in Taiwan. In this species the eggs are laid in a hole dug in the pith of a woody branch and the larvae are fed with small amounts of pollen by the female regurgitating each provision (Fig. 5.5 top left). This is one of the rare examples of progressive provisioning in bees. As provisioning and oviposition proceed at the same time the emerged young (both sexes) from the cells built early co-exist with the mother in the nest while those hatching late are still small. In this 'public ward' system the eggs, larvae and pupae are arranged from the end to the entrance (the most vulnerable eggs and small instar larvae are placed in the end part). If this order is changed the mother carries the displaced larvae or eggs by the mouth, like ants, and re-arranges them. The daughters have not been seen to work in this species but in *Allodapula* of Australia the emerged daughters are known to collect a large part of the pollen needed in the nest. According to Michener (1962) they are not fertilized and their ovaries are not developed, i.e. caste differentiation is already present.

Here I want to touch on Euglossinae and Bombinae of Apidae. Members of the former, restricted to tropical America in distribution, are mostly shiny blue or green and have a long proboscis. Most are solitary, each female building a nest independently in a tree hollow. The cells are similar to those of bumblebees and are made of resin, mud, etc. and are mass-provisioned. Although there are large nests with many cells in this group, some nests (e.g. of *Euglossa*) contain more females than there are cells, suggesting joining of females from other nests (Sakagami *et al.*, 1967). In *Eulaema nigrita* the sisters expand the mother's nest after her death and provide communal caring for the young (quasisocial) (Zucchi *et al.*, 1969). Emerged females of this species showed dimorphism; the large individuals left the nest (possibly to nest singly) and the small individuals oviposited. Thus, in spite of morphological differences caste differentiation is not present – a phenomenon somewhat akin to the migratory locust.

Bombus and *Psithyrus* constitute the Bombinae. The latter genus is labour parasitic whereas the former gives progressive provisioning and is eusocial (with caste differentiation). Wilson (1971) and Michener (1974) recognized two modes of life in this group: 'the pollen-storer' and 'the pocket-maker'. In both, the fertilized and overwintered females build nests in mouse holes, etc. in the temperate region and supply pollen pellets before laying a few eggs. They also make special cells to store honey (Fig. 5.5 bottom left). The larvae compete for food and differences in nutrition produce size differences in the emerged females. However, they all become workers. In the case of the pollen-storer they store pollen in pots made of old cocoons and wax. They remove the wax lid of larval

Fig. 5.6. Arrangement of larvae in cells of the European bumblebee *Bombus agrorum* (pocket-maker) (drawn from Michener, 1974). The hatched area shows the consolidated pollen above which the larvae make their own cells with silk. The peripheral larvae obtain less food and grow to smaller workers. The outside line is the wax cover built by the mother.

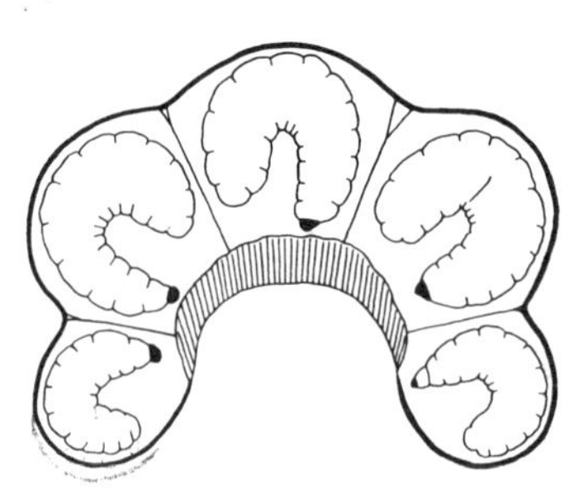

cells and pour liquid from these honey pots into the cells (progressive provisioning). In this group the difference between the queen and the workers is small. The queen emerges in autumn probably from the larva that has been given a large quantity of food.

In the case of the pocket-maker the collected pollen is pressed against a corner of the nest and eggs are laid on the pile (mass provisioning). The larvae when they hatch produce silk to contain themselves in at the place of hatching. The larvae occupying poor sites obtain less food and grow to small adults (Fig. 5.6). Only the larva destined to become the queen is fed by workers as in the pollen-storer.

The size of workers varies greatly in bumblebees. Their queen is capable of founding a nest alone unlike the queen of the honeybee or the stingless bees.

R. Zucchi (cited by Sakagami, 1975) demonstrated that the Brazilian bumblebee *Bombus atratus* in the tropics has a perennial nest. When the death of the queen is imminent a few prospective queens of the next generation return to the nest and start laying. They fight amongst themselves until all but one queen are killed. Zucchi's colony lasted ten years repeating this process of multi-female to single female systems. Even in this situation single females have been shown to be able to found new independent colonies (haplometrotic). The queen maintains the ability to work whereas workers can mate and produce eggs in the absence of the queen. In the cold region where the summer growing season is short, bumblebees produce only a few workers. Their caste differentiation is still not complete.

Returning to the Halictidae, we find various stages, including the matrifilial (mother–daughter) system, of social evolution in this primitive group. In the temperate region most halictid bees are solitary with one generation a year. The overwintered, fertilized females build nests in spring and after mass provisioning lay eggs, i.e. I(LO)C. Both sexes hatch from these eggs and the females overwinter after copulation.

There are two ways towards social evolution from this basic reproductive pattern. One way is the foundation of communal nests and the other the prolongation of longevity to permit co-existence of two generations (mother and daughters).

Occasionally two females are found in spring in the nest of the first generation. One of the females has degenerate ovaries but the two females belong to the same generation. This is seen in *Lasioglossum problematicum*. Sakagami (1970) interprets this as meaning that a female, instead of building her own nest and ovipositing, which she is capable of doing, enters a nest of the mother bee and one of the bees becomes the functional queen while the other functions as a worker.

The halictid bees often nest in aggregation and sometimes share the same burrow for nesting. Fig. 5.5 (top centre) represents a nest of *Lasioglossum ohei* as depicted by Sakagami *et al.* (1966). In this figure each gallery within the burrow is occupied by one female except for F with two females. This sharing of the gallery might indicate rudimentary cooperative caring as a step further from communal nesting (such an example is rare in this species).

A similar example is found in *Augochloropsis sparsilis* in Brazil (Michener & Lange, 1958, 1959). This species is basically solitary but often builds communal nests. The first emergences after the nest is founded include both sexes and they also produce both sexes (second generation). The second generation of adults mate and overwinter. The females of one nest are mostly sisters but occasionally include females of other nests (overlapping generations are extremely rare). Females are almost identical in their morphology but vary in the development of ovaries. As there are some unfertilized females ending their life as provisioners, we may recognize a very early differentiation of functional castes in this species.

However, there is as yet no example known of morphological caste differentiation from one generation of females alone. Differentiation of functional castes among the females of the same generation (particularly among the sisters) may be considered as a preadaptation to caste differentiation between two overlapping generations, which Michener appears to stress without the clarification of his logic. Sakagami (personal communication, 1978) is sympathetic to Michener's view as he considers that differentiation of castes among co-existing sisters would give the matrifilial caste system a selective advantage. I think it is probably impossible to proceed from semisocial (D′) to eusocial (D) in Fig. 5.2.

As for the prolongation of longevity, most halictid bees (solitary – infrasocial bees) have one generation per year but some definitely have two generations a year (Sakagami, 1970). The reason for saying 'definitely' is that there are two peaks of adult activities a year, seemingly of different generations, which quite often results from two laying periods of the same females, once in spring and once in summer with the help of daughter bees. This is typically seen in *Lasioglossum duplex*, which has been studied by Sakagami & Hayashida (1958, 1968). The fertilized and overwintered females of this species build an underground nest with several single cells in spring and after mass provisioning lay eggs and seal the nest. This is the same as in other infrasocial bees but when the new generation emerges a new phenomenon, seldom seen in hunting wasps, appears. That is, most of the newly emerged bees are females and smaller than the mother (caste differentiation). When they emerge the mother, who has been waiting at the entrance of the burrow since the sealing of the nest, starts laying for the second time. At least some members of the daughter generation then help the mother. The new offspring emerging out of the eggs of the second laying are as large as the mother and half of them are males. They mate during autumn and only the females overwinter.

Thus in *Lasioglossum duplex* two generations co-exist and functional castes have differentiated between them. This differentiation, however, is only rudimentary, for the offspring from the initial eggs include males (about 10%) with which some females mate before leaving the nest and produce offspring in autumn. Although both sexes are possible, in practice these offspring are mostly males (Sakagami, personal communication, 1977). Thus the reproductive ability of workers has not been lost.

On the other hand, when the mother was removed experimentally 65% of the females developed ovaries, suggesting that the presence of the mother had suppressed the development of gonads. There is, already at this stage of evolution, the possibility that some pheromone is operating as seen in *L. zephyrum* (Michener, 1974).

In *Lasioglossum duplex* caste differentiation was mostly functional but in *Halictus tumulorum* in Hokkaido (different from the European race), in which the old mother lays again in summer as in *L. duplex*, the size difference between the mother and daughters was negligible and functional cast differentiation was less developed than in *L. duplex*. The best-developed caste differentiation is seen in *L. malachurum* which breeds three times a year. The overwintered female (the queen) produces only females in spring and summer and the latter grow into small adults, clearly distinguishable from the queen morphologically. Thereafter the queen concentrates on laying and produces both sexes in autumn. The new generation of bees mate and overwinter. Morphologically distinguishable castes are found also in other halictid bees; in this group the genuine caste system of mother–daughters (eusociality) has been established.

Plateaux–Quénu (1959, 1962) in France made a significant observation of the extraordinary life history of *Lasioglossum marginatum* which provides a clue to the social evolution of the Halictidae. According to her this species has a life history as shown in Fig. 5.7. It breeds once a year in summer and produces females. Both the mother and the daughters (unfertilized) overwinter in their nest. In the following year only the mother reproduces and the daughters expand the nest and collect food (functional castes). As a result seven or eight females will be reared and the old daughters will die in autumn. The mother will enter the second winter with new daughters. This cycle lasts five or six years, during which time only the mother survives. In her last year (fifth year) more than 500 cells are constructed and for the first time she produces males and females that will found new nests. They mate and disperse from the nest as the aged mother finally ends her life (its adult life is longer than that of any other insect except social insects).

The significant point about this life cycle is that although the mother and daughters show such clear caste differentiation the castes are not fixed. Plateaux-Quénu dug out a nest in its third year and marked a new female (worker) and transferred her to another nest in its fifth year. The female mated with a male from the nest and after overwintering founded

Fig. 5.7. Life cycle of a long-lived halictid bee, *Lasioglossum marginatum* (from Plateaux-Quénu, 1959, in Sakagami, 1970).

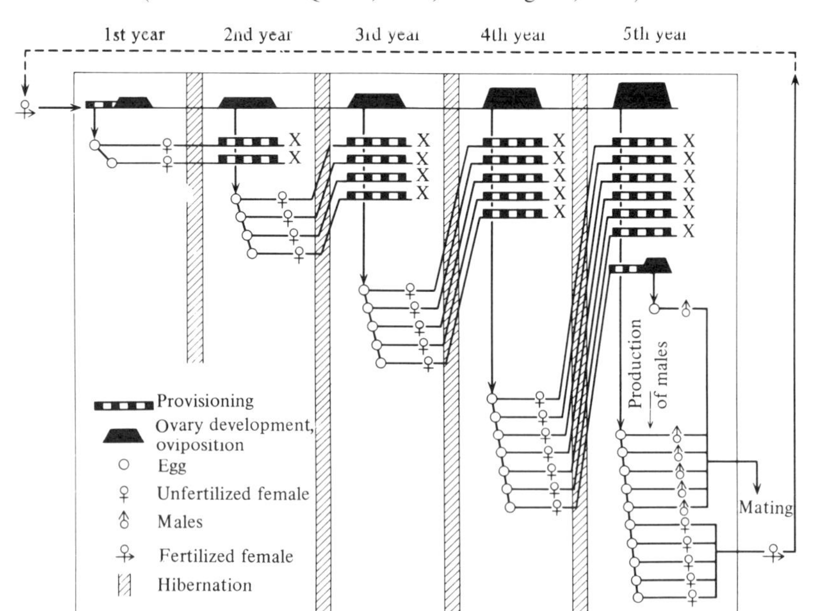

a new nest alone. When a female from a fifth-year nest (expected to become a queen) was transferred to a third-year nest she was found as a worker in the following year. They have given us splendid examples of early flexibility in the process of caste differentiation.

Once again we return to the Apidae and examine stingless bees (Meliponini). This group consists of a large number of species, varying in size from a few millimetres in length to the size of a honeybee. All maintain the matrifilial caste system. The number of individuals in a colony ranges from 500 to 4000 in *Melipona* and 300 to 80 000 in *Trigona* (Wilson, 1971).

The nest of the subgenus *Frieseomelitta* (*Trigona*) resembles that of bumblebees and contains honey pots and a cluster of cells in the burrow. Unlike other stingless bees the cells in this group do not constitute a comb but lie about in a cluster. Even in such a group the castes are distinct and the queen founds a new nest by swarming (in contrast to the honeybees in which the old queen takes some workers with her when she leaves, the stingless bees shift to a new nest after workers have prepared it for the queen). More advanced stingless bees build a huge nest with a comb. In *Dactyrulina staudingeri* of Africa the comb is of a vertical type with two sides like that of honeybees.

In spite of their highly developed social organization the stingless bees have mass provisioning without exception. Thus workers provision the cells and seal them as soon as eggs are laid in them. There is no trophallaxis or other close relations between the adult and young. How is the caste determined, then? The stingless bees other than *Melipona* have two types of cells: the large cell is provided with a large amount of food and a queen or a male emerges from it, whereas the small cells produce workers only.

The caste is thus determined by the amount of nutrition. If additional food is given to one of the young that was destined to become a worker it will grow to the size of a queen. If she is placed in the nest from which the queen is removed, she will mate and lay eggs functionally as the queen (C. A. de Camargo, 1972, cited by Michener, 1974). By contrast, at least some members of *Melipona* have genetically determined, as well as nutritionally determined, castes.

How do stingless bees maintain such a large colony with mass provisioning? In the honeybee the workers build cells and the queen oviposits in them; when the young hatch the workers mouth-feed them. When does the queen of stingless bees lay eggs? According to Sakagami and his colleagues (Sakagami *et al*., 1965; Sakagami & Zucchi, 1967,

1968) the egg laying of stingless bees is a result of a tense, dynamic relation between the queen and the workers, which may be interpreted as being hostile. The original papers should be consulted for details. A typical example might be that at the time of laying the workers would surround the queen and excitedly move backwards and forwards. Suddenly a worker would put its head in the cell, regurgitate food, then withdraw quickly. The queen would oviposit in this cell without delay and workers would make a lid to put on the cell. Though it is mass provisioning the method of provisioning presents a very special case.

There is no need to describe details of honeybees in this book. In all four species groups, *Apis dorsata*, *A. cerana* (Asian honeybee), *A. mellifera* complex and *A. florea*, honeybees provide progressive provisioning in a perennial colony of several thousand to tens of thousand individuals, and the colony multiplies by swarming. The queen cannot survive independently while the workers cannot regain their reproductive abilities completely even in the absence of the queen. The honeybees and the stingless bees, though different in their methods of provisioning, both show complete differentiation of the queen, which only lays eggs through her life (type O), and the workers, which do not reproduce.

In effect, the social evolution of bees is very complicated. The diagram opposite shows the main pathways of evolution thought to have been taken by the major groups of Apoidea.

Within each group the thin lines indicate the pathways that involve mother–daughter co-existence (overlapping generation) and the bold lines indicate the pathways that involve communal nests shared among sisters after the mother's death and contributed to by the joining of females from other nests. The pathway to communal nesting with the co-existence of sisters and joined females is certainly common among the bees (among the hunting wasps a surprising number have been seen in tropical Polistinae). I find it difficult to believe, however, that this could develop into morphological castes beyond the degree seen in *Augochloripsis sparsilis* or *Lasioglossum problematicum* without going through the phase of mother–daughter co-existence. There is no clear evidence, at least in our present knowledge, that caste differentiation has arisen among the sisters of the same generation in the absence of the mother.

This line of argument casts some doubt on the general applicability of the semisocial route suggested by Michener. The social evolution culminating in the complete caste system in which the queen is incapable of independent life and the workers cannot produce a queen may have been

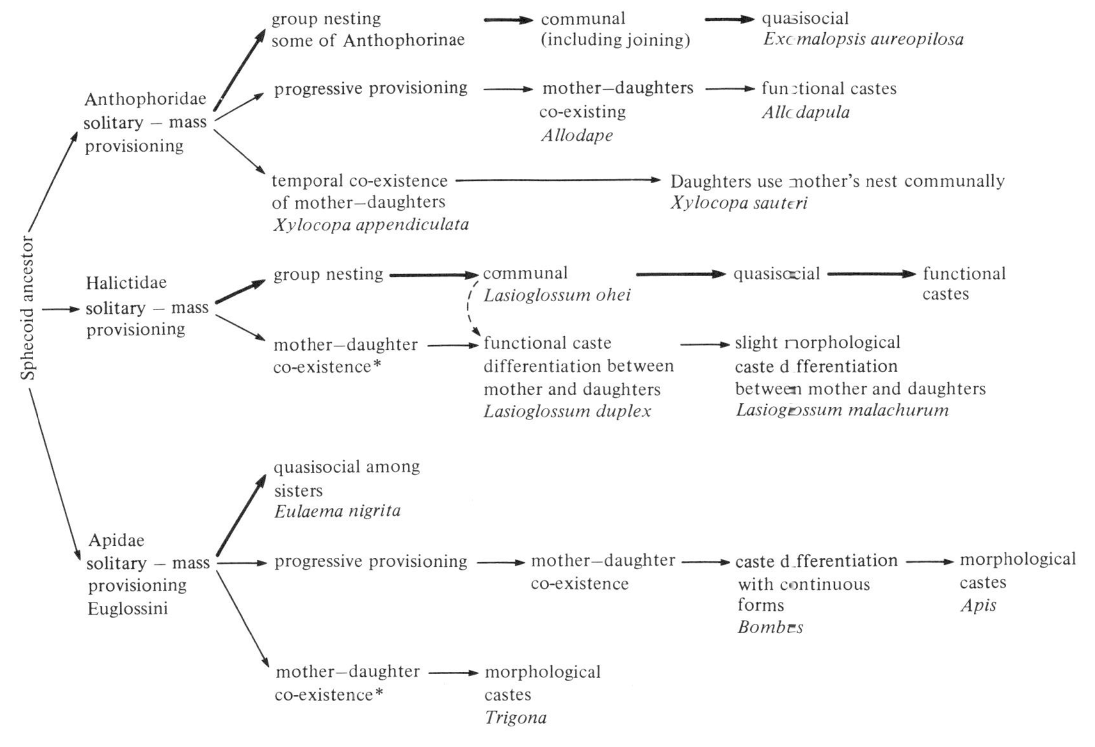

*Long-lived mother meets her daughters at the stage of mass provisioning

ultimately derived from the pathway of mother–daughter co-existence as proposed by Wheeler. The evolution of halictid bees to this stage with only mass provisioning could be explained if it started with overlapping generations of mother and daughters in the same nest as found in carpenter bees.

Social evolution of ants

One of the most successful groups of animals on the earth is the Formicidae (ants), which is thought to have evolved from tiphiid-like ancestors of Scolioidea. The oldest known ant is the worker belonging to *Sphecomyrma freyi* from a mid-Cretaceous deposit of some 100 million years ago. The female of this species resembles the apterous form of a tiphiid wasp, *Methocha* (Fig. 5.8) which hunts the larvae of tiger beetles (Wilson, 1971). Since the worker is known in *Sphecomyrma* the origin of sociality in ants goes back to at least 100 million years ago. All living ants are eusocial with true castes, making it extremely difficult to trace the origin of sociality in this group. As in the discussion of other groups I exclude social parasitism (in the case of ants it is slavery) from the following discussion.

Fig. 5.9 is a phylogenetic tree of ants given by Wilson (1971). Here, two major branches are separated: the Myrmecioid complex, which includes *Formica*, also differentiated into the primitive Myrmeciinae, Dolichoderinae, etc. through Sphecomyrminae (known only from fossils including the above-mentioned species); and the Poneroid complex differentiated from primitive ponerid ants. The life histories of Myrmeciinae and Amblyoponini render support to the above relationships.

The most primitive living ants known are *Myrmecia* and *Nothomyrmecia* of Myrmeciinae found in Australia and New Caledonia,

Fig. 5.8. The tiphiid wasp *Methocha fimbricornis* from the Philippines (redrawn from Wilson, 1971).

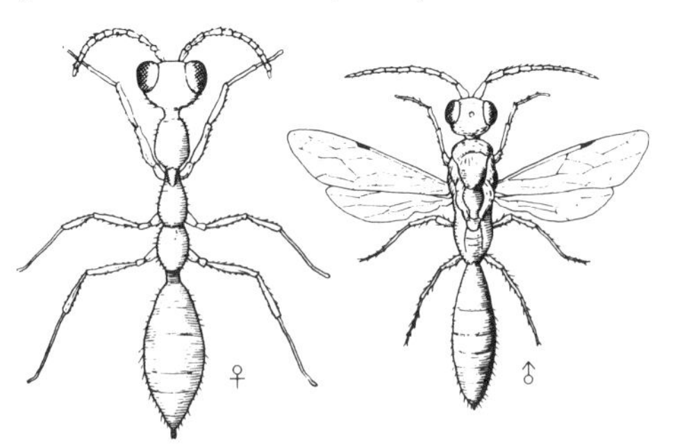

of which *Myrmecia* has been studied in detail by the Haskinses (Haskins & Haskins, 1950; Haskins, 1970, etc.) and *Nothomyrmecia* by Taylor (1978).

As in *Polistes* and *Vespa* the females of most ant species found a nest singly after mating. The female after constructing a nest waits for the development of eggs while fasting. When the eggs hatch the founding female (the queen) feeds the larvae with saliva. The wings of the female have fallen by then and the energy for this care comes from the wing bud muscles (Wheeler, 1923; Schneirla, 1952). The first lot of larvae nourished by part of the starved mother's body become infertile females – workers. By the time the workers emerge the queen is no longer building or feeding but concentrating on laying. This is the foundation behaviour of most ants. In the bulldog ants, *Myrmecia*, the queen collects food herself initially and cares for the young. If both honey and insects are provided in a cage the queen coming out of the nest takes honey until the eggs hatch but after hatching she carries insects to the nest (Haskins, 1970). The queen normally tears the insect apart and gives it to larvae. Other primitive characters of *Myrmecia* are listed in Table 5.1. The characteristics that are in contrast to the primitive traits given in the table (Wilson,

Fig. 5.9. A phylogenetic tree of ants proposed by Wilson (1971).

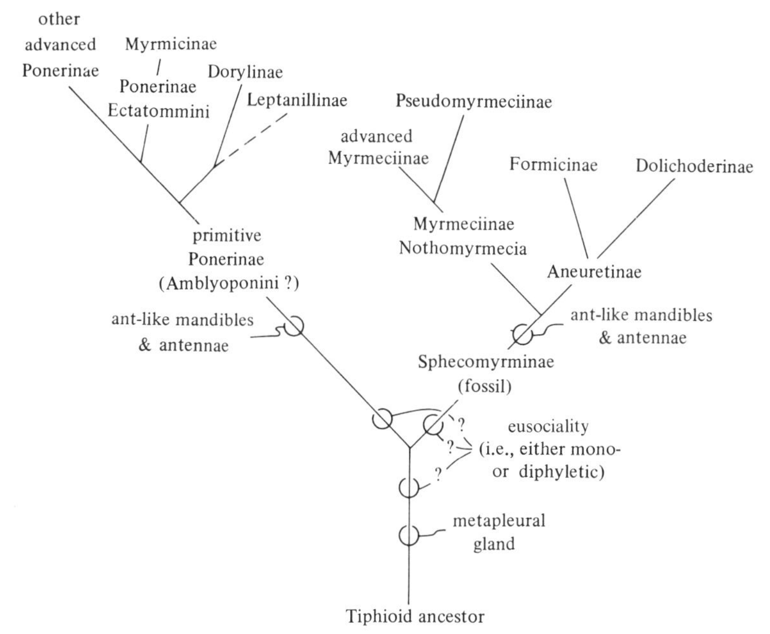

1971) are not necessarily the advanced ones. That the adult mainly takes nectar and the larvae depend on insects clearly indicates the tiphiid-like life of *Myrmecia* (the adults of tiphiid wasps take nectar). The role of regurgitation in feeding in *Myrmecia* is almost negligible compared with other ants. This and the fact that the queen is capable of working suggest that *Myrmecia* is in an early stage of social development. Some species may produce the ergatogyne – the apterous or brachypterous repro-

Table 5.1. *Primitive and advanced traits in behavioural and other characters of the bull-dog ants* Myrmecia

Primitive traits	
1	Multiple queens occur in many nests
2	The eggs are spherical and lie apart from one another on the nest floor
3	The larvae are fed directly with fresh insect fragments
4	The larvae are able to crawl short distances unaided
5	The adults are highly nectarivorous and collect insects mainly as food for the larvae
6	Transport of one adult by another is rare, awkward in execution, and not accompanied by tonic immobility on the part of the one being transported
7	There is neither recruitment among workers to food sources nor any other apparent form of cooperation during foraging
8	Alarm communication is slow and inefficient; the nature of the signal is still unknown
9	Colony founding is only partially claustral
10	When deprived of workers, nest queens can revert to colony founding behaviour, including foraging above ground
Advanced traits also found in higher ants:	
1	The queen and sterile worker castes are very distinct from each other, and intermediates are rare
2	Worker polymorphism occurs in many species, manifested as the coexistence of two well-defined worker subcastes
3	The colonies are moderately large and the nests regular and fairly elaborate in construction
4	Regurgitation occurs both among adults and between adults and larvae
5	Adults groom each other as well as the brood
6	Trophic eggs are laid by the workers and fed to other workers and the queen
7	The workers cover the larvae with soil just prior to pupation, thus aiding them in spinning cocoons; and they assist the newly eclosed adults in emerging from the cocoons
8	Nest odours exist and territorial behaviour among colonies is well developed

After Wilson (1971).

ductive individual with working ability, which may or may not indicate primitiveness (this also happens in higher ants). The fact that workers do not attract other individuals to the food source reflects the primitive stage of their social life. However, caste differentiation between the queen and the worker is clear and among the workers size differences appear depending on the species. Production of trophic eggs, laid by workers and given to larvae, which is wide-spread among ants has already occurred in this group.

Multiple queens are common in *Myrmecia* nests. This is due to the return of young queens to the nest after their mating flights. This character is not necessarily primitive. For example, among the races of *Formica rufa* found in Europe the following differences are known (Sakagami, 1958): *F. r. rufa* is haplometrotic (with one female) and constructs a single nest; *F. r. rufopratensis* (major) has up to 100 queens in a colony connecting some twenty nests, each established by swarming; *F. r. rufopratensis* (minor) is pleometrotic (up to 5000 queens per colony) and several hundred nests are connected by corridors in one colony, sometimes covering a whole wood.

Another group which retains many primitive characters of social evolution as in *Myrmecia* is *Amblyopone* (Ponerinae) considered to be primitive on morphological grounds also. The colony of this species is small and loosely organized. The nest structure is simple and the difference between the queen and the worker is small. A single queen founds a colony and as in *Myrmecia* she forages outside the nest. She has been observed to forage even after the emergence of worker ants. An interesting point about food collecting is that when a prey animal too big to be carried is located workers carry the larvae to the prey. Foraging of army ants may be an extreme development of such a habit. The workers of *Onychomyrmex* (tribe Amblyoponini) are known to attack a large insect together and feed the queen and larvae that accompany them in foraging.

According to Haskins & Haskins (1950) the pupa of Amblyoponini leaves its exuvium inside the cocoon when emerging. The reproductive individual then unfolds its wings and bites its way out of the cocoon. Haskins (1970) found that the queen often helps the emerging young though workers of *Amblyopone* can emerge without help. The newly-emerged, light-coloured adult is called the callow. The callow starts working immediately in this group and the male is already pigmented at emergence. In *Myrmecia*, however, the adult can hardly emerge on its own though the exuvium is shed in the cocoon. When there is a sign of emergence workers come to its aid and pull at the cocoon. When the emerging adult working inside protrudes its mandibles out of a hole workers grab the mandibles and pull out the ant. In many species of higher ants the adult cannot emerge without workers opening the cocoon from outside. In *Formica* even the exuvium is removed by workers. The newly emerged adult is pale and fragile and it takes many days for it to become functional as an adult. Schneirla (1952) wrote that this was neoteny that accompanied social evolution, analogous to what was typically seen in the evolution of man, and that ants learned many of their habits in this period.

Within the Myrmecioid complex the Pseudomyrmeciinae has a symbiotic existence with plants and defends territory. In the Dolichoderinae chemical communication is developed and some form a huge colony. *Leptomyrmex* in this subfamily has a degenerate sting which still functions but in Dolichoderini and Tapinomini the sting has been converted for the secretion of chemical repellents. In Formicinae the sting has degenerated and many show differentiation of workers. In *Oecophylla* which weaves leaf nests, the large workers collect food while the small workers care for larvae and eggs in the nest. However, even in the advanced forms the workers are often monomorphic.

In the poneroid ants there is much social differentiation in Ponerinae. As already mentioned Ponerini, in parallel with the primitive Amblyoponini, shows a strikingly high degree of prey specialization within the arthropod-eating habits (e.g. species specialized in hunting millipedes or termites). In this group there are also taxa with foraging behaviour similar to that of army ants (e.g. some of the Amblyoponini and Ponerini, and Leptanillinae). Myrmicini, a prosperous group in the temperate regions, has well-developed chemical communication.

The development of large colonies in ants is connected with the change from the carnivorous habit of primitive ants to the omnivorous habit including storage of seeds or to hunting of social insects. The large colonies found in Formicini, Dolichoderinae, Myrmicinae and others may have been established as a result of adopting herbivorous habits (though many still eat animal food). Changes in the above two directions are accomplished typically by the army ants of Dorylinae and Ecitoninae (they are very closely related) and the fungus-growing ants of Attini, in the poneroid complex.

Dorylus (Dorylinae) is a driver ant (an army ant of the Old World), which hunts termites among other animals. Its colony is the largest among all social insects and contains up to 20 million ants. Another Old World genus, *Aenictus*, hunts *Vespa* (wasp) and other ants.

The Ecitoninae comprises the New World army ants and has been studied in detail by Schneirla and his colleagues (Schneirla, 1933, 1938, 1940, 1949, 1952; Schneirla & Piel, 1948; Schneirla *et al.*, 1954; and others). These army ants form a colossal colony with hundreds of thousands of ants (the largest known number is 700 000 for *Eciton burchelli*), not nesting underground but wandering inside tropical rain forest. They take any prey they come across but particularly important as prey are social wasps, termites and other species of ants. Some are specialized for raiding the colonies of termites or ants. There is no fixed leader when they march. They have a certain cycle of reproduction. The queen starts laying while the young from the previous cycle are in pupal stage. At this time they form a bivouac in a tree hollow or other places but when the workers emerge they enter the foraging cycle for two weeks or so. They carry larvae in their jaws for this period. During the night they form a compact group with the larvae and place the queen in the centre. They depart in the morning for foraging. As a typical example the cycle of *Eciton burchelli* is shown in Fig. 5.10 from Schneirla & Piel

Fig. 5.10. Cycles of reproduction and behaviour in the army ant *Eciton burchelli* (modified from Schneirla & Piel, 1948). In this example the period at the bivouac (stationary: St) is 20–21 days, after which nomadism (Nm) begins. They form a 'nest' with their own bodies for each night (dots in St columns). Fine dots shown in the first Nm column (omitted from the second) indicate foraging behaviour.

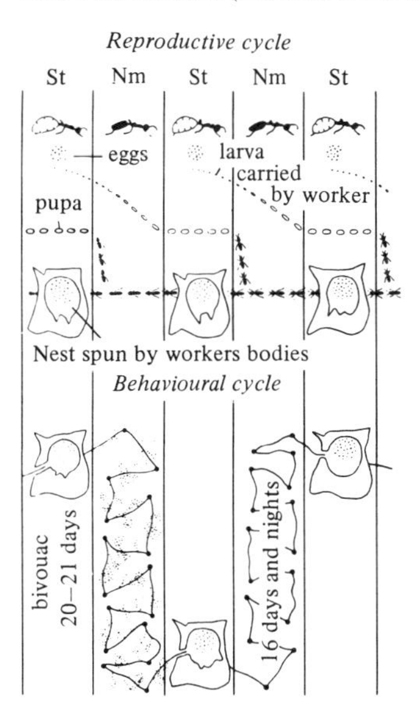

(1948) and Schneirla (1949). Schneirla (1952) suggested that the contradictory habits of carnivory and the maintenance of large colonies in Dorylinae and Ecitoninae were harmonized by the nature of nomadism. I would add as another important factor the exploitation of other social insects that form large colonies.

The fungus-growing ants (some are also known as leaf-cutting ants), which are distributed only in the New World, mostly cut leaves and ferment them with fungi in the nest (*Macrotermes* in Isoptera shows convergence in Asia and Africa). The species of symbiotic fungi on which they feed is different for each species of ant (why other fungi do not grow is an enigma). In the region of their distribution the leaf-cutters are the dominant group and often become very important agricultural pests. The colony size reaches hundreds of thousands.

Social evolution of termites

The termites (Isoptera) are the only social insects outside the Hymenoptera. Wheeler (1923) thought that the Isoptera evolved from Protoblattoidea – ancestors of cockroaches – and this idea was supported by the study of Cleveland (1934) and others on the symbiotic microbes of cockroaches and by the broad palaeontological study of Martynov (1938).

Fig. 5.11. A phylogenetic tree of termites (from Wilson, 1971).

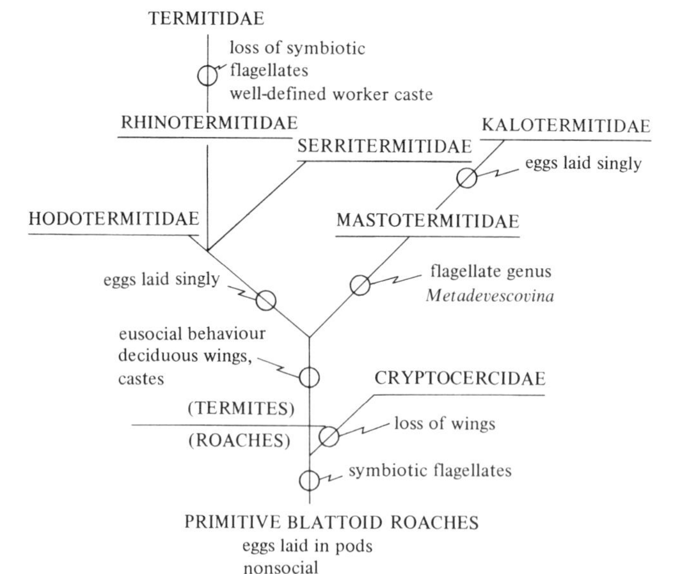

In the Blattodea the wood-eating cockroach *Cryptocercus* seems to be close to the ancestor of termites. The members of this genus live in family groups in rotten wood and digest wood with the help of specialized symbiotic protozoans in the gut. The newly hatched larvae and the larvae that have lost the protozoans by ecdysis cannot digest wood. They gain the symbionts by eating faeces of other individuals. Thus group life is indispensable for this group of cockroaches so long as they are adapted to wood-eating life. According to Cleveland (1934) and Emerson (1938), *Cryptocercus punctulatus* excavates many galleries in parallel with the pith of a tree. In the breeding season the female makes a large nursery and lays an ootheca with a dozen or so eggs strung together in a special cockroach fashion. She covers this with her body.

As with ants the oldest termite fossil is found in mid-Cretaceous. This is *Cretatermes carpenteri*, of which only the wings are known. It appears to be well advanced, suggesting that the history of termites goes back at least 100 million years. The phylogenetic tree of termites given by Wilson (1971) is presented in Fig. 5.11. There are six families in the Isoptera today, of which one family (Serritermitidae) is monospecific (*Serritermes serrifer*). The modes of life of the five remaining families were compared by Emerson (1938) and Goetsch (1941). A summary is given in Table 5.2.

Morphologically the most primitive family in this table is Mastotermitidae. Though many fossils belonging to this family are known there is only one living species, *Mastotermes darwiniensis*, which has symbiotic micro-organisms as in wood-eating cockroaches and which lays an egg mass almost identical to the ootheca of cockroaches. This species has prospered in the past and has a large colony size. Emerson (1938) considered it to be specialized rather than primitive. The next most primitive family is Kalotermitidae. The so-called 'lower' termites include the above families and Hodotermitidae and Rhinotermitidae. The 'higher' termites refer to the Termitidae. The common characteristics of the lower termites are that they have symbiotic flagellates and lack true workers. [Rhinotermitidae may be found to have workers (Miller, 1969).]

The growth of the lower termites is peculiar; the case of *Kalotermes flavicollis* (Kalotermitidae) is shown in Fig. 5.12. Generally, the individuals produced after five to eight ecdyses are called the pseudergates which substitute workers. Including this stage the larvae of fourth instar or older may become soldiers via the pre-soldier stage or may take part in reproduction as the supplementary reproductive members. Both the pseudergate and the soldier include both sexes. The pseudergate becomes

Table 5.2. *Evolution of behaviour and polymorphism in termites*

Taxon	Soldiers	Worker-like nymphs	Adult workers	Enlarged queen	Nest construction	Use of saliva or excretions for nest construction	Division of cells in nest	Shelves and supporting columns	Temperature–humidity control with ventilation pores	Food gathering	Food storage	Cultivation of fungi	Intestinal flagellates
Criptocercus					+								+
Kalotermitidae[a]	+	+			+								+
Mastotermitidae	+	+			+	+	+			+			+
Hodotermitidae	+	+			+	+	+	+		+	+		+
Rhinotermitidae	+	+	+?	+	+	+	+	+					+
Termitidae													
Subfamilies other than Macrotermitinae	+	+	+	+	+	+	+	+	+	+	+		
Macrotermitinae (broad sense)	+	+	+	+	+	+	+	+	+	+	+	+	
Anoplotermes (Amitermitinae)		+	+	+	+	+	+	+	+	+	?		

Based on Emerson (1938), Goetsch (1941) and Wilson (1971).

[a]Emerson (1938) described *Zootermopsis*, which uses salivary secretion for nest construction, as a member of Kalotermitidae, but this species was put in Hodotermitidae according to Snyder (1949).

the nymph after ecdysis (the termites are hemimetabolous; the young without wing buds are called larvae and the ones with wing buds are called nymphs), but the first-instar nymph may return to the pseudergate following the next ecdysis or the second-instar nymph may return to the form of the first-instar nymph. The only time they can attain the alate adults after two positive ecdyses is when the inhibitory pheromone of the king and the queen has been removed. So long as there is this inhibition the termites remain as juveniles regardless of the number of ecdyses. The termite society is heterosexual; there are both sexes in workers, soldiers and alates.

In the higher termites 'adult' workers that do not change further are produced (Noirot, 1961). They contain both sexes but depending on the species the size and functions of the two sexes may differ. There are also species whose soldiers belong to one or the other sex (female in *Noditermes* but male in *Trinervitermes*). Some are sexually

Fig. 5.12. Caste differentiation in *Kalotermes flavicollis* (Kalotermitidae) (from Lüscher, 1961). Each moult is indicated by an arrow; pseudergates appear after 5–7 moults (ecdyses).

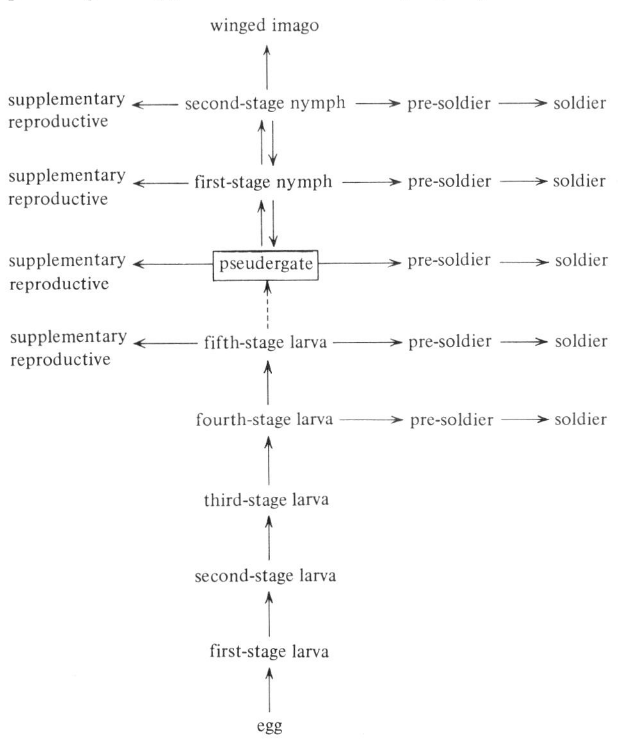

dimorphic. The higher termites lack symbiotic flagellates, whose place is taken by the symbiotic bacteria. In Macrotermitinae, *Termitomyces* is cultured.

As mentioned earlier, the termites seem to have emerged from an ancestor of wood-eating cockroaches as a result of their group life which began when they started using symbiotic protozoans for the digestion of cellulose. Perhaps the high humidity and darkness inside the wood retarded the development of this group and, aided further by the constant temperature conditions of the tropics, the co-existence of many individuals would have been made possible. Additional ecdyses may have been acquired under such conditions. Although the reasons are not clear the insects living in plant tissue of high humidity have a highly variable number of ecdyses. As a rule, termites belonging to Kalotermitidae nest only inside wood or, strictly speaking, in galleries eaten out by the larvae. Thus not only the pseudergates but also other immatures are, in effect, building the nest. They do not go out of the timber for foraging. Expansion of distribution is achieved only by the flight of the alate form.

Once such colony life is established the structure of the nest becomes gradually complicated; the cells and the corridors are differentiated, secretions and excreta are used for the construction of a proper underground nest and, when further advanced, they have come out to the ground surface. In the Rhinotermitidae they construct a roof of special structure outside and in the Termitidae they produce large mounds above the ground. The number of termites in a colony increases with such changes in nest structure: several hundreds to two thousands in the Kalotermitidae to several million in the Termitidae. In the Hodotermitidae and above they store food and in the Macrotermitinae they collect leaves and culture fungi. In some members of the Termitidae complete caste differentiation emerged as the soldiers disappeared and only the king and the queen were permitted to reproduce.

Unlike ants, the termite soldier is more basic than the worker. Even the family lacking adult workers has a soldier caste that does not change further. The fragile body structure and limited migratory ability of termites, living in an environment of abundant food supply, probably raised the survival value of the soldiers more than the workers.

As in ants and bees, trophallaxis occurs in termites. This is rather more important for the termites than for others. They exchange food by mouth and eat each other's faeces. In so doing they have cultivated their way into wood-eating.

Determination of castes and integration of colonies

Various theories have been advanced to explain the mechanisms of caste differentiation and co-existence of different castes in social insects. I leave the physiological and biochemical aspects of these questions to other books (Ishii, 1971; Birch, 1974; Yushima, 1976) and touch only on the ecological aspect in this section. Specially influential among the many theories put forward are the theories of trophallaxis (Wheeler, 1923, 1928), dominance hierarchy, pheromone determination and kin selection (Hamilton, 1964). The first three deal with differentiation of castes and integrating mechanisms among the castes while Hamilton's theory tries to explain the genetic problem of how the characters of workers which cannot leave progeny have been inherited. This last question has been explained in relation to the haploid nature of the hymenopteran male. I exclude this question from the following and refer to Sakagami (1975) who has discussed it in detail in Japanese.

Wheeler started with the observation of mouth-to-mouth exchange of saliva between ants and licking of salivary secretion of larvae (trophallaxis) by social wasps. According to Wheeler the workers receive saliva as a 'reward' from the larvae which they feed and to receive more reward they have to construct more cells. The theory of trophallaxis is that as a result of this reward system each larvae received less food and its nutritional level is lowered, leading to the production of the sterile caste. Thus this theory is coupled with the theory of nutritional castration where undernourished females become sterile workers.

Trophallaxis has been demonstrated widely among ants, Vespoidea and termites and is known to play a decisive role in the survival of a colony, specially in the case of some *Vespa* species. Ikan *et al.* (1968) in Israel discovered that in *Vespa orientalis* if the cells with larvae are isolated from the queen by a nylon film the queen dies. Further Ishay & Ikan (1969) proved, using carbon-14, that in this species only the larvae have the proteolytic enzyme and can provide adults with sugar which they produce from protein received from adults. For termites the acquisition of symbiotic microbes by eating faeces (trophallaxis in a broad sense) is indispensable. In some species of Rhinotermitidae the flagellates are absent in the first- and second-instar larvae, whose survival is impossible without regurgitated food of older larvae. We have already seen that the higher ants nourish larvae only with saliva at the time of colony foundation.

Although the above facts demonstrate the importance of trophallaxis in the integration of a colony, they do not provide evidence for Wheeler's original idea of the expanding colony (in order to obtain more saliva) and nutritional castration. For example, it has been established by means of a radioisotope that adult ants are involved in active trophallaxis (Wilson & Eisner, 1957) but this is no proof that the queen receives saliva of the larvae at the time of colony foundation, let alone that she lays more eggs and rears more workers as a result of receiving larval saliva. Further, the bees mostly furnish the nest with mass provisioning and there is no contact between the larvae and the workers. As Sakagami (1975) suggested we should retreat as far as the proposition that the contact between the mother and her daughters has influenced caste differentiation in some way or possibly in various ways, and start again from this point.

Since Pardi (1948, 1952) of Italy reported on dominance hierarchy in *Polistes gallicus* in 1946, dominance behaviour among adults has been considered to have a role in caste differentiation. This species is pleometrotic, i.e. many overwintered females together found one nest, and forms a linear dominance hierarchy (see next chapter for dominance hierarchy) among the females. The low-ranking females have small ovaries to start with but their ovaries degenerate further until they stop ovipositing and start providing labour for nest building and food collecting (differentiation of functional castes). There is a hierarchy also among the workers and the top-ranking worker starts ovipositing (though producing males only) if the queen is removed. Further observations showing a hierarchy among pleometrotic females and conversion into the functional worker caste of low-ranking individuals have been reported for *Polistes fuscatus* (Eberhardt, 1969), *Parischnogaster* sp. (Yoshikawa *et al.*, 1969) and *Belonogaster griseus* (Piccioli & Pardi, 1970). [The last mentioned species was described by E. Roubaud (1916, cited by Wheeler, 1928) as having entirely equal females and puzzled Imanishi (1951, pp. 111–12).]

In bumblebees *Bombus*, dominance behaviour has been noted for a long time. [Wilson (1971) refers to E. Hoffer (1882–83), who found a stable dominance relation of bumblebees, as the proposer of the idea of dominance prior to Schjelderup-Ebbe (see p. 303).] According to E. Lindhard (1912, cited from Wilson, 1971) the queen of *Bombus lapidarius* prods workers to induce laying and gives the eggs to the larva destined to become the queen. Brian (1952) also observed in *B. agrorum* that a high-ranking worker can lay eggs in the absence of the queen (the worker of bumblebees can mate and produce female eggs). As mentioned earlier many queens are killed through aggression, leading to the haplometrotic colony in *Bombus atratus*.

However, dominance hierarchy cannot be a general integrating principle; there is no hierarchy in the ants, stingless bees (except for the dominance behaviour of the queen towards worker bees found in *Melipona quadrifasciata* by Sakagami *et al.*, 1965) and honeybees, which have the largest colonies among social insects. The colony would be too large to permit individual recognition among the members in these groups.

In 1954 Butler proposed a 'queen substance' as a factor inhibiting the construction of royal cells and in 1959 he and his colleagues determined this substance as 9-oxodec-*trans*-2-enoic acid. Since then a number of pheromones concerned with caste differentiation and communication in social insects have been discovered (Table 5.3). In particular, at least four different pheromones have been found to inhibit or stimulate ecdysis of the pseudergate into a king or a queen in the termite *Kalotermes flavicollis* (Lüscher, 1961). In terrestrial ants and termites pheromones play important roles not only in caste differentiation but also in mobilizing worker ants to a food source (trail pheromone); this corresponds to the tactile and sound signals given by the waggle dance of honeybees leading an aerial mode of life (Frisch, 1950; and others).

It is obvious from the above that there is no simple explanation to caste differentiation. At the low levels of sociality, as seen in some members of halictid bees, bumblebees, stingless bees, paper wasps and lower ants, caste differentiation may have developed only as a result of undernourishment at the time of colony foundation (there are many reports showing the capability of what was destined to be a worker to become a fertile female by changing its nutritional condition during the larval stage). In the pleometrotic species and communal nesting species there are cases of dominance hierarchy being the major factor for the differentiation of functional castes. The hierarchical system would also be involved in the determination of the replacement queen. However, such nutritional and behavioural determination of castes did not last long. The pheromones appeared. In species that construct royal cells or provide royal jelly to specific larvae, the mechanism of caste determination is the quality and quantity of food but what sort of food is given is determined through the behavioural regulation of workers by means of pheromones. In the case of termites the pheromones themselves are involved in caste determination. This method of behavioural regulation in the social life of insects clearly separates insect societies from vertebrate societies. This is probably not unrelated to the limitation imposed on insects by the exoskeleton, which prevents enlargement of the body

Table 5.3. *Regulation of behaviour and physiology in social insects by pheromones and other secreted substances (e.g. venoms)*

Reaction	Type of chemical stimulus	Group
Primers		
Caste determination	Reproductive-caste inhibitory pheromone	Bees, wasps, termites
Ovarian inhibition	Queen substance	Ants, bees, termites
Inhibition or stimulation of queen-cell construction	Queen substance	Ants, bees, termites
Releasers		
Regurgitation and solid food exchange	Solicitation pheromone	Ants, bees, wasps, termites
Alarm behaviour	Alarm pheromone; trail pheromone; defence substance; toxic substance	Ants, bees, wasps, termites
Recruitment and emigration	Trail pheromone; alarm pheromone; specialized attractant pheromones	Ants, bees, termites
Aggregative (=clustering) behaviour	Aggregation pheromone	Ants, bees, termites
Attraction of alate reproductives	Territorial pheromone; sex attractant	Ants, bees, wasps, termites
Flight induction of sexual alates	Flight induction pheromone	Ants
Repellency	Alarm pheromone	Ants, bees
Abdominal pumping by adults	Thermoregulatory pheromone	Wasps

Modified by Yushima (1976) from the original table of Blum (1974).

and, as a result, also the central nervous system. Finally, castes are known today to be determined by genes for only one genus, *Melipona* (Meliponini) (Kerr, 1950). However, this is expressed when the stored food is plentiful (the segregation ratio of the queen to the workers is one to three); if the nutritional conditions are bad all eggs turn into workers (W. E. Kerr, A. C. Stort, M. J. Montenegro, 1966, cited by Wilson, 1971).

Communication of the honeybee, by means of the waggle dance, to transmit information on the direction and distance to the food source is so well known today that there is no need to repeat it here, except to mention that, in contrast to the Nobel-prize winning study of von Frisch on honeybees, A. M. Wenner and others suggest odour as the main means of discovering food sources (see Sakagami, 1975, for introduction in Japanese). Honeybees spray a pheromone when a honey source is discovered, suggesting that the waggle dance is not the only means of communication in this regard. Here I would introduce the observation of Esch (1967) on a stingless bee, *Melipona quadrifasciata*, which is indicative of the origin of the highly sophisticated communication in the honeybee. This stingless bee, when a good honey source is discovered, returns to the nest and tries to lead others there by performing a conspicuous zigzag flight in the direction of the food source. Although the following flight of other bees is limited to 30–50 m, they eventually fly in the right direction and many of them find the food source after the performance of the informer is repeated many times. Esch also discovered that the workers of this species and *M. merillae* stridulate with their wings and its duration is related to the distance to the food source as is the time spent by the honeybee in the straight portion of the path taken during the waggle dance. Stridulation occurs also in *Trigona* but no distance information is loaded. Such a behavioural chain is thought to have evolved into a more precise pattern in the honeybees that nest in dark hollows.

By comparison the trail pheromones used by ants and termites are little developed and cannot even indicate the originating or terminating point. In this case, also, the origin of communication may have been in the excitement of individuals in response to chemical stimulation (Schneirla, 1952), as found in *Camponotus herculeanus*, in which the excited individual runs around and discovers food without the ability to follow the trail. As mentioned earlier there is hardly any such communication in the Myrmeciinae.

The origin of sociality and the strategies of survival

Since Wheeler's work there have been many publications on the origin of sociality in insects but these are all concerned with the comparison of sociality and attempt to establish an order by which the various stages of sociality have evolved – that is, the *result* of evolution and not the *cause* of evolution, or why such evolutionary series could gain new survival values under natural selection. Only the kin selection theory of Hamilton (1964) and the ergonomics model of Wilson (1968) have provided mathematical explanations of the existence of the worker caste in the former and of polymorphism within it in the latter. Their discussions are based on the premise that the existence of the worker caste is advantageous to the species and they have ignored the mechanisms of colony size determination.

However, to discuss the survival values of the life-habits in social insects with the life-table parameters is almost impossible at present if actual values are to be used instead of imaginary ones. The following is a presentation of merely a few ideas for such discussion in the future.

In the evolution of sociality in insects we must distinguish between the Hymenoptera and the Isoptera. Both build nests of complex structure and establish large colonies accompanied by the caste system. The Hymenoptera with the exception of ants are aerial insects; the adults and the larvae have different modes of life so that the co-existence of two generations was made possible only after the establishment of a special behavioural sequence necessary for the protection of young. In the case of the Isoptera parental care was consequential to the wood-eating habit in aggregation because ingestion of wood itself was nest building. In the moist wood of the tropical environment generations overlap; young live in the galleries eaten out by adults and obtain symbiotic microbes from faeces of the parents. On this point the life of termites is closer to the life of ambrosia beetles (*Xyleborus* and *Platypus*) than to hymenopterans. Do ambrosia beetles lack castes? If they do it may be due to the fact that they are holometabolous and generations do not easily overlap. In the evolution of termites it was not particularly necessary to provide parental protection accompanying low fecundity. In the social evolution of hymenopterans, parental protection evolved and the lowering of fecundity reached its extreme when suddenly the course was reversed towards high fecundity.

In the first stage of lowering fecundity it is clear that the same process as seen in other low-fecundity animals was involved; namely, it was associated with colonization of a particular environment, or an adoption of a particular mode of life, in which food was not easily obtainable by the young. For many sawflies food was always abundant and so were predators. When parasitoids emerged from them new food items were usually eggs and larvae of ground insects, but so long as adults could find them and oviposit in them there was no difficulty for the hatched larvae to obtain food. On the debit side the risk of predation for the host (hence for the parasitoid) was great, particularly for those that parasitized the host in the early stages of development when the host suffered high mortality.

Among the parasitic wasps there are also species that evolved towards high fecundity. They are found in Trigonalidae, *Euceros* (Ichneumonoidea), *Eucharis* and *Chelonogastra* (Chalcidoidea), etc., which do not oviposit directly in the host. Members of the Trigonalidae are all secondary parasites and lay minute eggs on the plant food of sawflies and lepidopteran larvae, which are the potential hosts of the parasitoids. The eggs eaten by the primary host hatch inside it and later move to the larva of a parasitoid or a hunting wasp. According to Iwata (1971) the number of mature eggs carried by a female is 5300–14 000 in *Poecilogonalos fasciata*, 4400–13 200 in *P. maga* and 4800 in *Satogonalos elongata* (one sample). The female of *Euceros pruinosus* carries 3000–5000 mature eggs in 450 or so ovarioles, the largest egg number recorded among the Japanese ichneumonoids. *Perilampus japonicus* also lays minute eggs on leaves and its recorded egg number, 2240, is the highest among the Chalcidoidea. As to the capacity to oviposit some species of *Eucharis* are said to lay as many as 10 000–15 000 eggs.

In contrast to such high-fecundity species, there are scolid wasps that search for scarabeid larvae deep underground and spider wasps (Pompiloidea) that hunt specially dangerous spiders. They have had to develop time-consuming or complex behavioural sequences to succeed in hunting and thus have become low-fecundity species. This was also demanded by the principle of allocation of energy. In the scolid wasps, low fecundity was balanced by the lowered early mortality which resulted from their subterranean life and likewise for the spider wasps, with their initial subterranean life and subsequent residence in carefully constructed nests. The long-lasting paralysis of the host was indispensable for the reduction of early mortality but at the same time the habit of nest construction induced additional risks. When the low-fecundity/parental-protection strategy was established the hunting wasps were able to utilize spiders, adult orthopterans, and large lepidopteran larvae, which opened up new possibilities for further evolution.

Most spider wasps nested in sandy soils; this was probably a habitat in which it was difficult to obtain food but in which only few predators occurred. This tendency was more pronounced in the Sphecoidea, many members of which were river-bed dwellers. Some groups began nesting in cliffs or trees but this would have required further reduction in fecundity.

As the egg number was reduced the few larvae that were born had to be better protected. The adoption of progressive provisioning also helped to protect young from predators – the mother would return to the nest many times and stay by the entrance between feeding excursions.

However, as the hunting wasps and primitive bees prospered, the number of specialized predators and labour parasites that depended on them would have increased. Theoretically there are two ways to cope with this: one way is to disperse a large number of small colonies with further protection of young; the other way is to reverse the direction of evolution and to improve the defence ability by aggregation and using the aculeus (the ovipositor/sting used to paralyse the host) as a weapon. However, the construction of a well-protected nest necessarily reduces fecundity. Besides, the time and effort required for nest building would be too big a load to select for the habit of building many small nests to house merely one or two larvae per nest. If this was the case the only practical way would be to form a gigantic colony with armament. Prior to the transfer to this strategy, when the low-fecundity/protection strategy was still in progress, territorial defence by the male was seen even in the Hymenoptera (see p. 208). After the transfer the hymenopteran male became a drone, with no role to play other than copulation.

We have seen many attempts at development of group nesting or pleometrotic characters in various bees and, in particular, in the polybiine wasps of the tropics. In the bees the initial demand for it might have been due to the limited availability of suitable substrates for nesting. But, at the same time, the swarming of adults may have improved their defence function against natural enemies. Pleometrotic foundation by *Polistes* is commoner than used to be thought and many new cases are being discovered even in temperate Japan. It is particularly widespread in the tropical rain forest where there are many natural enemies. This fact seems to point to its significance in defence functions.

Enlargement of the nest may be realized in two ways. First, by the quasisocial aggregation of females belonging to the same generation, either of the same colony (sisters) or of different colonies. Secondly, the aggregation of a mother and daughters. The first type of aggregation appeared continually in social evolution and is specially widespread among the bees, as we have seen. In the insects that are constrained in size by the exoskeleton there is no assurance for the development of individualism that permits individuals of equal ability to assume different functions in a semi-permanent aggregation. It was also difficult to develop morphological (physical) castes by allowing the co-existence of individuals *only* of one generation. As explained by Hamilton (1964), it was not possible to transmit the character that produced an infertile caste without the mechanism to produce the haploid male parthenogenetically, unless the relatedness of individuals was very high as in the colony of termites in wood. For the first time in overlapping generations of the mother and her daughters the establishment of a large colony with caste differentiation seems to have been assured. So long as the queen is fecund, the bigger the colony the stronger the power of defence. This defence function or device against natural enemies seems to provide the stimulus for the establishment of large colonies.

Why, then, do not the colonies of all social hymenopterans become large? There appear to be at least two reasons for this. First, the arrival of winter in the temperate region and, secondly, the storage of food. We have seen that small colonies with only a few workers have evolved in the bumblebees of the arctic circle and that most of the social hymenopterans with large colonies are distributed in the tropics or are of tropical orgin. These facts indicate the importance of seasonal factors in determining the colony size.

The benefits from predatory life and those of living in large colonies do not often coincide. There is no doubt that the transfer from carnivorous to omnivorous habits in ants and from carnivorous to herbivorous habits in bees made it possible for them to form large colonies. The fact that colonial nests are seldom found in the sphecid wasps but are common in bees is probably related to the availability of food. Thus the ants, stingless bees and bees may be considered as high-fecundity strategists that have re-exploited an environment, abounding in food and predators, with large colonies that accompany the caste system. The sting of hunting wasps and bees degenerated in ants. The ants, instead, acquired strong mandibles and venomous substances, which they could use with advantage when living in large colonies. Besides, a special caste for defence – the soldier ants – emerged from them. The ants established large colonies with these two defence mechanisms and grasped unsurpassed prosperity in the tropical region. Their prosperity inevitably caused the evolution of specialized ant predators. [Ants and termites are eaten by animals of many different phylogenetic groups: in mammals, anteaters (Edentata), scaly anteaters (Pholidota), echidna (Monotremata); in birds, wrynecks (Jyngidae); in insects, army ants, etc.] To counter increased predation the ants had to intensify their high-fecundity strategy and defence function. The fire-ants, leaf-cutting ants and army ants are probably the end products of such trends. It is also interesting to surmise that the army ants (in the broad sense) have evolved by attacking other social insects. The colony of social insects provides a rich source of food if there is a means to attack it.

The origin of large colonies in hunting wasps is not clear. In *Vespa*, as in army ants, the habit of attacking other social wasps and bees makes it possible to procure a large quantity of food. How *Vespula*, while preying on a variety of insects, can maintain the largest colony in the Vespoidea, cannot be explained today. Detailed studies of colony metabolism in *Polistes* and *Vespula* are strongly needed.

I said that the termites took a different pathway, which did not go through the low-fecundity/parental-protection strategy. However, when they were prospering in and near abundant decaying logs there also appeared specialized predator groups. As food was abundant the problem for the termites was defence against predators. Thus they survived in tropical rain forest by producing large colonies with increased fecundity and the defence caste. It seems reasonable that they should have a large proportion of soldiers in the colony.

6

Group life

Dominance hierarchy

(1) Dominance, as a rule, refers to all dominance–subordination relationships found among conspecific individuals. It includes cases where dominance of an individual over another is revealed statistically as well as the relation in which one dominates another at all times. The term dominance order or dominance hierarchy is used only if an animal society is organized on the basis of dominance, that is when dominance has been shown to play a certain role in the adjustment of the society of the species concerned.

It has been known for a long time in Europe that certain hierarchical relations existed among a herd of grazing cattle or horses. In Japan aggressive behaviour of the domestic fowl and racial differences in aggressiveness were noted long ago and the fighting cock has been genetically selected for since the sixteenth century.

What marked the beginning of modern animal sociology, which provided methods for quantitative analysis of differences in the role of individuals in a group, was the study of dominance behaviour in chickens and other domesticated birds reported by Schjelderup-Ebbe in Norway. [Schjelderup-Ebbe (1922) is generally taken as the first report on dominance hierarchy, but according to Allee (1952) the paper written in Norwegian in 1913 was the earliest. As pointed out on p. 294 there is an earlier report of dominance behaviour in insects but Schjelderup-Ebbe's study is still considered as the starting point of modern animal sociology.]

In a flock of chickens kept in a pen one bird will suddenly peck at another. The one pecked at runs away calling. Such a common phenomenon in a poultry farm has some regularity though it appears to occur at random to a casual eye. If pecking occurs between the sexes it is the cock that pecks at the hen and the hen always retreats. In a unisexual flock there is one that always pecks at others and another that always runs away.

The following is an example of dominance relations among the hens reported by Schjelderup-Ebbe:

Of seven hens (A–G),

A pecks at	B, C, G, D, E, F	(total 6 birds)
B pecks at	C, D, E, F	(total 4 birds)
C pecks at	G, D, E, F	(total 4 birds)
G pecks at	B, D, E, F	(total 4 birds)
D pecks at	E, F	(total 2 birds)
E pecks at	F	(total 1 bird)
F pecks at		(none)

Here we have a triangular relation:

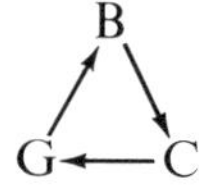

Apart from this, the dominant (A) pecks at all others; the subordinate bird (F) pecks at no one; and the individuals in between peck at certain members of the flock depending on their position in the hierarchy. The most typical case is where each individual pecks at those birds below its position and there is no triangular relation. In such a case the dominance hierarchy is said to be linear.

This study was followed up by Allee and his school in America (Allee, 1938, 1951, 1952). According to Masure & Allee (1934) the dominance hierarchy in the chicken is usually linear but triangular relations are not rare. Once established the dominance relation lasts for some months and it is not reversed unless the new individual sexually matures or the dominant bird becomes ill. In any case pecking is in one direction in the chicken; it is always the high-ranking bird that attacks the low-ranking bird and the one pecked at runs away. Generally there is no pecking at each other. In the case of pigeons, both the domestic rock pigeon *Columba livia* studied by Masure & Allee (1934) and the wood pigeon studied by Bennett (1939) differ from the chicken. As in the example given in Table 6.1 the low-ranking bird often pecks at a high-ranking bird so that dominance can only be determined statistically. In the example given WT, the second-ranking bird on the 32nd day, delivered more pecks than O, the most dominant bird. Yet because WT pecked at O only four times while O pecked at WT twice as many, O is placed above WT. Allee (1938, 1951) distinguished the dominance hierarchies of the chickens and the pigeons as the peck-right and the peck-dominance respectively.

The dominance hierarchy of birds is usually linear but in rodents or fish the most dominant individual is clearly distinguishable while the others are of the same rank. This is despotism and the dominant individual is the despot.

Despotic dominance is often found in animals with little intelligence or

Table 6.1. *Sociograms describing the peck-order among ring doves* Columba palumbus

1. Fourteenth day

Order	WT	O	P	RR	RT	Loss total
WT	—	7	0	0	0	7
O	1	—	1	0	0	2
P	1	1	—	0	0	2
RR	9	0	0	—	0	9
RT	2	0	0	0	—	2
Win total	13	8	1	0	0	22
No. of defeated opponents	3	$1\frac{1}{2}$	$\frac{1}{2}$	0	0	—

2. Thirty-second day

Order	O	WT	RT	P	RR	Loss total
O	—	4	1	1	2	8
WT	8	—	0	0	0	8
RT	1	7	—	0	0	8
P	2	5	1	—	1	9
RR	3	13	0	9	—	25
Win total	14	29	2	10	3	58
No. of defeated opponents	$3\frac{1}{2}$	3	$1\frac{1}{2}$	1	0	—

After Bennett (1939).

little behavioural ability. To establish a linear hierarchy individual recognition within the group is necessary but in despotism recognition of the despot is sufficient.

From the above there are three types of dominance:

(I) Linear dominance

(a) peck-right (A→B→C→...)

(b) peck-dominance (A↔B↔C↔...)

(II) Despotism [A→(B – C – D ...)]

(2) Since the evaluation by Allee and others in the 1930s of the significance of dominance behaviour many studies have been conducted to date on this topic. However, many dominance relations discovered may not have a real function in nature. Most experiments conducted in the laboratory have dealt with species which maintain territories in the field but which have been enclosed in a small space; e.g. the experiment of Noble (1939b) on the Mexican swordtail *Xiphophorus helleri* conducted in an aquarium. In this case inevitable fighting ensues but the heavier individual or the one with greater sexual activity naturally wins the fight. In some cases unnecessary fighting may be avoided through learning but this does not mean that dominance has functions in the field. Also, a flock of small birds gathering at a feeding station or of water birds coming to a pond may show dominance but the role played by such dominance in population dynamics is probably not so great. Dominance behaviour plays an important role in the survival strategy only when it is connected to the life of a group and expressed, for example, as a function of leadership in adjusting the behaviour of the individual group. The introduction of dominance hierarchy in this section is for convenience; in the following sections individual relations in wild groups will be compared and the relationship between group life and dispersed solitary life or family life will be examined.

Group life of poikilothermal vertebrates

Animals of all major phylogenetic groups form aggregations. Barnacles are attached densely on rocks whereas aphids breed in thousands on a host plant. However, groups of animals composed of leaders and followers are not found in the invertebrates (apart from colonies of social insects). There is no leader among several thousand cucumber beetles gathering in one place or in an overwintering aggregation of coccinelid beetles, let alone in the gregarious larvae of the fall webworm *Hyphantria cunea* (Arctiidae) or the caterpillars of *Malacosoma neustra*

(Lasiocampidae) moving in their collectively formed webs. In a related species of the latter moth *M. pluviale*, active and inactive larvae are found (Wellington, 1957) but they merely reflect individual differences in physiology and no definite leader exists among many active larvae.

Even in the spiders which are known as solitary predators, some species have recently been found to use a communal web though there is no leader in the commune. Adults kill the prey caught in the web and young appear later (Kullmann, 1972; Burgess, 1976).

Similarly, schools of fishes do not constitute groups in the above sense. Observing individuals in a school one may notice that a few fish in the front row come to the flank as the school changes its course (Fig. 6.1). Wilson (1975) contends that dominance hierarchy does not exist or has very little effect, and there is no firm leadership, in schooling fishes.

Nevertheless, there is adaptive significance in schooling as many fishes do form schools and this is essentially related to the origin of group life in birds and mammals. Neill & Cullen (1974) cite an experiment of D. V. Radakov (1958) in the U.S.S.R. showing that a cod, *Gadus morhua*, catching a young schooling coalfish, *Pollachius virens*, requires five times the amount of time needed to catch a solitary individual. Neill & Cullen released predatory fish into tanks containing one, six and twenty prey fish respectively and found that the success rate (number of captures/number of contacts) was greater when the prey was not schooling.

No dominance relations having functions in the field have been found in the amphibians. In the reptiles only some lizards provide such examples. Greenberg & Noble (1944) studied the behaviour of *Anolis carolinensis* inside a large glasshouse. When males were put in a small container ($8 \times 5 \times 6$ cm) a despot appeared and the rest became subordinate (despotism). In a medium-sized container ($12 \times 14 \times 14$ cm) with plants, two males established territories, with the larger individual

Fig. 6.1. Change of orientation in schooling fish (redrawn from Wilson, 1975). The fish in the front row (hatched) do not necessarily come to the leading edge after a change of direction.

occupying the better site, and the rest remained subordinate, hiding in inconspicuous places such as under-logs within the territories of the dominants. Under such conditions only the territory holders could mate. When the middle-sized containers were opened to allow free passage of some individuals inside the glasshouse more interesting relations developed. Fig. 6.2 shows this arrangement, in which three males established territories around two containers and a tree. Within each territory there were subordinate individuals wandering in the periphery of the territory. The container seemed to be too small for a territory of large individuals and M_7 freely visited the territories of M_1 and M_3 while M_1 entered the territory of M_3, forming 'territories within a hierarchy'. M_3, the lowest ranking of the three, occupied the container H as his territory and defended it against intruders other than M_7 and M_1.

Within such territories of males, several females formed a dominance hierarchy. When M_2 later occupied the containers E and H there were two females (F_5, F_4) in E and three females (F_1, F_{12}, F_{18}) in H, not shown in the figure. In E, F_5 was dominant over F_4 and always associated with M_2, whereas in H, F_1 was dominant and F_{12} and F_{18} were subordinate though they dominated F_5 which occasionally visited the container. The dominance hierarchy in this container was thus linear.

Among the males the higher the rank the larger the territory he occupies. This fact and the characteristic display of the throat pouch by

Fig. 6.2. Arrangement of territories with dominance hierarchy in *Anolis carolinensis* observed in a large glasshouse (from Greenberg & Noble, 1944). M, male; F, female. Large and small circles indicate dominant and subordinate individuals respectively, at the site where the circle is placed. A pair, M_{44} and F_{33}, were introduced later.

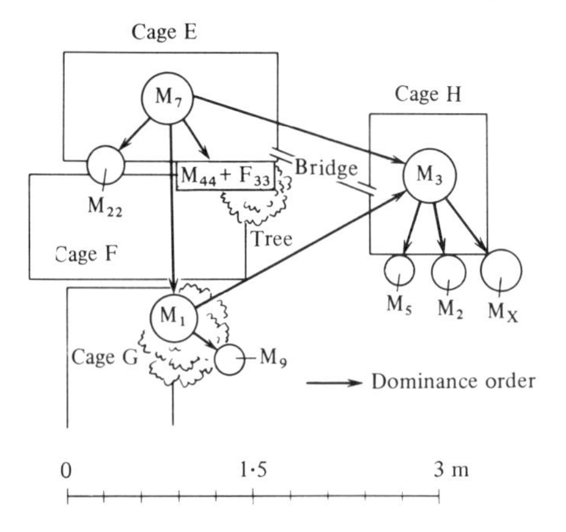

the dominant male attract many females. As a result of the dominance hierarchy among the females the large females with conspicuous secondary sexual characters mate more often than others. Thus in this species dominance behaviour has a role in sexual selection for either sex.

As mentioned in Chapter 4, the male of *Anolis sagrei* forms a harem in the field (Evans, 1938). In a desert species, *Sceloporus occidentalis*, males occupy almost non-overlapping territories in the flat country but, if there is a rock suitable for a sentry post, a large male establishes a large territory containing this rock and a subordinate male may occupy a small territory centred around the slope of the rock within the territory of the dominant male (Brattstrom, 1974). Thus we find, among the lizards of dry areas, not only territoriality but also exclusive possession of many females by dominant males as seen in grouse – this necessarily limits participation of young males in breeding. However, there is no report of group life with leadership in the reptiles.

Group life of birds

Specially conspicuous among the groups of birds are the seasonal flocks outside the breeding season (e.g. roosting flocks of crows, flocks of water birds) and the breeding colonies of herons, gulls, etc. Within such groups there may be dominance hierarchies, over the occupation of good feeding or nesting sites, which may in some cases prevent breeding of low-ranking males. The avian groups are not a permanent organization; they are formed either in the breeding season or in the non-breeding season. One of the reasons for group formation is that suitable sites are limited. Secondly, it is to increase defence against predators and, thirdly, it functions as an information centre for detecting sources of food (Itô & Murai, 1977). However, no special leader has been recognized in most flocks of the non-breeding season, breeding colonies, or groups of herons, gulls or albatrosses.

In contrast a family group with the father as the leader in monogamous ducks and geese may last several months after the hatching of eggs (Jenkins, 1944). In geese the post-breeding migration is by the family unit and is led by the adult male. When a mixed flock of water birds gather at a pond there is no leader who leads the entire flock though a family of geese may maintain its integration within such a flock. Cranes probably behave like geese in this respect.

The best developed society is seen among the corvid birds. The carrion crows *Corvus corone* in Nagano Prefecture gather at seven roosting sites (there are only seven within the Prefecture) and overwinter in enormous flocks (Yamagishi, 1976). According to Yamagishi these groups (without leaders) split into small groups when the breeding season approaches. Then they divide themselves into pairs. After breeding, small groups are formed at scattered sites and they gradually amalgamate into larger groups by winter. They return to the same roosting sites accompanied by their young. Within this large group the family stay together until about March and each morning when the group disperses from the roosting site family groups may be identified (Haneda & Iida, 1966). The family defends the territory of the group during the day (young birds also take part in defence). However, when the breeding season comes smaller territories are formed around the nests within the territory of the group. The young of the previous year leave the parents' territory by the time laying starts and the two-tier structure of group and family territories disappears. The female roosts in the nest during incubation and the male either stays near the nest or returns to the roost of the group. Thus in the carrion crow the parent–offspring relation remains within a large group but how the dominance hierarchy functions within the group is not clear.

Among the jays of the New World a variety of breeding systems are known; the monogamous unit with type A territory, colonial nesting with small territories around the nests within the colony, and two or more pairs sharing a common territory. In the last case juveniles assist

Fig. 6.3. Brown's (1974) scheme showing two pathways of social evolution in the New World jays. Hexagon with solid line, territory or, area of dominance; hexagon and circle with broken line, home range; P, pair; H, helper; NH, non-helper floater (with an arrow, being driven away).

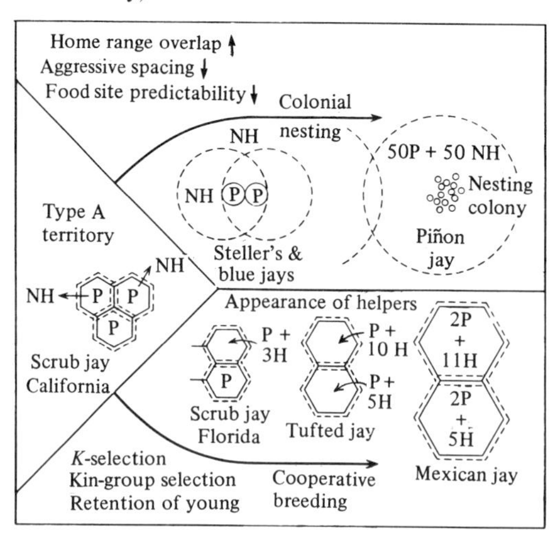

parents as helpers in feeding nestlings (analogous to social insects!). Brown (1974) examined such variations of social morphology and proposed a scheme of evolution from the ancestral monogamous system to the colonial and communal reproductive systems (Fig. 6.3).

The basis of this evolution is a species with type A territory and it is seen in the scrub jay *Aphelocoma coerulescens* of California. This species is monogamous and its home range coincides with its territory. Since non-territorial birds are chased away to the forest edge, etc., there is no helper in this system. One pathway of evolution is seen in the overlapping of home ranges. In the blue jay *Cyanocitta cristata* and the Steller's jay *C. stelleri* the distance between nests is short while home ranges are large and overlapping. Each pair defends a certain area around the nest (Fig. 6.4). Those that fail to establish territories leave the area or live inside home ranges but outside the territories of other birds. The piñon jay *Gymnorhinus cyanocephalus* is a species in which the proximity between nests has become such that 50–300 pairs nest in aggregation and each pair defends only around the nest. The female stays in the nest during incubation while the male feeds in a flock and delivers food to the female. The group has a large home range of about 20 km^2, in which there are also a few dozen non-breeders. These floaters occasionally feed the young of other birds but the first-year birds never feed young.

The other evolutionary sequence ends in cooperative breeding. While the scrub jay of California defends the type A territory the Florida scrub jay permits young birds to stay in the type A territory of parents. These

Fig. 6.4. Arrangement of territories (unbroken lines) and home ranges (broken line, pair A; dotted line, pair B) of two pairs of the Steller's jay *Cyanocitta stelleri* (from Brown, 1974). Home range is widely overlapping.

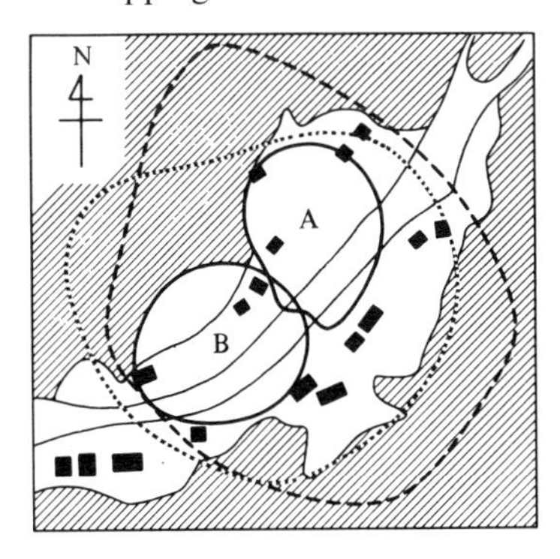

juveniles feed the nestlings of the territory-holding pair – they become helpers. According to Woolfenden (1975) about half the pairs have helpers. The helpers do not build a nest or incubate, but take part in the feeding of young and the defence of territory. When the parent–offspring relation was investigated by individual marking it was found that two-thirds of the helpers were helping their parents. Thus the young remain mostly in the territory of their parents and work for the parents (the territory does not shift from year to year). As Wilson pointed out this resembles the workers of social insects, specially halictid bees. Further, in the tufted jay *Cyanocorax dickeyi* the presence of helpers is normal and in the Mexican jay *Aphelocoma ultramarina* two or more pairs defend a large type A territory in cooperation. There are always 8–20 helpers cohabiting in a territory of the Mexican jay and they contribute about half the food taken to the nest. Outside the breeding season the pair disbands but the territory of the group is maintained. The helpers are mainly young birds (the Mexican jay does not form pairs until it is three or four years old, thus the young birds include birds of adult age in other species). Many individuals spend their entire lives within the family territory and new breeding grounds in this species may be established by budding off. There is a dominance hierarchy among the helpers of the tufted jay and the Mexican jay; when the breeder dies it is replaced by the dominant helper.

The evolution of sociality in jays stops here. Communal breeding as seen in the promiscuous cuckoos (p. 243) is not known in jays.

What then is the cause of social evolution? Brown (1974) states that it is the paucity of nesting sites in the arid region and increased kinship in this environment – this would gradually reduce antagonism between individuals. In the species with helpers maturation is necessarily retarded. This is a direction of low fecundity/protection and as in social insects this would have induced evolution of corresponding predators, which in turn further the evolution of special defence mechanisms.

Group life of mammals other than primates

Discussion of the group life in cetaceans, which includes the very interesting example of dolphins, is omitted here (see Wilson, 1975).

(1) Large marsupials of grassland lead a loosely organized group life. The whiptail wallaby *Macropus parryi* living in open woodland of the Dividing Range in Australia forms a group of 30–50, in which linear dominance hierarchy has been observed among the males (Kaufmann,

1974). Although the group organization of marsupials is loose and individuals often shift between groups, sometimes groups merging, even this degree of social organization is not known among the reptiles.

Among the rodents, lemmings migrate in groups but there does not appear to be any leader in the group. Some of the middle-sized rodents, such as the marmot which lives in the arid zone, form harems (see Chapter 4). The development of groups from the harem structure has culminated in the prairie-dog. According to King (1955) the black-tailed prairie-dog *Cynomys ludovicianus*, living in the open habitat of western United States, formed a town of about 1000 individuals. Natural barriers such as ridges and streams separated the wards within the town and each ward consisted of many coteries. The coterie was the true social unit – the group – whose mean composition was 1.7 adult males, 2.5 adult females, 3.6 immature males and 2.4 immature females (mean total 10.2). The largest coterie was composed of 38 individuals. The members of a coterie used the same burrows and defended a certain area around the burrows by means of vocalization, after which the animal was named. The dominant male in territorial defence is most active. The home range of the group is much larger (just under 400 m in diameter) than the territory.

Anthony (1955) observed the dominance hierarchy of this species in captivity. Dominance could be determined by the mouth contact (kissing) between two individuals and by the submissive give-way when food was introduced into the cage. The prairie-dog has a well differentiated vocal signal system, by which defence and collective escape into the deep burrow are signalled and unnecessary fighting between individuals of the same group is avoided.

Smith *et al.* (1973) confirmed the observation of King and speculated that the main factor responsible for the evolution of sociality in the prairie-dog was the defence function against predators. If this was the case the question would arise as to how overcrowding was prevented. Perhaps some regulatory mechanism is operating within the group – through dominance hierarchy – but this is yet to be discovered.

(2) The history of group life in ungulates is old. The restoration of the bedground of a Miocene camel *Stenomylus hitchcocki* in the American Natural History Museum depicts the life of this species in groups of several to ten or more individuals. The polygynous bond of the vicuña and the wildebeest has already been mentioned (Chapter 4). In either species the males without territories could not breed and this might help

regulate the population (birth control and dispersal of females). While the family of vicuñas is semipermanent the wildebeest leads group life only outside the breeding season. The herd may be a group of a female and her young or a male group. In the former territorial defence is known but the large groups formed out of many such herds during the dry season with limited feeding stations have no leadership, suggesting that this is merely an aggregation. In the Uganda kob each polygynous family leads an independent life when food is abundant; since their density is high in the flat grassland many individuals may be in sight at the same time. However, these wildebeests tolerate others within their home range and aggregations are formed in the grazing area and at the waterhole. In the Cervidae the caribou forms a large herd but it is probably loosely organized. On the other hand, deer in the temperate regions form small groups in which leadership has been recognized. Darling (1937) discovered leadership in herds of the red deer *Cervus elaphus* in Scotland. Only the groups consisting of the females and their young are organized by leadership and the male herds have no special leaders. The leader in a matriarchal group is usually the largest individual and the followers are usually her offspring. The leader leads the group on feeding excursions and utters an alarm call to warn others of approaching danger. According to Kawamura (1952) the leader of a female group (mother, daughters and grandchildren) in the sika deer *Cervus nippon* is also the oldest female and gives alarm calls. If the leader is absent the second-ranking female utters warning. If the leader is present, the second-ranking female with such an ability only retreats without giving a call. Dasmann & Taber (1965) found that the yearlings

Fig. 6.5. A male group of bighorn sheep *Ovis canadensis* in procession (drawn from a photograph in Geist, 1971). The individual with the largest horns is in front and others follow in decreasing order of horn-size.

and small young of the black-tailed deer *Odocoileus hemionus columbianus* do not escape from danger unless the adult female (only one in the group) gives an alarm call. There is no dominance hierarchy in the male group.

In the bighorn sheep and the Dall sheep both the male group and the female–offspring group are led by the largest individuals (Geist, 1971). Particularly in the male group the dominance hierarchy is linear according to the size of the horn and the group proceeds in this order (Fig. 6.5). The male with small horns adopts the submissive posture of a female to the male with large horns. There are very few such examples of male herds having leadership, but as yet little equals the detailed study of Geist to permit generalization. The young (even the male) can stay in the mother's herd for a long time. In the breeding season the male becomes the leader in mixed groups.

The basic unit of ungulate society may have been the mother–offspring group. The adult female may have been either solitary or in a large group consisting of many females, each accompanied by their respective offspring. The polygynous group of the vicuña seems to have been formed by a male joining such a group for an extended period in arid areas or mountains.

The culminating point of matriarchy in the ungulates is reached by the African elephant *Elephas* (*Loxodonta*) *africana*. African elephants usually form a herd of 10–20 adult females accompanied by their offspring. In this herd females of several generations co-exist, the female-to-female bond lasting as long as fifty years (Wilson, 1975). The leader is the oldest female (Fig. 6.6) and the young elephants can obtain milk from *any female*! The old female controls movements of the herd and helps injured

Fig. 6.6. Groups of African elephants *Elephas africanus* (drawn from Wilson, 1975). A, a matriarchal herd led by an old female with torn ears (marked ×); B, a loosely organized male herd.

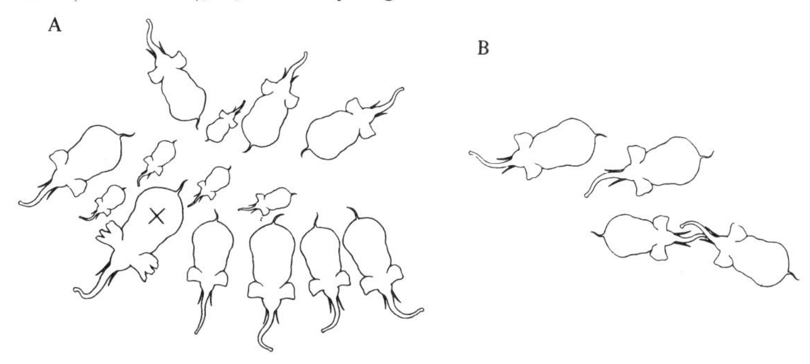

animals to their feet (a huge elephant fallen in the African savannah would be overheated in no time). It is the old female that takes sacrificial defence action when facing a big danger. This is the opposite of social insects. When a matriarchal group becomes too large some members of different ages bud off in a group as in the swarming of bees.

(3) Many aquatic species of the Carnivora form large groups. These are breeding colonies, which are different from those of sea-birds in that they are not monogamous and strict dominance hierarchies appear, limiting the number of breeding males. Unlike the group of ungulates the male entirely dominates. According to Kenyon *et al.* (1954), several thousand fur seals *Callorhinus ursina* gathering on an island in the northern Pacific consist of pups, bachelors (immature males of one to seven years of age), bulls (harem bulls and idle bulls) and cows, representing five social statuses (Fig. 6.7). Only the harem bulls can mate and they may possess 40–50 females each. The bachelors congregate away from the harem area and the idle bulls position themselves near the harems. The idle bulls are young adults and old, exhausted males. Thus there are different classes of animals in relation to mating but no leader or leader group can be recognized.

(4) Let us move to the terrestrial carnivores. The Feloidea contains Viverridae (mongoose, etc.), Hyaenidae and Felidae, of which the

Fig. 6.7. A drawing showing group structure of the fur seal *Callorhinus ursina* in the breeding season (drawn from Kenyon *et al.*, 1954). Large animals in the centre are harem bulls and those in the periphery are idle bulls.

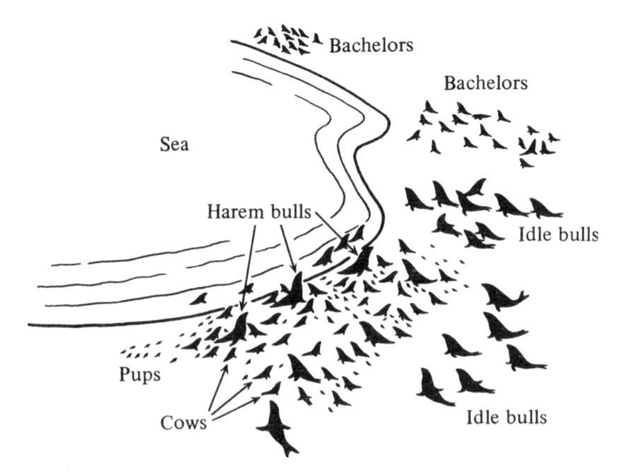

Viverridae has no species with group life. The members of the Felidae (except the cheetah and the lion) are all solitary. The cheetah *Acinonyx jubatus* forms a family group of one female and her offspring and a loose aggregation of a few non-breeding males; there is no lasting mixed group (Kleiman & Eisenberg, 1973).

As mentioned in Chapter 4 lions form a matriarchal group (pride) of a mother and her daughters, joined usually by two males (Schaller, 1972; Bertram, 1975). The matriarchal group contains not only the mother and immature young but also several mature females (sisters) and defends the same territory for generations. The size of a pride varies from four to thirty-seven (Schaller, 1972). They hunt in cooperation, sometimes attacking prey from different angles in relay. The female in a pride usually nurses her own young but occasionally suckles other young. The male, on the other hand, is merely a parasite and takes a share of the prey killed by females. He also mates indiscriminately with the females of the pride. The young male leaves his mother's pride at about three years of age and becomes a wanderer. The wanderer's behaviour has already been mentioned (Chapter 4). The significance of group life in lions is: first, the protection of young in a pride and, secondly, the improvement in hunting efficiency. The mortality of pups due to leopards, spotted hyaena and other predators is surprisingly high. Schaller has shown that the survival-rate of pups in a pride is higher than that of the young born to a solitary mother and that the success-rate of co-operative hunting is twice as good as solitary hunting. Probably for this reason the wandering males often live in pairs – usually said to be brothers.

The hyaenas in the Feloidea, analogous to the Canidae, have been thought to be scavengers until recently but it is becoming clear that they are in fact predators. The spotted hyaena *Crocuta crocuta*, the best adapted to open habitats among the hyaenas, has been studied in detail by Kruuk (1972). According to this study the hyaenas on the Serengeti plains and Ngorongoro Crater of East Africa ate their own kill 70 to 96% of the occasions in which they were observed in feeding. The spotted hyaena forms a group (clan) of 10–60 and hunts socially. A clan typically consists of more than one female, more than one male and young; a large female takes the role of the leader. On this point only does the hyaena represent the traditional matriarchal family of the Felidae. The clan usually has a territory and the captured prey is distributed among the clan members. Old individuals receive their shares after the prey is brought down. No member of the clan is attacked even when a small prey is being shared. Thus the number of hyaenas gathering around a prey animal sometimes reaches thirty. The young receive regurgitated food. The striped hyaena *Hyaena hyaena* and *H. brunnea* living in the forest zone are not 'social' (gregarious) (Eaton, 1973; Kleiman & Eisenberg, 1973).

In the Canidae the racoon dog *Nyctereutes* and the racoon *Procyon* are solitary and the foxes are monogamous (often solitary outside the breeding season) whereas the dogs – coyote, wolf, jackal and lycaon – living in the open country have the best developed group integration among the mammals other than the primates.

According to Estes & Goddard (1967) and Lawick & Lawick-Goodall (1971), *Lycaon pictus* forms a mixed pack of five to fifty or so with male leaders and hunts by the pack. Like the spotted hyaena they attack and kill ungulates larger than themselves. The pack leader (male) decides on the prey and other individuals do not move at the sight of prey until the leader takes action. In the absence of males a female is known to have acted as a leader. Dominance hierarchy, though it exists within each sex, is weak and there is little overt aggression within the pack. Thus dominance relations other than the distinction between the leader and the pack members are not clear. Estes & Goddard found no relay hunting of wolves in the lycaon in chasing prey animals. When the prey is captured as a result of group pursuit, all individuals pounce on the prey. Even when the prey is small it is distributed and young animals are given precedence. A further interesting fact is that though several females are found in a pack only one or two become pregnant each year and non-breeding females (probably low-ranking) help in caring for the young (again this is analogous to social insects). They provide food not only for young but also for suckling females, by regurgitation.

The wolf *Canis lupus* too is famous for its highly organized group. According to Etkin (1964) the pack of the wolf is centred round the pair-bond and contains offspring (both sexes) born over several years. The male that has founded the group – the oldest member – is usually the leader. As tradition says wolves hunt ungulates by relay (one chases prey towards another lying low) and the fallen prey is eaten on the spot by all individuals of the pack. If the mother, being pregnant or soon after parturition, did not take part in hunting, many individuals of the pack would take a mouthful of meat back to her. [Feeding of a female or young by a male is already developed in the monogamous red fox.] The most dominant male is the leader in prey pursuit and also confronts invaders. He mates with most females in the pack but other high-ranking males may also mate. The formation of a new pack is by budding off;

that is, a pair leaves the pack to produce young and form a new pack with the young. A pack has a territory but, as it is large like the lion's, other packs may often penetrate into it.

Thus the lycaon and the wolf possess the most developed social organization among the mammals except for some primates (more developed perhaps than most primates).

Group living versus solitary living

What, then, are the factors that select some animals for solitary or monogamous family living and others for polygynous harem or group living? Solitary animals in the following refer to those in which the male and the female lead separate lives except for a brief period of mating.

Fig. 6.8. Antelopes of Africa. A, royal antelope *Neotragus pygamaeus* (a solitary neotragin, its small size is in proportion to the eland shown in B); B, eland *Taurotragus oryx*; C, gemsbok *Oryx gazella*; D, gerenuk *Litocranius walleri* (living in small herds in woodland and adapted to browsing); E, greater kudu *Tragelaphus* (*Strepsiceros*) *strepsiceros*; B, C and E graze in grassland, and B and C form specially large herds.

Jarman (1964) compared social organizations of African antelopes (Fig. 6.8) and arrived at the conclusion that the group size is determined usually in relation to the body size and vegetation-type. He first examined vegetation-types in the habitats of antelopes and the feeding habits of each species, and found that the steppe had the most uniform vegetation while the forest was the most heterogeneous habitat in species composition and that antelopes selected certain parts of plants and this preference was strongest in the foliage-eating species. He, then, grouped their modes of feeding into five categories and classified the types of social organization mainly on the basis of group size, and examined the relation between these two. These are summarized below according to the type of social organization.

A: Duikers (Cephalophinae), pygmy antelopes (Neotraginae), etc. These are mainly small forest-dwelling antelopes living a solitary or monogamous life, occasionally accompanied by young. They eat only the special parts (e.g new buds) of a variety of plants but have small home ranges throughout the year.

B: Reedbucks (*Redunca*), *Pelea*, gerenuk (*Litocranius*), oribi (*Ourebia*, an aberrant member of Neotraginae), etc. They are adapted to forest and low woodland and usually form a group of two to twelve. The adult male is often accompanied by one to several adult females. They consume grasses and special parts (e.g. lower foliage of trees eaten by the long-necked gerenuk) of plants. They maintain home ranges throughout the year within their areas of specific vegetation. Sitatunga *Tragelaphus* (*Limnotragus*) *spekei* of swamps may also belong to this type.

C: Waterbuck (*Kobus*), greater kudu *Tragelaphus* (*Strepsiceros*) *strepsiceros*, kob (*Adenota*), gazelles (*Gazella*), impala (*Aepyceros*), etc. Females and young males form herds of four to sixty, sometimes up to 200; adult males are solitary. They live in flood plains and selectively feed on grass and broadleaf plants in several vegetation-types.

D: Wildebeest (*Connochaetes*), hartebeest (*Alcelaphus*) and other members of Alcelaphinae. Males are solitary but females and young males live in herds of four to one or two hundreds, occasionally thousands (superherds) by amalgamation of herds. They live in savannah, steppe and woodland, and feed selectively on certain parts or developmental stages of grasses. They migrate in search of suitable food.

E: African buffalo (*Syncerus caffer*), eland (*Taurotragus*) and probably some oryx (*Oryx*). The buffalo belongs to Bovine but is often treated with antelopes including Tragelaphini. This group lives in most open habitats of savannah, steppe and semi-desert and forms herds of a few hundreds to 2000 under favourable conditions. These herds are not temporary superherds formed by coalescence of small herds but of permanent nature and contain males as well. They eat grass indiscriminately in a variety of vegetation-types and migrate seasonally within an extensive home range.

Thus the size of a herd increases from A to E. Jarman compared the mean body weight among the antelopes representing these five types of social organization. The results shown in Fig. 6.9 indicate that small antelopes are solitary and the larger the animal the larger the herd formed. Also, as already mentioned, the small species consume only the particular parts of certain plants in the forest whereas the large species feed indiscriminately in grassland where vegetation is both structurally and floristically simple.

According to Jarman, the types of behaviour used to avoid predation are:

1. to avoid being discovered by predators (hiding);
2. to run away when discovered, before being attacked;
3. to run away when attacked; and
4. to threaten the predator or even counter-attack.

The small antelopes with type A organization mainly use (1) above, species with type B organization use mainly (1) but some take (2), species

Fig. 6.9. Mean body weights of antelopes according to the type of social organization (see text) (after Jarman, 1974). Group size increases in the order of type A to type E.

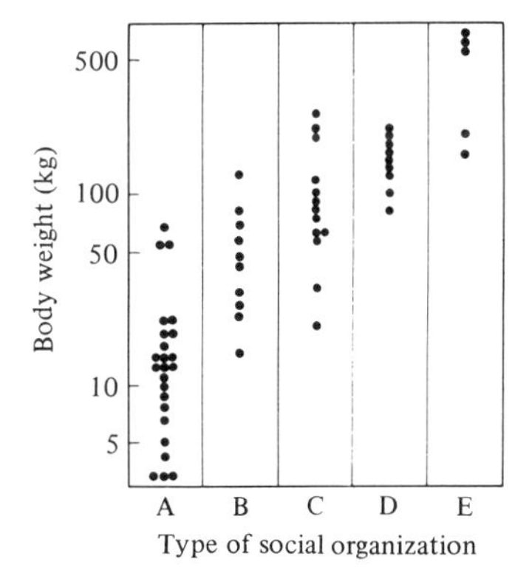

of type C organization use (2) and (3). The antelopes of type D, though occasionally run away from predators, often form a tight group to face enemies and make it difficult for them to select a victim. They sometimes counter-attack the predator. In the African buffalo (type E) an attack on one of the young is countered not only by the mother but also by all the other members of the herd coming to its aid.

Such group defence probably developed in the environment of open country where animals cannot hide – for hiding, group life is more disadvantageous than solitary life – and in this environment food is distributed uniformly and is easily obtainable. Thus the main point of Jarman's conclusions is that the group life of antelopes has developed as a result of the improved efficiency of food utilization which accompanies increased body size.

Estes (1974) also compared social organization of antelopes. His view is more lucid than Jarman's: the species is forced to select either the closed habitat or the open habitat in which to live. Those advancing into the field are selected for group life, large body size and running away from danger. According to Estes all species of solitary antelopes live in the forest habitat whereas only a dozen or so out of seventy species living in this habitat form herds. Of these very few are restricted to inside the forest.

The formation of groups by predatory Felidae and Canidae has been discussed by Kleiman & Eisenberg (1973). As already mentioned, almost all members of the Felidae are solitary animals, only the cheetah and the lion forming small groups. Only the spotted hyaena among the members of the Feloidea forms large groups. In the Canidae the racoon dog is solitary and foxes have monogamous families. The wolf, lycaon and few other wild dogs lead group life. Kleiman & Eisenberg (1973) considered that the pair-bond of foxes was the basis for the feeding of young and led to the establishment of large groups seen in the wolves, etc. in which food was distributed among the members. In the Felidae there was no habit of male feeding young. This probably led to matriarchal groups of the cheetah and the lion in which only the females fed young. In either case the question remains as to why group life has developed only in a few species. This may be considered as an adaptation to living in an open habitat which does not permit single predators to approach prey animals undetected. The grassland, however, provided assurance for a large quantity of food – enormous numbers of large herbivores – necessary for the group life of predators. There, instead of solitary stalking, hunting by relay pursuit was necessary and hunting by packs evolved. The reason why group life evolved further in the Canidae than in the Felidae is probably that the role of the father and hunting by pursuit had already existed in the Canidae before group life began.

Recently, Crook *et al.* (1976) examined the social organization of mammals including primates. The criteria used for comparison are the group size and the bonding of males to the group (separate, monogamous, multi-male co-existence, etc.). The factors affecting these are the characteristics of a particular species (they consider that primitive species are solitary animals, with several females having small home ranges within a large home range of a male), the distribution of resources (food and shelter) and the response to predators. Rather than drawing specific conclusions, Crook *et al.* emphasized the necessity for comparative studies of broad mammal societies by means of models incorporating these factors. This is certainly a most reasonable and unmistaken opinion, but can we not have a working hypothesis in our study? Various theories of Jarman and others described above contain many correct indications in parts but appear to lack applicability as universal theories. For example, in the theory of Jarman, if we expand it to the entire Artiodactyla the question arises as to why the argali *Ovis ammon* and the elk (moose) *Alces alces* do not form large herds. It is not clear why small wild cats and the tiger are solitary but the lion is social among the Felidae. In the theory of Kleiman and Eisenberg no clear explanation has been put forward for the development of group life only in the species best adapted to open country among the strongly solitary feline animals.

I have mentioned in Chapter 1 that the grassland as opposed to the forest is a continuous environment in which it is easy for young animals to procure food but the risk of predation is high. I have also pointed out that forests, mountains and the polar regions are similar in one respect even though they are very different in the total amount of organic energy – it is difficult for the young animals to obtain food. Also, the data on reproductive capacity and survival-rate have shown, when the high-fecundity strategy and the low-fecundity/parental-protection strategy are dichotomized in the survival strategies of organisms, that the former strategy developed in open environments like grassland and the latter evolved in forests and mountains.

Espinas (1877) said in his book on animal societies 100 years ago that 'Même à l'origine, la famille et la peuplade, sont antagoniques; elles se développent en raison inverse l'une de l'autre.'

Today, as the social morphology of many animals has become known we can see that this remark was very true. The formation of a pair and group life are generally antagonistic to each other. The birds and mammals that form lasting pair-bonds do not form organized groups; if they breed in a colony this is only an aggregation resulting from a shortage of nesting sites. On the other hand, in the group life with leadership, specific pair relations dissolve and promiscuity appears (see next section on primates). [Here the word promiscuity follows the usage of animal ecologists and refers to the tendency that some males mate with more than one female during the same breeding season and females, though there may not be as many cases as in the males, receive more than one male. This does not preclude particularly intimate relations of a specific male and a specific female for a short period. It does not include incest mating with parents as in the definition of Morgan-Engels. At least there has been no report of mating between mother and a son in the field among higher mammals except rare cases in provisionized troops of Japanese macaques. Though its mechanism is not clear the incest taboo has long been established.]

Some species form pairs in the breeding season and change to group life after breeding. What is the bifurcating point between family life and group life? A broad perspective of all birds and mammals including the primates to be discussed gives us a clue, as recognized by Estes (1974), that the grassland made the animals evolve group life and the forest habitat caused the evolution of monogamous families and solitary life. This is well demonstrated by a correlation table of habitats and social morphology (Table 6.2), constructed from reviews of Jarman and others and information given in Wilson (1975). Ungulates and carnivores both contain many species with group life in the grassland and of solitary and monogamous family life in the forest habitat. The reason for solitary living in some species today might be a result of population decline due to hunting, etc. If so, some of the species marked by indicators d and e in Table 6.2 might have to be raised to a higher social status though this would probably not change the overall tendency greatly. The species with the largest group size live in the most open habitat. Even in the species with gigantic body size, such as the rhinoceros and hippopotamus, or in the strongly solitary family of Felidae there is some social tendency in the grassland species. That the duikers and dik-dik are solitary is not because they are small but because they are forest-dwellers, like the tapir, elk and okapi, which are larger but also solitary. Smith *et al.* (1973) who discussed the evolution of group life in the blacktail prairie-dog *Cynomys ludovicianus* also thought that the very

Table 6.2. *Relationships between habitats and social structures among major groups of mammals (except primates); male groups are omitted*

Taxon	Grassland (savannah, steppe, semi-desert)	Transition zone (sparse forest, gallery forest)	Forest	Rocky mountain and arctic
Marsupialia	Wallaby[a], Red kangaroo[b]		Ringtail possum[c], koala[d]	
Lagomorpha		Rabbit[b]	Brush rabbit[d]	Pika[c]
Rodentia	Prairie-dog[a], Marmot[b], Kangaroo rat[e]	Beaver[c], ground squirrel[d], Microtus[d]	Chipmunk[d], Apodemus[d]	Lemming[c]
Edentata			Sloth[d], Giant anteater[d]	
Equidae	Grant's zebra[a] Mongolian wild horse[a]	Mountain zebra[c]		
Rhino. & Tapir	White rhinoceros[b]	Black rhino.[c], Indian rhino.[c]	Javanese rhino.[d], Tapir[d]	
Suidae & Tayassuidae	Warthog[c], Peccary[a]	Bush-pig[a]	Japanese wild boar[c]	
Hippopotamidae		Hippopotamus[a]	Pygmy hippopotamus[d]	
Camelidae				Vicuña[c]
Cervidae	Caribou[a]	Red deer[b], Black-tailed deer[b]	Moose[d], Roe deer[d], Mouse deer[d]	
Giraffidae	Giraffe[b]		Okapi[d]	
Bovinae	American bison[a], African buffalo[a]		True cattle (Gaur, Kouprey, Banteng)[d]	
Antelopes	Eland[a], Wildebeest[a], Impala[a],	Gerenuk[c], Kob[c]	Duiker[d], Pygmy antelope[d]	
Caprinae	Saiga[a]		Japanese serow[d]	Mountain goat[c], Bighorn[c]
Proboscidae	African elephant[a]	Indian elephant[b]		
Canidae	Wolf,[a] Lycaon[a]		Red fox[d], Racoon dog[d], Racoons[d], Bears[d]	Arctic fox[d]
Feloidea	Spotted hyaena[a], Lion[b], Cheetah[b]	Striped hyaena[c]	Tiger[d], Leopard[d], Mongoose[d]	

Social structure:

[a] Large groups consisting of several dozens or more individuals (mixed-sex group or matriarchal group).

[b] Matriarchal groups consisting of *ca* 10–20 individuals.

[c] Polygynous groups or relatively long-lasting groups of mother and her progeny.

[d] Monogamous families (juveniles leave the family) or solitary.

[e] Aggregation of monogamous or polygynous families.

(This table was constructed, based on literature cited in the text and personal communication of H. Obara.)

open habitat of grassland was responsible for the natural selection of group life in this species. Although the spotted hyaenas living in savannah form large groups, the striped hyaena *Hyaena hyaena* not found in savannah is not 'social' (Kleiman & Eisenberg, 1973). In the highly fecund pigs the large groups are formed by the bush-pig *Potamochoerus* of savannah and the peccary *Tayassu* of grassland; the wild pig adapted to the dense forest of South-East Asia may be found to be solitary.

We have already seen in the section on birds that, when the breeding system of one group differentiates into a monogamous pair system with territorial defence and a polygynous or promiscuous breeding system, the former is found in the forest habitat and the latter in the grassland. Crook (1965) compared the habitat-type, distribution of food and the type of social organization for all living families of birds. His main point is also that grassland- and ground-living induced group breeding whereas the arboreal life – particularly insectivorous and carnivorous – brought about the defence of territory by a pair. Only the frugivores of tropical forests are the exception (already touched on in p. 244 but also see below).

Group size is often different depending on the habitat even within the same species. Dasmann & Taber (1956) and Linsdale & Tomich (1953), who observed the black-tailed deer *Odocoileus hemionus* in forest and scrub, reported on the small group size with territorial defence but Palmer (1951) and McLean (1940) observed large groups of this species in the open habitat. Darling's (1937) herds of red deer were observed in Scotland where the primary forest had been destroyed but those living in the deep forest did not appear to form such a large herd. From these facts Dasmann & Taber (1956) concluded that in the discussion of social evolution in deer the herd is a characteristic of those living in grassland or tundra and that the species living in closed forest are solitary or form small groups and defend territories. Similar intraspecific variation appears to occur in some African antelopes such as the impala. Estes (1974) states that *Oreotragus oreotragus* and *Ourebia ourebia* which form small herds of up to ten in woodland form large herds and run away from danger rather than hide, when they come to grassland to feed on new sprouts after fire.

What characteristic of grassland or forest, then, have brought about such changes in animals? Many workers recognize that the forest is an environment in which animals can hide from predators and, for this, solitary life is advantageous. Grassland is an environment in which large animals cannot hide and thus the habit of defending by group would be more easily selected for. For prey catching also, the forest is suitable for stalking or lying in wait singly and the grassland for pursuit by a group.

However, I think that the factor behind all this is what caused the bifurcation of the high-fecundity and low-fecundity/protection strategies; in other words the procurability of food by the young and the continuity–discontinuity of the environment. Forest has more dimensions than grassland and it is only after a long period of care that young animals can obtain food in it (on this point large ground dwellers of the forest are the same, specially in tropical rain forest). Here the evolution of the low-fecundity/protection strategy is inevitable but the social structure that is most efficient in protecting young in this environment with patchily distributed food is the monogamous family system in which parents provide for the young. In the forest the danger of predation is relatively small and prey species can hide. This is probably the reason why small species have been selected for in the forest habitat. It is not the large size of species that induced herd formation as Jarman would postulate but that the formation of herds and the enlargement of body size are both selected for by the same factor of grassland. In grassland (for herbivorous animals) young birds after hatching or young mammals after weaning can immediately start feeding themselves from under their feet and the grass on which they feed is distributed continuously. For the carnivores the prey density is very high in grassland. Thus grassland is suitable for maintaining large groups from the point of view of food supply alone.

Thus formation of groups in grassland is concerned with food conditions and predation. The common factor to both is the continuity of the habitat. Looking at the animal kingdom as a whole it is then the procurability of food by the young that is the important factor. Colonial nesting with the monogamous system is seen in herons and oceanic birds; both the shortage of nesting sites and predation were involved in the case of herons while only the shortage of nesting sites was involved in the oceanic birds. In either case the difficulty of procuring food by the young would be responsible for maintaining the monogamous system. [The nidicolous birds have evolved in such a way as to require a long period of feeding of nestlings and they cannot return or change to nidifugous habit by advancing into an open habitat. This Catch-22 situation was resolved by jays which acquired helpers at the nest as social insects did – an incredible conversion.] However, the conditions have changed since the advent of large mammals in grassland. Both carnivores and herbivores increased their speed of locomotion in the course of co-evolution in grassland. This process, at the same time, seems to have resulted in enlargement of the body size, a small litter consisting of large young with well-developed locomotive ability and adaptation to group life in herbivorous mammals. In this view not only the high fecundity of nidifugous birds in grassland but also the large mammals with a litter size of one or two may be considered as an extension of the high-fecundity strategy. I should think that in the last stage of evolution they have overcome the confrontation of high fecundity versus low fecundity/protection and come to face the choice of solitary life or group life within the framework of the low-fecundity strategy.

Societies of primates

(1) The last section of this book is devoted to the social structure of the primates – the group from which man has emerged. At the time of publication of the first edition of this book data on the social structure of primates were available only for nine species, of which four had been studied by one pioneer worker in this field – C. R. Carpenter. Since then, however, we have witnessed an incredible advance in this field. An impetus was given by the formation of the Japanese primate study group with the strong backing of Kinji Imanishi. In the 1960s many workers in America and Europe also tackled this problem and today there are many centres using modern techniques to study primates. The number of species studied in recent years exceeds fifty and most of the works published before 1971 were cited in the review of Itani (1972) and Jolly (1972). Some of the more recent work may be seen in Wilson (1975). Thus in this section I shall limit citation to only a few relevant papers and give a bird's-eye view of the social structure of primates disregarding variations within species.

The classification of the primates is still unsettled (and said to be generally much more 'split' than other animal groups); the following classification is taken from Kawai *et al.* (1968).

Prosimii
- Tupaiformes
  - Tupaiidae
- Tarsiiformes
  - Tarsiidae
- Lorisiformes
  - Lorisidae
- Lemuriformes
  - Lemuridae

Indriidae
Daubentoniidae
Anthropoidea
Ceboidea
Cebidae
Callithricidae
Cercopithecoidea
Cercopithecidae
Hominoidea
Pongidae*
Hominidae

For a long time the Tupaiidae has been treated as a family of the Insectivora (Jolly's book does not include this family in the primates). This is a group showing intermediate features between the ancestral (insectivore) type and the primates.

According to Kawamichi (1976) and Kawamichi & Kawamichi (1979) the greater tupaia *Tupaia glis* of South-East Asia is solitary and diurnal, running around among trees and on the ground and eating insects and fruits. Territorial defence is male against male and female against female. The territory of a female (occasionally more than one female) is often included in the territory of a male, suggesting the existence of pair relations in this group. [I have interpreted Kawamichi's 'defence of home range' as territorial defence. Greater home range (or territory) of males is also found in rodents (see Chapter 4).] Young individuals generally have their home ranges within the mother's territory and they appear to maintain a loose association with parents for a time after weaning. The 'group' size given by Kawamichi is 2–6, which includes a sexual pair. In one case there were two adult males and two adult females in the 'group'. However, the pair relations and parent–offspring relations are not such close associations as seen in the pair relation of birds or in a family of gibbons. Movements are not organized but reveal much the same range among the 'group' members, and the sexual relation of a specific male and a specific female seems to be weak. Thus the tupaia may be considered basically as a solitary animal.

The tarsiers and lorises are nocturnal animals. Although there is still no systematic observation made on the Tarsidae, they are probably either solitary or in pairs. Of the Lorisidae all four species of the

* Jolly (1972) separates Hylobatidae (gibbons) from Pongidae (orangutan, gorilla and two species of chimpanzee).

Lorisinae are solitary. Their social life is similar to that of the tupaia in that males defend territories and females have narrow home ranges within them.

The life of the Galaginae (Lorisidae) has been investigated by A. J. Haddow & J. M. Ellice and P. A. Jewell & J. F. Oates (cited by Jolly, 1972). According to their observations the lesser galago *Galago senegalensis* in Uganda was either solitary or in a group of up to four (average two in twenty-five observations) while the Demidoff's galago *G. demidovii* was seen either singly or in twos (mostly sexual pairs) in fifty-nine observations. Jolly (1972) suspects that there may also be some small polygynous groups in the galago.

The Lemuriformes has been studied by Petter (1962, 1965) of France and Jolly (1966). According to Petter the aye-aye *Daubentonia madagascariensis* is solitary. The Indriidae forms small groups. Jolly found that the white sifaka *Propithecus verreauxi* lived in groups of two to eight (average five) and in one case one adult male and in another case two or three adult males were identified in the group. In the diurnal species of *Indri indri* Petter found monogamous groups with or without accompanying young in three out of four cases of observation and a combination of an adult female and a young in one case. The combination of a male and a female appears to last throughout a year. *Avahi laniger* is also found in family groups. Petter gives the following group structure:

One adult male, one adult female and three immature young ... two cases;

One adult male, one adult female, three immature young and one infant ... one case;

One young female, two immature young (sexes unknown) ... three cases.

The species mentioned above, except the tupaia, sifaka and indri, are nocturnal. The nocturnal species are either solitary or found in sexual pairs as a rule. The avahi is an exception; in the Lemuridae other nocturnal species, the mouse lemurs (*Microcebus*), dwarf lemurs (*Cheirogaleus*) and the weasel lemur *Lepilemur mustelinus* are solitary or in pairs.

Among the Lemurinae, *Lemur*, which is diurnal or crepuscular in activity, and *Hapalemur* form groups. Petter (1962) gives ages and sexes of group members observed in the black lemur *Lemur macaco* (Table 6.3). All groups contain more than one male and usually more males than females. The reason for this is not clear. Jolly (1966) gives the composition of three groups observed of the ring-tailed lemur *Lemur catta* (Table 6.4), which has a strong tendency for terrestrial life. Other diurnal

Table 6.3. *Group composition of the black lemur* Lemur macaco *at Nosy Komba Island near the Malagasy Republic*

Group no.	May 1956	November 1956	April 1957
1	7♂+3♀	5♂+3♀+2J	6♂+4♀
2	7♂+3♀	6♂+4♀+3J	6♂+7♀
3	2♂+2♀	4♂+2♀	4♂+2♀
4		5♂+2–3♀+2J	5♂+4♀
4′		5♂+5♀+2J	8♂+7♀
5	5♂+4♀	4♂+4♀+1J	
6	5♂+5♀	5♂+5♀+1J	
7		3–4♂+3♀	
7′		5♂+3♀+1J	5♂+3♀
8		5♂+4♀+1J	
Total	26♂+17♀	47–48♂+35–36♀+13J	34♂+27♀

After Petter (1962).
♂, ♀ and J refer to adult males, adult females and juveniles, respectively.

species of lemur seem to have similar group compositions. In no species was there any strong territorial defence by the group or any strained relation observed when two groups met. Although no hierarchical relations have been found within a group, adult males probably have a greater role to play than others. Males mark the territory, with a pheromone from the odour gland, more frequently than females.

(2) All New World monkeys, the Ceboidea, are arboreal. There are over

Table 6.4. *Group composition of the semi-terrestrial ring-tailed lemur* Lemur catta

Group no. and date	Adult males	Adult females	Subadult males	Subadult females	Juveniles	Infants	Total
1							
Sept. 1963	6	9	1	—	3	(4)[a]	(23)
Apr. 1964	5	9	1	1	7	—	23
2							
Sept. 1963	8	3	1	1	3	(2)	(18)
Apr. 1964	8	5	2	1	5	—	21
3							
Sept. 1963	4	7	—	—	1	?	12+

After Jolly (1966).
[a]Cases of uncertain number are put in parentheses.

sixty species (Napier & Napier, 1967), of which the night monkey *Aotus trivirgatus* is the only nocturnal species. All diurnal species live in families or groups. The Cebidae includes two species, the study of which paved the way to modern primatology. They are the mantled howler monkey *Alouatta palliata* (=*villosa*) and the black-handed spider monkey *Ateles geoffroyi* that appear in the pioneer work of Carpenter. Carpenter (1934) worked on the howler on Barro Colorado Island of Panama Canal,

Table 6.5. *Group composition of the mantled howler monkey* Alouatta palliata *on Barro Colorado Island*

Group number	♂	♀	♀ with an infant	Immature young	Total
1	1	2	1	1	4
2	2	2		1	5
3	2	2	2	2	6
4	2	4		2	8
5	1	4	3	5	10
6	1	4	2	6	11
7	2	6	2	4	12
8	3	4		6	13
9	2	5	3	7	14
10	2	6	4	7	15
11	2	5	3	8	15
12	3	6	5	8	17
13	3	8	3	7	18
14	3	7	5	8	18
15	3	7	4	9	19
16	3	7	4	9	19
17	4	10	2	6	20
18	3	9	5	8	20
19	3	6	5	12	21
20	3	9	3	9	21
21	4	8	4	11	23
22	3	8	5	12	23
23	4	9	4	11	24
24	5	9	7	11	25
25	5	11	6	9	25
26	5	13	3	9	27
27	3	9	9	15	27
28	5	12	4	12	29
Total	82	192	98	215	489

Based on Carpenter (1934).

where he counted some 500 in twenty-eight groups (average of seventeen monkeys per group) in 1933. Their composition shown in Table 6.5 indicates that the group usually consisted of a small number of males and twice as many females with some immature young. Each group had a home range in which they had several roosts and foraged in between. However the boundary of a home range was not fixed and only the familiar part of the range was frequently defended against other groups. In foraging a male tended to lead the group. Although group composition has been shown to vary these social characteristics have been confirmed by subsequent studies (see Carpenter, 1965).

When there is more than one male in the troop they share in leading and defence of the troop. As the socionomic sex-ratio, the ratio of the adult males to the adult females in a troop, is about one to two, there are surplus males outside the troops. They do not form male groups but live

Table 6.6. *Subgroup composition of the black-handed spider monkey* Ateles geoffroyi *in the Panama Canal Zone*

Subgroup	♂	♀	♀ with an infant	Immature young	Total
1		1	1	2	3
2	1	1		1	3
3	3	2	1	1	6
4	2	3		1	6
5	3	2	1	2	7
6		4	1	3	7
7	5	1		1	7
8	1	6		1	8
9	1	3	1	4	8
10		5	3	3	8
11	1	5	1	4	10
12	3	4		3	10
13	7	2		1	10
14	1	6	2	4	11
15	2	5	2	5	12
16	5	7	3	3	15
17	2	7	2	7	16
18	5	6	2	6	17
19	4	6	2	7	17
Total	46	76	22	59	181

Based on Carpenter (1935).

as solitary animals. Carpenter described the sexual relation of the howlers as 'communal', not monogamous or polygynous. However, there is often a consort period in which one male and one female live together.

An example of subgroup composition found in the spider monkey (Table 6.6) is taken from Carpenter's (1935) study in the Canal Zone. He found some 200 spider monkeys living in four troops. Each troop was divided into subgroups, each consisting of nine individuals on the average. In the table the number of males varied from one to seven per subgroup, five subgroups with single males out of sixteen subgroups in which at least one male occurred. There were, in addition, some male groups. Carpenter gave an example of a troop with three polygynous subgroups. I followed his idea of the polygynous family as a basic unit of social organization in the spider monkey (Itô, 1959, 1966, etc.), but judging from the way subgroups amalgamated or dissolved such a unit was not likely to be very fixed.

Nevertheless, it is doubtful if these four large groups could be called social groups. On the whole they do not seem to be exclusive in membership, as in troops of Japanese macaques, let alone have any leadership or status differentiation within the group; at least no such observation has been reported to date. According to Izawa (1976a, b) who studied New World monkeys in the upper reaches of the Amazon, the black-handed spider monkey *Ateles geoffroyi* and the long-haired spider monkey *A. belzebuth* were found either singly or in small groups of up to four. Larger groups of 10–30 were seen at a salt station provided. Izawa thought that these larger groups maintained the same membership, thus forming basic units in the long-haired spider monkey, but was this really the case? He found a Humboldt's woolly monkey *Lagothrix lagotricha* accompanying a troop of more than ten spider monkeys. As I shall discuss below the tropical rain forest is an environment which favours formation of large, loosely-organized groups with mixed species. In spite of Itani's (1972) criticism, I would maintain that the subgroup, rather than the larger group, is the social unit of spider monkeys even when there are a certain number of membership changes between subgroups. The red howler *Alouatta seniculus* which Izawa (1976c) also studied had troops consisting of 3–11 individuals, with a pair or a polygynous family in most cases. Only in three out of twenty-three cases did more than one male (two or three) occur in a troop.

Among other New World monkeys, the night monkey is solitary or in a monogamous family, the titis *Callicebus* spp. are in groups of 2–4, probably based on the monogamous group [six cases cited by Jolly (1972) of *C. moloch*, seven cases of *C. torquatus* in Izawa (1976c)], and the monk saki *Pithecia monachus* and the pygmy marmoset *Cebuella pygmea* are monogamous (Izawa, 1976a). Among the tamarins, Jolly reported groups of 2–6 containing a monogamous or bigynous family for the red-handed tamarin *Saguinus midas* and Izawa found groups of 6–12, based on monogamous relations, for the brown-headed tamarin *S. fuscicollis*. Izawa also observed larger groups of 30–40 containing more than one male, but for the reason given for the spider monkey I would consider the smaller group, said to be seen in the dry season, as the basic unit. Izawa describes the large groups seen in the wet season as loosely organized for a unit group. Thus large groups seen in tamarin may consist of smaller units (for the black-mantled tamarin *S. nigricollis*, see Izawa, 1978).

Woolly monkeys and squirrel monkeys are different from the above species in their social organization. According to Izawa, the Humboldt's woolly monkey lives in integrated troops of 20–70, containing adult males and adult females. When they are encountered in the forest several well-built adult males appear above the observer and shake branches or look at the observer. While dispersing there are frequent exchanges of calls between moving adult males, suggesting that they take the role of leaders. Such differentiation of status within a troop has not been found in other large troops of New World monkeys (see Izawa, 1976c; Nishimura & Izawa, 1975). The common squirrel monkey *Saimiri sciureus* has been seen foraging and moving in early morning and late afternoon in large troops of about forty (sometimes 200) but during the day the troop is dispersed over a large area. The sex and age compositions suggest that the socionomic sex-ratio is about one to one in large troops. Thorington (1968) observed a troop consisting of eighteen members, which was divided into smaller groups of males, females and immature young, separately, during movements.

In effect, my view is that most of the New World monkeys living in the canopy of tropical rain forest form small groups based on a monogamous or polygynous association or combinations of a few males and a slightly larger number of females. However, their group integration is not very strong and they often form loosely organized large troops. The Humboldt's woolly monkey remains enigmatic; this species may belong to a group with the largest troop size among the New World monkeys.

(3) As for the Old World monkeys, the Cercopithecidae, there is a wealth of information, not to mention the many studies on the Japanese macaque *Macaca fuscata*, which brought about the revival of field

primatology after the war. This family may be divided into Cercopithecinae and Colobinae (specialized for leaf-eating). The former includes guenons (*Cercopithecus*), macaques (*Macaca*), and terrestrial baboons (*Papio*), whereas the latter includes mainly arboreal *Colobus* and langurs (*Presbytis*), containing semi-terrestrial species.

The Cercopithecidae, as in the Ceboidea, contains no solitary species. *Cercopithecus* is a large genus with twenty-three species (Napier & Napier, 1967) but most of them live in trees of tropical rain forest (e.g. greater white-nosed guenon *C. nictitans*, crowned guenon *C. pogonias*, red-eared guenon *C. erythrotis*, moustached guenon *C. cephus*, talapoin *C. talapoin*, red-tailed monkey *C. ascanius*, mona monkey *C. mona*). The red-tailed monkey prefers the forest edge whereas the mona monkey lives in the middle to the lower strata of the forest. Struhsaker (1969) described the mona monkey as an arboreal and terrestrial forest-dweller and the Preuss' guenon *C. l'hoesti preussi* as a terrestrial forest species. The only species not found in the forests is the vervet (savannah) monkey *C. aethiops*, which is semi-terrestrial in riverine woodland and savannah with many trees.

Table 6.7 shows the social structures of some of these monkeys, which are relatively well known, together with those of other Old World monkeys. Arboreal species of tropical rain forest, such as the greater white-nosed guenon, crowned guenon and mona monkey, form small single-male troops. [In the following the 'single-male' troop refers to a troop based on a monogamous or polygynous association whereas the 'multi-male' troop refers to a troop with more than one adult male and more than one adult female.] The basic social unit of the guenon is probably a polygynous group led by the male. Struhsaker (1969) observed single-male troops also in the moustached guenon (three cases) and the red-eared guenon (four cases). However, Gautier & Gautier-Hion (1969) found multi-male troops in the greater white-nosed guenon and the mona monkey while Bourlière *et al.* (1970) found them in the Lowe's guenon *C. campbelli lowei*.

There are detailed studies of the red-tailed monkey by Lumsden (1951) and Buxton (1952). This species lives in the lower strata of forest edges and swamp forest, and ranges in troop size from several to more than 100. When they are in a large group they divide into subgroups (resting bands) at night and join together in the morning. The data in Table 6.7 are taken from the resting bands. Struhsaker (1969) criticized Buxton's treatment of the data but his criticism strengthened the view that the basic social unit of the red-tailed monkey is a small group. The subgroup appears to be centred round the mother–infant relation and contains one

Table 6.7. *Size and structure of unit groups of monkeys and apes belonging to Cercopithecoidea and Hominoidea; species for which one or only few countings were made were omitted*

Species	Group size $\bar{x}$ (range)	No. of groups counted	No. of males per group	Main habitat	Author	Notes
Greater white-nosed guenon *Cercopithecus nictitans*	9(5–13)	8	1[a]	Forest–Arboreal	Struhsaker (1969)	
Mona monkey *C. mona*	9(3–13)	6	1	Forest–Arboreal	Struhsaker (1969)	
Crowned guenon *C. pogonias*	13(9–19)	7	1[a]	Forest–Arboreal	Struhsaker (1969)	
Red-tailed monkey *C. ascanius*	4(2 – 11 +)	81	1 or many	Forest–Arboreal	Lumsden (1951)	night group
L'hoest's guenon *C. l'hoesti*	4(2 – 7)	15	1 or many	Mountain forest–Semi-terrestrial?	Struhsaker (1969)	
Savannah monkey *C. aethiops*	11(6–21)	46	many	Savannah–Semi-terrestrial	Gartlan (1968)[e]	
Patas *Erythrocebus patas*	18(5–25)	8	1	Grassland–Terrestrial	Hall (1965)	♂-group exists
White-collared mangabey *Cercocebus torquatus*	(3 – 18 +)	3	—	Forest–Semi-terrestrial	Struhsaker (1969)	
Black mangabey *C. albigena*	(10 – 25 +)	4	1 or many	Forest–Arboreal	Struhsaker (1969) & Chalmers (1968)[e]	
Crab-eating macaque *Macaca irus*	43(13–72)	5	many	Seaside forest–Semi-terrestrial	Furuya (1965)	
Bonnet macaque *M. radiata*	35(6–58)	4	many	Village–Semi-terrestrial	Simonds (1965)	

Table 6.7 continued

Species	Group size $\bar{x}$ (range)	No. of groups counted	No. of males per group	Main habitat	Author	Notes
Rhesus monkey *M. mulatta*	(16–78)	many	many	Village–Semi-terrestrial	Southwick *et al.* (1965)	
Japanese macaque *M. fuscata*	(17–120)	9	many	Forest–Semi-terrestrial	M. Kawai (personal communication)	
M. fuscata	35(12–70)	4	many	Forest–Semi-terrestrial	Suzuki (1965)	
Chacma baboon *Papio ursinus*	46(12–109)	18	many	Grassland–Terrestrial	DeVore & Hall (1965)	
Savannah baboon *Paio doguera*	41(12–87)	9	many	Grassland–Terrestrial	DeVore & Hall (1965)	
Hamadryas baboon *P. hamadryas*	5(2–13)	8	1[b]	Rocky mountain–Terrestrial	Kummer & Kurt (1963)	♂-group exists
Drill *P. leucophaeus*	24(9 – 55)	12	many	Forest–Terrestrial	Struhsaker (1969)	
Gelada baboon *Theropithecus gelada*	9(5–35)	30	1	Rocky mountain–Terrestrial	Crook (1966)	♂-group exists
Abyssinian colobus *Colobus guereza*	10(6 – 15)	4	many[c]	Forest–Arboreal	Schenkel (1967)[e]	
Satan colobus *C. satanus*	6(2–8)	5	1	Forest–Arboreal	Struhsaker (1969)	

Silvered lutong *Presbytis cristatus*	35(22–48)	2	many	Forest–Arboreal	Furuya (1962)	
Nilgiri langur *P. johnii*	14(10–25)	5	1	Forest–Arboreal	Tanaka (1965)	
Hanuman langur *P. entellus*	13(9–24)	38	generally 1	Village and Forest–Semi-terrestrial	Sugiyama (1964)	♂-group exists
Hominoidea[d]						
White-handed gibbon *Hylobates lar*	4(2–6)	21	1	Forest–Arboreal	Carpenter (1940)	
Agile gibbon *H. agilis*	3(2–4)	8	1	Forest–Arboreal	Ellefson (1967)[e]	
Siamang *H. syndactylus*	3(2–4)	22	1	Forest–Arboreal	Koyama (1976)	
Mountain gorilla *Pan gorilla*	10(5–18)	4	1, 2	Forest–Terrestrial	Kawai & Mizuhara (1962)	
Pan gorilla	15(5–27)	10	1 or many	Forest–Terrestrial	Schaller (1963)	

Mainly based on Jolly (1972) and Itani (1972) with modifications according to the originals.
[a]Gautier & Gautier–Hion (1969) found a group of more than ten including several males (in Itani, 1972).
[b]Here 'one-male unit' in Kummer & Kurt (1963) is taken as a group.
[c]Struhsaker (1969) observed two groups, both of which were one-male groups.
[d]The orangutan is a solitary species. In both the chimpanzee and the pigmy chimpanzee social structure is flexible.
[e]Figures are cited from Jolly (1972) and the reference is omitted from the list.

adult male in small groups or two in large groups. The socionomic sex-ratio is 1 to 1.5. This sex-ratio and the fact that the solitary individuals which constitute 10% of the population are males (Haddow, 1952) suggest that this species is polygynous. I think it is the subgroup that forms a basic unit in the cercopithecine society in tropical Africa. Lumsden (1951) presented a table showing the sizes of the resting band observed over six months in forest edges in Uganda (Table 6.8). According to this information many species including the red-tailed monkey rest in groups of four on average. Since the male group is not known for arboreal Old World monkeys, small groups of less than six listed in the table may be considered to consist of one male and several females with accompanying young. Although Buxton (1952) found no leader in the group of red-tailed monkeys, Haddow (1952) observed the largest individual giving alarm calls from outside his troop when danger approached. Because alarm calls were never given by two males at the same time there might be some social order among the males though such order, if it existed, would be less rigid than in the semi-terrestrial monkeys.

Itani's (1972) table of social morphology gives figures of 100 and 200 for the maximum numbers recorded in the troops of the red-tailed monkey and 115 for the talapoin, but these may represent aggregations around large crops of fruit on some specific trees in tropical rain forests. There seem to be other records of large troops in the talapoin, and M. Kawai (personal communication, 1978) considers that the talapoin and the red colobus *Colobus badius* may have large social units. As in South America large aggregations in the canopy of tropical rain forest in Africa often contain more than one species of monkey. Itani (1958) recorded thirteen large aggregations, of which four were mixed troops. There may be a few species with a large basic social unit living in the forest habitat, but most of the arboreal species are likely to have a small troop size as the basic unit.

By contrast, the savannah monkey with semi-terrestrial life in savannah is more commonly seen in multi-male troops than otherwise, and its troop size is large. Struhsaker listed the figures of 7–53 (1–4 adult males) obtained by himself, 7–20 (2–6 adult males) from Hall & DeVore (1965) and 4–22 from Hall (1965). In forest-dwelling *Cercocebus*, the black mangabey *C. albigena*, an arboreal species, and the white-collared mangabey *C. torquatus*, a semi-terrestrial species, form relatively small troops. The former seems to form both single-male and multi-male troops.

Table 6.8. *Size of night-resting groups of some monkeys in Uganda*

	Number of individuals										
Species	1[a]	2	3	4	5	6	7	8	9	10	11–
Black mangabey *Cercocebus albigena johnstoni*	4	17	44	53	28	33	6	4	1	—	3
Blue monkey *Cercopithecus mitis stuhlmani*	2	2	3	5	2	—	—	1	—	—	—
C. m. denti	—	—	—	2	—	—	—	—	—	—	—
De Brazza's monkey *Cercopithecus neglectus*	—	1	1	—	1	—	1	—	—	—	—
Red-tailed monkey *Cercopithecus ascanius schmidti*	—	8	11	25	12	11	5	5	2	1	1
Abyssinian colobus *Colobus guereza (abyssinicus) ituricus*	41	26	24	42	36	48	11	2	2	2	—

After Lumsden (1951).

[a] Solitary.

Most of the macaques (*Macaca*) are semi-terrestrial and their troop life is typified by the Japanese species.

Discovered originally by Itani (1952, 1961) and confirmed by many subsequent workers in Japan, the Japanese macaque *Macaca fuscata* lives in troops of 10–120 head and forages for leaves and other food along fixed routes (the troop size of 170 at the beginning of the study at Takasakiyama is exceptional). The foraging range is fairly well defined, depending on the troop, and forms a home range (Itani & Tokuda, 1954). If there are two or more troops having contiguous ranges territorial defence may be seen but the boundary is not always fixed. As a rule a troop contains more than one adult male and about twice as many adult females, and immature young. The troop is not divided into subgroups or family groups but is differentiated functionally. Itani (1954, 1961) distinguished the following categories of status of individuals in the troop (Fig. 6.10):

> *Leaders* (bosu) – they are positioned in the centre of the troop and lead the troop in movement; the position is usually occupied by one or more males but rarely joined by an adult female (in artificially constituted troops); there is a rigid linear hierarchy among them.

Fig. 6.10. A diagramatic representation of troop structure in the Japanese macaque *Macaca fuscata*. Left, Takasakiyama troop (after Itani, 1954, 1961); right, Kōshima troop (after Tokuda, 1956). Arrows indicate directions of movement between different classes; the numerals in the left figure indicate the numbers of individuals and those of the right figure the individual numbers (all members are shown).

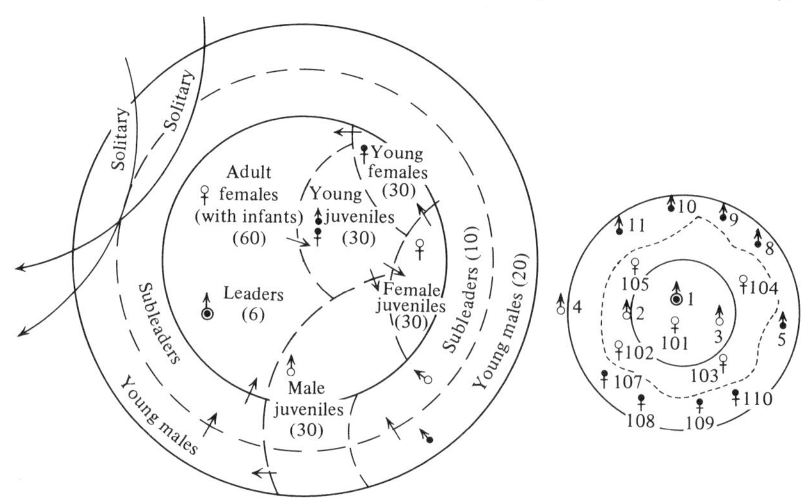

Subleaders (bosu-minarai) – they are adult males not admitted to the centre of the troop by the leaders; there is a social hierarchy among the individuals of this category.

Young males (wakamono) – they are immature males (2–3 years old) and found in the periphery of the troop standing on guard. Social hierarchy exists.

Adult females (otona-mesu) – they live with the leaders in the centre. Social hierarchy exists but it is not so clear-cut as in males.

Young females (musume) – they are immature females (2–3 years old) positioned away from the centre; social hierarchy is obscure.

Young juveniles (kodomo, 'children') – they consist of 'boys' and 'girls' of one- or two-year-old juveniles independent of their mothers; they are permitted to stay close to the centre but 'boys' gradually move to the periphery of the troop and eventually join the young males; they learn the culture of the troop through play in which they are often engaged; social hierarchy develops in this period.

Infants (akambo) – they are less than a year old and dependent on mother in the centre of the troop; there is no hierarchy among them.

There are *solitary males* (hitorizaru) outside the organization of troops.

The composition of relatively well-known natural troops is shown in Table 6.9. Of the troops analysed one at Takasakiyama in 1958 is unnatural in that population increase has begun as a result of 'provisionization'. The socionomic sex-ratio of the nine remaining troops combined is 1 to 2.5 (94 miles : 253 females). Since then, many natural troops, including one in Shimokita Peninsula of Aomori Perfecture at the northern limit of the distribution of monkeys in the world (Izawa & Nishida, 1963; Suzuki, 1965), have been investigated and the results showed much the same social structure. That is, in the troop of Japanese macaques the leaders, females and juveniles are found in the centre, and young males with a social hierarchy guard the troop at the periphery. The differentiation of the status within a troop is accomplished by the difference of social positions between different status-groups and a linear social hierarchy within each status-group. The hierarchial relations are revealed by feeding a 'provisionized' troop or by observing the behaviour during movement – low-ranking individuals avoid high-ranking individuals. Some of the functions served by a social hierarchy may be observed in the field. When a troop is foraging young males would stand

Table 6.9. *Group composition of wild Japanese macaques*

Locality	Male				Female			Juvenile					
	Leaders	Sub-leaders	Young males	Sub-total	Mature females	Young females	Sub-total	♂		♀	In-fants	Sol-itary ♂	Total
Takasakiyama 1953	6	10	20	36	60	30	90	30		30	30	2	*ca* 218
Takasakiyama 1958	4	10	*ca* 90	*ca* 104	150	*ca* 30	*ca* 180	*ca* 100		*ca* 100	86	20	*ca* 590
Kôshima 1954	2	2	3	7	7	4	11	4		4	5	2	33
Taishakukyô 1956	1	1[a]	3	5			10	5		9	6	4	39
Takahashi 1956	4	5[b]	6	15			*ca* 35		45		*ca* 25	0	*ca* 120
Shôdoshima (K)[c] 1956				19			41					(9)	
Shôdoshima (T) 1956				11			15		20		5	0	51
Shôdoshima (O) 1956				10			18		14		8	0	50
Tsubaki 1956[d]	2	1	0	3	4	3	7	3		1	2	1	17
Arashiyama 1956	1	16[e]	2	19	13	3	16		10		7	1	53
Shirahama 1956	1	7	1[f]	9			10	5		1	5	0	30

Mainly based on Reityôrui Kenkyû Gruup (Primate Research Group) (1957).

[a] An old male considered to have been a leader and called a 'declined leader' by Mizuhara. It is put in here for convenience.

[b] Reported by the observer as being high-ranking among the young males.

[c] This troop is little differentiated and social hierarchy is also obscure. Nine 'solitary' males formed a male group and accompanied the troop.

[d] Composition of this troop was described in terms of age classes and not by the social position.

[e] Only one was the subleader, the rest formed a male group in the periphery.

[f] Recorded as two years old.

on guard and an individual discovering approaching danger would climb a tree and give alarm calls. As in the red-tailed monkey the alarm call is never given by two individuals at the same time. If the danger remains this function is given to larger individuals in turn. The solitary monkey is usually a male, which has left the troop through the hierarchical struggle regardless of whether or not direct fighting has taken place. The leaders dominate in copulation but the subleaders may take chances, usually for low-ranking females on heat.

Recent studies have elucidated the origin of solitary monkeys and provided a framework of male life history by which to explain the wanderers and male groups. It now seems clear that most, if not all, males leave their natal troops when three to seven years old and become wanderers (Sugiyama, 1976). These troop-leavers (*hanarezaru*) appear to join other troops after a few years of wandering.

Other species of *Macaca* studied, the rhesus monkey *M. mulatta* (Southwick *et al.*, 1965) and the bonnet monkey *M. radiata* (Simonds, 1965), have a social structure similar to that of the Japanese macaque (an intensive study by Simonds provided descriptions of hierarchical relations within troops and behaviour concerned with the hierarchy). Southwick *et al.* investigated over 400 troops of rhesus monkeys in India and found, among other things, that troops near villages consisted of 10–20 whereas those in temple yards and forest consisted of 40–50 monkeys, and that the overall socionomic sex-ratio was about 1 : 2. In India monkeys, including the Hanuman langur which will be mentioned later, are fully protected for religious reasons and they attain very high densities. Thus it is somewhat doubtful if the large troop size was not an artifact. Findings on other species of *Macaca*, e.g. the crab-eating macaque *M. irus* (in Table 6.7), the lion-tailed macaque *M. silenus* and the pig-tailed macaque *M. nemestrina*, indicate that the troops of these species are similar to the medium-sized or small troops of the Japanese macaque in size and structure, and they are multi-male troops except for the very small troops.

Before discussing baboons I shall touch on colobine monkeys which show adaptations to living in the canopy of tropical rain forest. The two species of *Colobus* listed in Table 6.7 have small troops and their basic social unit is probably a single-male troop. Among the arboreal species of langur, the silvered lutong *Presbytis cristatus* and the Nilgiri langur *P. johnii* are listed in Table 6.7. The former forms multi-male troops of size similar to the Japanese macaque's while the latter forms single-male troops.

Table 6.10. *Group composition of two species of leaf monkeys (lutongs) in the canopy of tropical rain forest*

	Presbytis obscurus					*P. melalophos*				
Group	♂	♀	J	I	Total	♂	♀	J	I	Total
1	2	4	2	2	10	2	5	2	0	9
2	2	3	3	1	9	1	3	1	2	7
3	1	3	4	1	9	1	5	0	1	7
4	2	4	1	1	8	1	3	2	0	6
5	1	4	0	1	6	1	3	1	0	5
6	1	3	1	1	6	1	3	0	1	5
7	1	3	0	2	6	1	2	1	1	5
8	2	1	1	0	4	1	3	0	0	4
9	1	4	0	0	5	1	2	1	0	4
10	1	3	0	1	5	1	2	0	0	3
11	1	3	0	1	5					
12	1	2	1	1	5					
13	1	2	1	1	5					
14	1	2	1	1	5					
15	1	2	1	1	5					
16	1	2	0	2	5					
17	1	1	2	1	5					
18	1	3	0	0	4					
19	1	3	0	0	4					
20	1	2	0	1	4					
21	1	1	1	0	3					
22	1	1	1	0	3					
Total	26	57	20	19	122	11	31	8	5	55
Mean	1.2	2.6	0.9	0.9	5.5	1.1	3.1	0.8	0.5	5.5

Based on Koyama (1976).
J = juveniles.
I = infants.

Recently, Koyama (1976) published accounts of troop structure for the dusky lutong *P. obscurus* and the banded leaf monkey *P. melalophos* (Table 6.10). In both species the basic unit is a single-male troop. Occasionally two males were observed in a troop but in such cases the troop showed signs of possible splitting (N. Koyama personal communication, 1978).

In the genus *Presbytis* the Hanuman langur *P. entellus* is well known. This species occurs in a wide range of habitats, i.e. rain forest, deciduous forest and woodland, and maintains high densities. It is semi-terrestrial and its troop structure is rich in variation; many troops are single-male but some are multi-male. Sugiyama (1964) observed thirty-eight troops, of which thirty-one were single-male and seven were multi-male. He found six male groups in addition. On the other hand, the troops observed by Jay (1963, 1965) in other parts of India were multi-male and larger than Sugiyama's.

In the Cercopithecidae six taxa became ground-living (terrestrial). Namely, the patas *Erythrocebus patas*, the hamadryas baboon *Papio hamadryas*, the mandrill *P. sphinx*, the drill *P. leucophaeus*, four species of savannah baboon (the guinea baboon *P. papio*, the doguera baboon *P. doguera*, the yellow baboon *P. cynocephalus* and the chacma baboon *P. ursinus* – these four are very closely related and may not all be proper species; they are collectively referred to as the savannah baboons or merely baboons by Washburn and DeVore) and the gelada baboon *Theropithecus gelada*. Of these the drill and the mandrill are ground-living forest animals and have been said to form multi-male troops of 10–60 in the former and 10–20 in the latter. However, Gartlan (1970) gave troop sizes of 14–179 for the drill and Sabater Pi (1972) recorded a large troop of more than 100 for the mandrill (cited by Itani, 1972).

The patas became ground-living independently of baboons and its basic social unit is a polygynous group, with separate male groups (Hall, 1965). Unlike other open field monkeys, sexual dimorphism of this species is not well developed. Jolly (1972) considers that the patas became adapted to grassland recently and its polygynous system is a relict of social structure from the days of arboreal life of a guenon (see p. 336).

When the first edition of this book was written many people believed that the social system of baboons was a polygynous harem, but I predicted that because they lived in savannah their social structure would resemble that of the Japanese macaque which is semi-terrestrial to terrestrial. This prediction was proved to be correct by the study of Washburn and his colleagues (Washburn & DeVore, 1961: Hall, 1962; DeVore & Hall, 1965). [Rowell (1966) examined the society of baboons in a habitat with more trees than the habitat in which the above work has been conducted, but his result showed the same social structure as others except that the troop size was generally small.] The basic structure of baboon society is that leaders of more than one adult male occupy the centre surrounded by females and juveniles, and low-ranking males and young males are at the periphery (Fig. 6.11). The socionomic sex-ratio is 1 : 2.5. Sexual relations are promiscuous in a broad sense in that the female mates with more than one male and most copulations involve central males though mating with peripheral males is possible. The troop has a fixed home range but conflict between troops is not so intense as in the Japanese macaque. In fact, during the dry season when the distribution of water and grass is limited many individuals congregate around the water hole. If the troop did not tolerate congregation the species would become extinct. The hierarchy within each status-group is linear (peck-right type) like the Japanese macaque. Males often move from troop to troop and young males also occasionally shift affiliation (Rowell, 1966).

The hamadryas baboon living in steep rocky mountains and semi-desert of Ethiopia has a different life. According to the study of Kummer & Kurt (1963) this species congregates in the evening on rocky shelves of cliffs for camping. The camping group is usually composed of dozens of animals but occasionally the number is as large as several hundreds. In the morning, however, the same polygynous groups as in the previous day are formed and each is led by the adult male of the group to a foraging area. The socionomic sex-ratio is about one to three; an example cited by Bourlière (1961) is one to five. According to Kummer (1968), the hamadryas baboon has three tiers of society: the minimum unit is a polygynous group (one-male unit), several of which form a band. A few bands in turn form a troop which allows free exchange of bands with other troops. The large aggregations for camping are troops of this

Fig. 6.11. A moving troop of baboons (redrawn from Hall & DeVore, 1965). The leaders (male sign with a dot) are in the centre of the troop, accompanied by nursing females and juveniles. Young males are found on either side of the troop. Other adult males (male sign) and females are located at the front and rear during procession. Two females on heat (double circle) and adult males are in consort relations.

type. Within a band the membership of one-male units is usually the same. In the daytime the band moves as a unit but within it each one-male unit maintains its independence. Surplus males, about 20% of the males, do not become solitary males but are accepted in bands or troops though not as members of one-male units. Females may copulate with young males outside the one-male unit but form consort relations only with the males of their units. Needless to say, such a male is the leader of the one-male unit.

The gelada baboon, belonging to a different genus, is also a terrestrial baboon adapted to the high mountains of Ethiopia. Its life history has been studied in detail by Crook (1966) and Kawai (1974, 1976, 1979). The two populations investigated in different regions had identical social structures. That is, a two-tier structure of the polygynous group as the basic unit (let us call it a one-male unit following the social unit of the hamadryas baboon) and the herd consisting of tens to hundreds of individuals. The size of the one-male unit is 5–35 head in Crook's study and 6–24 head in Kawai's report. The socionomic sex-ratio for the latter is 1 : 2.6. According to Kawai (1974) the number of individuals in a herd varied from 27 to 350 and one of these was composed of eight one-male units, one male group and three 'free-lancing' males (individuals attached to a herd). The herd is well integrated as a social organization (in this sense it corresponds to the band rather than the troop of hamadryas baboons), but it is an open system, permitting temporary affiliation of one-male units from other herds (Fig. 6.12). There is neither leadership nor territoriality in the herd. Although a large herd is often formed by

Fig. 6.12. A scheme of social structure in the gelada baboon *Theropithecus gelada* (after Kawai, 1974). The units shown in the same pattern have close association.

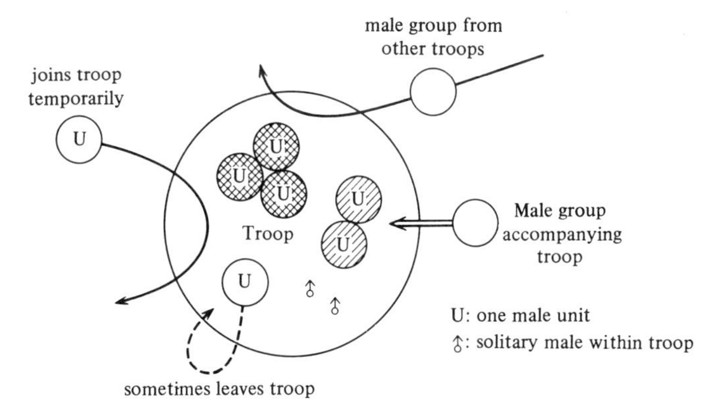

amalgamation its composition is much the same. From these observations Kawai concluded that the basic social unit of the gelada baboon is the one-male unit. In the herd which he studied there were some units with two or three adult males, but in these cases one specific male was always the leader and the second- and third-ranking males had no chance of mating. Thus even in the multi-male units the reproductive unit was still the 'one-male'.

The origins of the two similar social structures in the hamadryas and the gelada seem to be different. Kummer observed that the young male hamadryas abducted a female infant (less than a year old) which had been reared in a one-male unit. Such a young male looks after the infant and adopts her. A year or two later the two animals move together, and Kummer considers that the relation of such two animals develops into one-male unit. As the period of observation is limited no definite answer can be obtained from his work. [Itani (1959) mentions that male parental care (not necessarily by the father; the father is not identified) appears in the Japanese macaque.] In the case of the gelada (probably as in the polygynous groups of spider monkeys and the hanuman langur), young males leave the group while females remain in the group. A male from a male group may take over a one-male unit. Although the leader is a male the group is maternally based.

I thought (Itô, 1966), as Itani (1967) did, that the basic social unit of the hamadryas is a polygynous group. Crook (1966) considered that the single-male group is an adaptation to the environment in which food is scarce and scattered. Itani (1972) claimed from the above fact that (1) the basic social organization of the hamadryas is not the one-male unit but probably the band in which infant abduction occurs (so assumed by Itani) and (2) the view of Crook in treating the polygynous system of the hamadryas and the gelada as analogous is mistaken. I would rather pay attention to the remarkable similarity of the social organization found in the two species in spite of their different origins. Their polygynous groups are, in fact, the basic social unit and in both species emerged as a form of adaptation to the low-productivity zone of semi-desert.

(4) Apart from preliminary reports of Nissen (1931) on chimpanzees and of Bingham (1932) on gorillas, the first species to be studied among the apes was the white-handed gibbon *Hylobates lar*, a member of the group best adapted to living in the canopy of tropical rain forest among the primates. Carpenter (1940) reported that this species had monogamous family life and caused a sensation in Christian countries where this finding was taken as a proof that man had been monogamous from the

beginning. As shown in Table 6.11, 20 out of 21 groups observed were monogamous. Koyama (1976) also found all seven groups observed to be monogamous. Each group – family – has a territory in the canopy of the forest and defends it by actual fighting and vocalization (Fig. 6.13). There is no difference in the hierarchical position between sexes. That there is hardly any secondary sexual character in anatomical features and that there is no definite breeding season seem to be the basis for the monogamous system in this species (Carpenter, 1952). One aberrant group which Carpenter found had one very old male completely suppressed by a young male. There are solitary animals in both sexes. According to Koyama (1976) and Ellefson (cited by Jolly, 1972), the

Table 6.11 *Group composition of two species of gibbons*

	Hylobates lar					*H. syndactylus*				
Group	♂	♀	J	I	Total	♂	♀	J	I	Total
1	1	1	2	1	5	1	1	1	1	4
2	1	1	2		4	1	1	1	1	4
3	1	1			2	1	1	1	1	4
4	1	1	2		4	1	1	0	1	3
5	1	1	3	1	6	1	1	2	0	4
6	1	1	2		4	1	1	2	0	4
7	1	1	3	1	6	1	1	1	0	3
8	1	1	2	1	5	1	1	1	0	3
9	1	1	3		5	1	1	0	0	2
10	1+1	1	2		5	1	1	1	0	3
11	1	1	3	1	6	1	1	0	1	3
12	1	1	2		4	1	1	0	0	2
13	1	1	2	1	5	1	1	1	0	3
14	1	1	3	1	6	1	1	0	0	2
15	1	1	1	1	4	1	1	0	0	2
16	1	1	1	1	4	1	1	0	0	2
17	1	1	2	1	5	1	1	0	1	3
18	1	1	1		3	1	1	0	1	3
19	1	1	1	1	4	1	1	1	0	3
20	1	1			2	1	1	0	1	3
21	1	1	2		4	1	1	1	0	3
22						1	1	0	0	2
Total	22	21	39	11	93	22	22	13	8	65
Mean	1.0	1	1.9	0.5	4.4	1	1	0.6	0.4	3.0

Based on Carpenter (1940) for *Hylobates lar* and on Koyama (1976) for *H. syndactylus*.
J = juvenile.
I = infants.

siamang *Hylobates syndactylus* is also monogamous and according to Ellefson the agile gibbon *H. agilis* is the same.

Carpenter (1938) suggested from his single observation in Sumatra that the orangutan *Pongo pygmaeus* might be bigynous. Schaller's (1961) observations after the war showed, however, that both monogamous and bigynous associations occur as well as solitary males and females. Schaller thought that the orangutan was basically solitary and does not form a specific group except the mother–infant bond. According to the recent investigations of Rodman (1973) and MacKinnon (1974), this species is solitary and the male has a large territory which is occupied by more than one female.

The gorillas *Pan gorilla* are divided into the lowland gorilla *P. g. gorilla* living in the forest zone of West Africa and the mountain gorilla *P. g. beringei* living in the mountains almost to the timber line in Zaire, Uganda, Kenya, etc. In the past it was the mountain gorilla that received the attention of field workers. Today, however, much intensive work is in progress on the lowland form whose life history has not been known for a long time.

Fig. 6.13. Maps showing overlaps of home ranges in the white-tailed gibbon *Hylobates lar* and in the mountain gorilla *Pan gorilla beringei* (from J. O. Ellefson & G. B. Schaller in Jolly, 1972). Only little overlap of home ranges is seen in the gibbon. Hatched areas are considered to form territories.

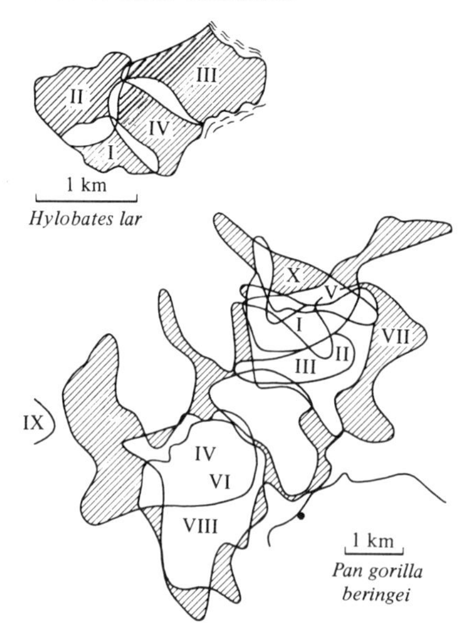

The gorilla has unique ontogenic characteristics. The period of infancy lasts up to three years during which time the infant is in close contact with its mother. Then the juvenile period lasts about three years (three to six years of age) in which the young have little contact with their mother. The female then matures sexually but the male is called the 'black-back' (the back is black) between the ages of six and ten and, as far as is known, does not perform sexual activities. In the male older than ten years the colour on the back changes to silvery white and only this 'silver-back' is the true adult male.

The socio-ecological study of gorillas was initiated by Japanese. The four groups that Kawai & Mizuhara (1962) observed in Virunga Volcanoes of Uganda in 1959 consisted of 5–18 head, only the largest containing two silver-backs. There were 2–4 adult females per group Kawai & Mizuhara considered that the polygynous group is the basic social unit in gorillas. The silver-back was the leader of each group and in the group with two silver-backs the younger one often left the group.

Schaller (1963) investigated ten groups in Kabara which were led by silver-backs. As can be seen in Table 6.12, seven out of ten groups in Kabara were led by one silver-back. Itani (1972) presented a table of group composition for twenty groups whose memberships were completely known. Of these, seventeen were single-male groups consisting of

Table 6.12. *Group composition of the mountain gorilla* Pan gorilla beringei *in Kabara, Uganda*

Group	♂	♂′	♀	J	I	Total
I	1		3	2	2	8
II	1	3	6	5	4	19
III	1		2	1	1	5
IV	4	1	10	3	6	24
V	2	2	3	2	2	11
VI	1	1	9	2	7	20
VII	1	2	6	4	5	18
VIII	1	2	8	3	7	21
IX	4	3	9	5	6	27
X	1	1	6	2	6	16
Total	17	15	62	29	46	169
Mean	1.7	1.5	6.2	2.9	4.6	16.9

From Schaller (1963).
♂, silverbacks; ♀, females; ♂′, blackbacks; J, juveniles and young; I, infants.

four monogamous groups and the rest with two to ten adult females, about half of which contained one to three black-backs. Thus the basic social unit of gorillas seems to be the polygynous group. Schaller also reported that there was no group territory and home ranges overlapped (Fig. 6.13); in the case of groups with more than one male, social rank existed between males but the low-ranking animal was not chased – there was even a case where the leader did not interfere with the copulation of a low-ranking silver-back; and addition of new members to a group, exchange of members between groups and separation from a group occurred frequently.

Recent studies by Fossey (1974) and Harcourt *et al.* (1976) show that the basic social unit of gorillas is a strict polygynous group. The male matures at the age of eleven and leaves the group then to become a solitary male. The female matures at the age of eight but leaves the group when she is about six and forms a new group with a solitary male. The solitary male thus gathers young females from different groups and forms a polygynous group. The female may shift affiliation from group to group. Both Fossey (1974) and Elliott (1976) confirmed that gorillas do not defend group territories but there is often a rival relation between groups. They also observed rare fightings between the leader of a group and a solitary male.

In effect, the society of gorillas is based on the polygynous family as a unit but its structure is fairly fluid. There may be neighbourhood relations among the groups of one region.

Finally the chimpanzee *Pan troglodytes* remains to be discussed. The social organization of the chimpanzee is unique among all the primates, for it is so flexible that it cannot be specified in terms of social morphology, such as monogamous, polygynous, Japanese macaque-type and so on (Sugiyama, 1969). There are recent reviews of Japanese studies on the subject (Sugiyama, 1973; Itani, 1977).

Kortlandt (1962) observed chimpanzees of rain forest and found two types of group: the nursing group composed of mothers and infants, and the sexual group comprising more than twenty members of both sexes (more than one adult male involved). According to the circumstances the same animal may become a member of either group.

Azuma & Toyoshima (1963) saw four types of party: (1) the mother–infant party, (2) the party of immature young, (3) the polygynous party, and (4) the large mixed-sex party, containing more than one adult male. However, the observations made by Reynolds & Reynolds (1965) in the Budongo forest in Uganda did not suggest any closed social unit in the party of chimpanzees, for the party constantly divided, amalgamated and replaced its members. Their field notes were filled with descriptions of peaceful sexual relations not found in the Japanese macaque; for example, a young male copulating in front of the leader and two adult males grooming the same female. These observations showed the existence of promiscuity at Budongo. Goodall (1965), who observed chimpanzees in the riverine forest of Gombe Stream to the south of Budongo, also found promiscuity; there are four cases in which the same female copulated with more than one male, in one of these cases one female was mounted by four males one after another. There was no closed party at the Gombe Stream. In this area a total of 498 temporary associations (same individuals recorded more than once) have been observed and the largest party consisted of twenty-three head, three cases of more than twenty and forty-four cases of more than ten. Solitary individuals were sighted sixty-four times, some of them females.

In spite of such loose associations there appears to be a dominance hierarchy among the adult males and male leaders may exist within a party. This was suggested by the movement pattern of a party observed by Itani & Suzuki (1967) in a defoliated forest of Tanzania (Fig. 6.14). This arrangement of procession is such that females with infants are in front, followed by adult males and then females with infants again at the rear. The probability of having such a combination by chance is

Fig. 6.14. Procession of a large party of the chimpanzee *Pan troglodytes* observed in dry, raingreen forest of Firabanga, Tanzania (after Itani, 1966). The party is headed by a female (1) with an infant (2).

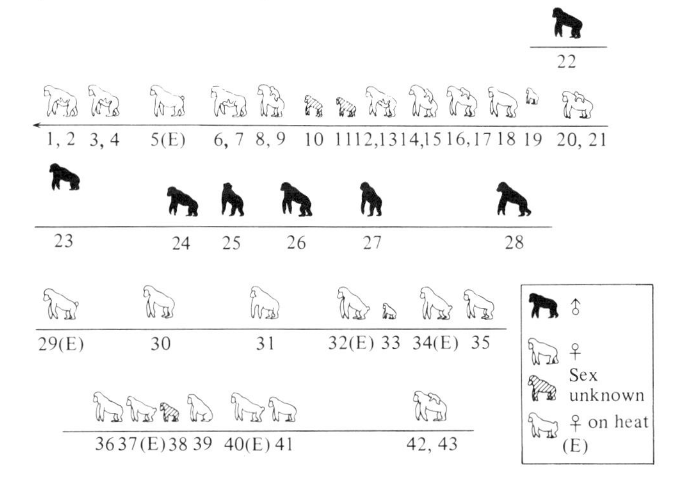

almost nil. This observation thus suggests some functional differentiation within the party. Earlier, Itani and Azuma recorded the passage of a party of twenty-seven head at another locality (Itani & Suzuki, 1967; Suzuki, 1969). The arrangement of this party was similar to one described in Fig. 6.14. Izawa's (1977) example also showed that the males were in the centre of the moving party.

Itani (1972) summarized the composition of seven parties investigated by the Japanese primate research group (Table 6.13). All parties listed here are multi-male groups; the socionomic sex-ratio being 1 : 1.6. Nishida (1968) observed movements of females between two neighbouring groups (under baited conditions). In spite of this, each group on the whole maintained approximately the same composition and some vocalizations suggested territorial display between groups. From these observations Itani (1972) modified his earlier view (Itani, 1966) that chimpanzees might have unit groups within a large party, and concluded that the basic unit of chimpanzee society is neither paternal nor maternal but is a group like a highly fluid commune. There is no fixed family relation within it and sexual relations are promiscuous. The only thing is that the males when they mature leave the group to 'observe a taboo against incest'. Because groups within a region interchange members and even mothers change their domicile, Itani considers that young males may travel very far beyond the boundary of the regional population. According to A. Suzuki (1975, 1976), in the forest habitat the proportions of mother–infant groups and male groups were high in the chimpanzee groups encountered whereas in savannah large groups containing both sexes and young were observed more often than not. Suzuki's interpretation seems to be that the latter group corresponds to Itani's commune-like group (in which males become leaders) and that the chimpanzees can dissolve or organize this group depending on circumstances. Recent findings on the pygmy chimpanzee *Pan paniscus* show that social structure of this species is also fluid. Kano (1979) considers that its basis lies in the constant receptivity of females.

(5) Finally in the discussion of primates let us examine how the different social systems (morphologies) have emerged in the primates. In the previous sections I surmised that for the ungulates and the carnivores the forest habitat caused the evolution of monogamous family and solitary life while the grassland caused the evolution of group life. The factor responsible for this difference is the mode of response to the procurability of food by the young and to the danger of predation. Operating behind

Table 6.13. *Structure of unit groups of the chimpanzee*

	Adult males	Young males	Adult females	Juveniles	Infants	Unknown	Total	Observer
Firabanga, Tanzania	6	1	21	5	9	1	43	Itani & Suzuki (1967)
Kabogo, Tanzania	5	0	13	9	4	α	31 + (α)	Itani & Azuma[a]
Kasoge, Tanzania	6	1	12	2	8		29	Nishida (1968)
Ishanda, Tanzania	4	0	11	2	6		23	Itani & Kano[a]
Kasakati (Z), Tanzania	9	4	20	6–7	7		46–47	Izawa (1970)
Budongo (A), Uganda	18	3	16	11	8		56	Sugiyama (1968)
Budongo (C), Uganda	8	1	14	2	7		32	Sugiyama (1968)
Total	56	10	107	37–38	49		260 +	
Mean	8.0	1.4	15.3	5.3	7.0		37 +	

From Itani (1972).

[a]Described in Itani & Suzuki (1967). α = a small unknown number.

this factor are the discontinuity of the forest evironment and the continuity of grassland. I argued in the first edition that 'the primates are no exception to this rule' and subsequently repeated this view with additional information (Itô, 1966, 1970). Kawai (1972, p. 33) and Itani (1972, p. 70) among the Japanese primatologists disagreed by pointing out that arboreal species, such as the colobus, red-tailed monkey and some South American species, are known to form troops of more than 100 individuals.

Other countries lagged behind in the comparative examination of primate social structure. DeVore (1963) may have been the first to put forward a general hypothesis. He proposed that in the anthropoids the arboreal species formed small groups and the terrestrial species formed large groups regardless of phylogenetic differences. At that time, however, the social structure was known only for a handful of species. Using the data accumulated in a short subsequent period, Crook & Gartlan (1966) produced a grading of primate societies, which has become a standard reference (see p. 246 for the argument used for birds by Crook, 1964). Their table is reproduced here in an abridged form (Table 6.14). Among many factors involved Crook & Gartlan emphasized the habitat and considered that the social structure reflected different levels of adaptation to forest, savannah with many trees, grassland and arid environments. In response to changes in the habitat the social structure changes from solitary to multi-male groups with status differentiation. This is considered to accompany other changes, e.g. nocturnal to diurnal, insectivorous (utilizing scattered food sources) through frugivorous to graminivorous.

To explain why insectivores and frugivores remain in small groups, they propose that the food supply is limited and scattered in the tropical rain forest. The conditions in savannah and semi-desert where the Grade IV and V animals live are different; apart from the region in which the Japanese macaque experiences cold winter, the rainy season and the dry season alternate, providing an abundant supply of continuously distributed food in one season and causing high mortality in the other. This condition and the fact that the risk of predation in savannah is greater than in the forest habitat made the animals form large groups and selected for sexual dimorphism. The Grade IV primates except the chimpanzee form multi-male groups with status differentiation.

Why should the patas, the hamadryas and the gelada, which are entirely terrestrial and living in the most open habitat, form polygynous groups? Crook & Gartlan consider that it would be advantageous to

Table 6.14. *Grades of social organization among primates*

Ecological and behavioural characteristics	Grade I	Grade II	Grade III	Grade IV	Grade V
Habitat	Forest	Forest	Forest–Forest fringe	Forest fringe, tree savannah	Grassland or arid savannah
Diet	Mostly insects	Fruit or leaves	Fruit or fruit and leaves. Stems, etc.	Vegetarian–omnivore. Occasionally carnivorous in *Papio* and *Pan*	Vegetarian–omnivore *Papio hamadryas* occasionally also carnivorous
Activity	Nocturnal	Crepuscular or diurnal	Diurnal	Diurnal	Diurnal
Size of group	Usually solitary	Very small groups	Small to occasionally large parties	Medium to large groups. *Pan* groups inconstant in size	Medium to large groups, variable size in *Theropithecus gelada* and probably *P. hamadryas*
Reproductive unit	Pairs where known	Small family parties based on single male	Multi-male groups	Multi-male groups	One-male groups
Sex dimorphism and social role differentiation	Slight	Slight	Slight. Size and behavioural dimorphism marked in *Pan gorilla*. Colour contrasts in *Lemur*.	Marked dimorphism and role differentiation in *Papio* and *Macaca*	Marked dimorphism Social role differentiation
Population dispersion	Limited information suggests territories	Territories with display, marking, etc.	Territories known in *Alouatta*, *Lemur*. Home ranges in *Pan gorilla* with some group avoidance probable	Territories with display in *Cercopithecus aethiops*. Home ranges with avoidance or group combat in others. Extensive group mixing in *Pan*	Home ranges in *Erythrocebus patas*, *P. hamadryas* and *T. gelada*. Show much congregation in feeding and sleeping. *T. gelada* in poor feeding conditions shows group dispersal

From Crook & Gartlan (1966).

forage in small groups in an environment in which the food supply is extremely short. Evidence for this is that the gelada formed a large group in areas of abundant grass but foraged scattered in polygynous groups in barren areas. [According to Kawai (1974) the gelada spend 70% of daylight eating grass, compared with 37% spent by the red colobus picking and eating leaves in the forest.] The hamadryas and the gelada form large aggregations on rock shelves when they camp for the night. The shortage of suitable footholds may have caused this large aggregation. The patas does not form a large group. Instead, they sleep individually in trees at night.

Jolly (1972) generally followed the idea of Crook & Gartlan, but she emphasized the difference of feeding habits in grading the primates: (A) nocturnal; (B) diurnal arboreal and herbivorous; (C) diurnal arboreal and omnivorous; (D) partly terrestrial and herbivorous; (E) partly terrestrial and omnivorous; (F) arid-zone inhabiting. She summarized that (A) is mainly solitary, (B) forms small multi-male groups, (C) forms monogamous to small multi-male groups, (D) forms polygynous or small multi-male groups (the gorilla and Hanuman langur), (E) forms small or large multi-male groups (l'Hoest's guenon is polygynous), and (F) forms large groups at night but polygynous groups by day. Here too, the degree of terrestrial tendency is considered important as a factor. Jolly thought that it would be advantageous to feed singly when food was dispersed in small patches and to feed in groups when food occurred in a large mass or spread evenly. The forest is the former habitat in which hiding from predators was also facilitated by dispersed living. Struhsaker (1969), on the other hand, thought that it would be impossible to predict the size of the group simply from the arboreal or terrestrial habit of the species (he grouped the savannah baboons and the hamadryas as terrestrial) and that we should place emphasis on the phylogenetic relations (for example, all members of *Cercopithecus* tend to form single-male groups).

The same problem was discussed by Eisenberg *et al.* (1972). They pointed out, as Itani and others did, that some arboreal species formed large groups, and emphasized the importance of phylogeny. They consider that some phylogenetic groups have a tendency to form large groups while others have a tendency to form small groups and that differences due to environment appear within this restriction. They also thought that in the arboreal primates the basic unit was originally a single-male group but with increase in density they would become multi-male groups, which would split into single-male groups if the density increased further. In spite of this argument they accept that in the medium-sized primates the co-existence of many males is an adaptation to terrestrial life whereas the single-male group is an adaptation to arboreal life.

Returning to the criticism of Itani and others, I agree that the phylogenetic origin necessarily influences social behaviour – this was also pointed out by Eisenberg *et al.* The genus *Macaca* has a tendency to form multi-male groups with status differentiation whereas *Presbytis* remains polygynous in spite of its strong terrestrial tendency. Nevertheless, the largest group of macaques is formed by the species or population that has contact with human settlement or other open environments. Also, the dusky lutong and Nilgiri langur, arboreal species, form monogamous groups, whereas the Hanuman langur, a semi-terrestrial species, is polygynous and occasionally forms large groups.

According to H. Obara (personal communication, 1978) the phylogenetically older types of mammals (e.g. tapir, rhinoceros) tend to be solitary, but the rhinoceros, being a grassland species, forms a herd though small. Even within the strongly solitary family of Felidae the lion living in open country forms a pride. Reptiles probably did not have large groups even at the height of their era, but it is interesting to note that the gigantic brontosaurs have recently been estimated to have lived terrestrial life in groups in the open habitat instead of semi-aquatic life as has been thought (Bakker, 1971). Even Itani accepts that with some exceptions the group size of open field species is generally greater than that of forest species.

The tendencies that we have seen in birds and mammals other than primates are associations of arboreal species with family life and terrestrial species with group life. The obstacle for the application of this rule to the primates was the large group found in several arboreal monkeys. Here we must recall the characteristics of the tropical rain forest which I have mentioned repeatedly in this book. The tropical rain forest, though resembling the temperate and subtropical forests in its three dimensional structure, is entirely different from these forests in the diversity of species. Here trees belonging to the same species are scattered and at one time their fruit may ripen in large quantity at one place while at another time fruiting may occur in another species at another place. The same applies to the sprouting of tender leaves. In the tropical rain forest fruit-eating birds also formed large mixed aggregations. Here it seems as though tolerance was necessary for life in the canopy not only among different families within the same species but also among different

Table 6.15. *Distribution of primate species based on Crook & Gartlan's grades and types of social organization*

	Solitary	1♂+1♀	1♂ +♀♀	♂♂+♀♀ Mean 4–30	♂♂+♀♀ Mean 31–150
Type I	Tarsiers[a]				
Forest–	Loris[a]				
Nocturnal	Aye-aye[a]				
	Potto[a]				
	Demidoff's galago[a] ↔	Demidoff's galago[a] ↔	(Demidoff's galago?)[a]		
	Dwarf lemurs[a]				
	Night monkey[b] ↔	Night monkey[b]			
		Avahi[a]			
Type II		Indri[a]			
Forest–		White sifaka[a] ↔	(White sifaka?)[a]		
Crepuscular		Grey gentle lemur[a]			
Type III				Black lemur[a]	
Forest–		Tamarins[b] ↔	Tamarins[b]		
Diurnal			Mantled howler[b] ↔	Mantled howler[b]...*	
			Black-handed spider m[b]. ↔	Black-handed spider m[b]...*	
		Titis[b]			
				Humboldt's woolly monkey[b] ↔	
			Four spp. of guenons[c]...*		
			Red-tailed monkey[c] ↔	Red-tailed monkey[c]...*	
			Colobus[c] ↔	Colobus[c]	
			Nilgiri langur[c]		
				Dusky lutong[c]	
		3 spp. of gibbons[c]			
	Orangutan[c] ↔	Orangutan[c]			

Type IV Forest fringe–semi-terrestrial			Ring-tailed lemur[a]	
			Vervet monkey[c]	
			White-collared mangabey[c]	
			Several spp. of macaques[c]	
			Japanese macaque[c] ↔	Japanese macaque[c]
			Rhesus monkey[c] ↔	Rhesus monkey[c]
		Hanuman langur[c] ↔	Hanuman langur[c]	
		Mountain gorilla[c] ↔	Mountain gorilla[c]	
	←	←	Chimpanzee[c] →	→
Type V Grassland, mountain–terrestrial				Three species of baboons[c]
		Patas[c] ↔		(Patas)[c]
		Gelada baboon[c] ↔		(Gelada baboon)[c]
		Hamadryas baboon[c] ↔		(Hamadryas baboon)[c]

[a] Prosimians.
[b] New World monkeys.
[c] Old World monkeys and apes.
* Temporary feeding groups.
( ) Night aggregations.

species, so much so that in monkeys hybrids between species are often found in the canopy (Itani, 1972, pp. 76–78). Therefore I would maintain that large groups of more than 100 individuals per group seen in the spider monkeys, squirrel monkey, red-tailed monkey and talapoin are temporary in nature and their basic unit is the size of spider monkey's subgroup or red-tailed monkey's camping group. If re-examined in this light Grade III in Table 6.15 will contain only a few species with multi-male groups as a basic unit and it becomes clearly different from Grade IV. In the canopy zone of the tropical rain forest primates developed two parallel structures of large groups and small groups, and the basis for the existence of large groups is the temporary concentration of abundant food supply. I consider that the apex of canopy adaptation is occupied by the monogamous family of gibbons and the solitary life of the orangutan.

Why should the patas, hamadryas and gelada form polygynous groups in terrestrial life? My view for the last two species is the same as that of Crook & Gartlan, that it is an adaptation to the habitat of semi-desert mountains where food is very difficult to obtain (the montane species had small groups in ungulates or in grouse – recall the polygyny of the vicuña). There is still a danger of predation (probably far more predators lived in Ethiopia in the past), which they avoid by camping on rock-shelves. Shortages of suitable rock-shelves and dispersion of food – with different origins as mentioned earlier – probably caused the development of dual structures (large groups and small groups) in the two species. Might it be the increased size of males, developed originally for defence, that kept the polygynous group as the basic unit of their groups?

As for the patas there is still much to learn but Jolly says that this species may have become terrestrial recently from a forest cercopithecid which lived originally in small groups. It has not abandoned trees completely, judging from the use of trees such as for sleeping. Yet this species is said to be able to run faster than any other species of primate on the ground (Hall, 1964). This characteristic may have made it possible for it to survive in the savannah without large groups or sexual dimorphism.

(6) Finally, let us consider the origin of man, limiting our discussion to only the ecological aspect. Washburn (1950) proposed that for the emergence of man two stages of locomotor adaptation were necessary. The first was the evolution from an insectivore-type ancestor to an arboreal primate during the early Tertiary period. This change has been accompanied by: (1) A conversion from olfaction-centred life to vision-

centred life. With this they acquired an enlarged brain and the ability to see in three dimensions. This is an advance from the two-dimensional life to the three-dimensional life (Itô, 1966). (2) The use of arms in locomotion among the trees led to the evolution of hands and arms which, along with accompanying changes in the pelvis, were preadapted to upright living. (3) Vocal communication developed in arboreal life and the linguistic ability of the future was founded. The second stage was the re-invasion of the ground from the mid- to late-Tertiary period. As a result of this change the arms were relieved of their locomotive function, and the production and use of tools could begin. At the same time the arboreal life meant the adoption of the low-fecundity/parental-protection strategy and this produced a long period of parental care (and thus an increase in the importance of learning behaviour), leading to the conspicuous neoteny of man and strengthening the family bond. It would have been impossible to produce the family system of man without the series of adaptations to arboreal life in the past.

Population pressure (food problems) and other factors caused attempted re-invasion of the ground by primates several times in the past (Fig. 6.15). The baboons and others that came down to the ground too early returned to quadropedal locomotion. On the other hand, the gibbons are too specialized for arboreal life (well adapted in brachiation, using only the arms to move from branch to branch, and the thumb has degenerated) to become ground-living again.

The occupation of the ground is an advance into a continuous environment, and this is associated with adaptation to the high-fecundity strategy. However, the primates did not take such an option. Instead, they formed large groups in response to the existence of a spatially continuous and abundant supply of food and to the danger of predation. Increasing sexual dimorphism – i.e. strengthening of the defence function in the male – is seen but the family-bond which would have existed during the arboreal life has been lost in the macaque and chimpanzee. It is interesting that among them the hamadryas and the gelada, of different origin, have kept the two-tier social structure of the polygynous group and the large group – the proposition of Espinas (p. 323) is maintained on integration ('*aufgehoben*').

However, even more interesting is the extremely fluid society of chimpanzees. Whereas gorillas are catholic in their herbivorous habit and live a quiet life in small groups, chimpanzees have a very flexible life, sometimes living in monogamous, sometimes polygynous, sometimes unisexual and sometimes large groups. It is not only these associations

Fig. 6.15. A model of evolutionary pathways among primates showing two types of ocomotor adaptation (after Itô 1966).

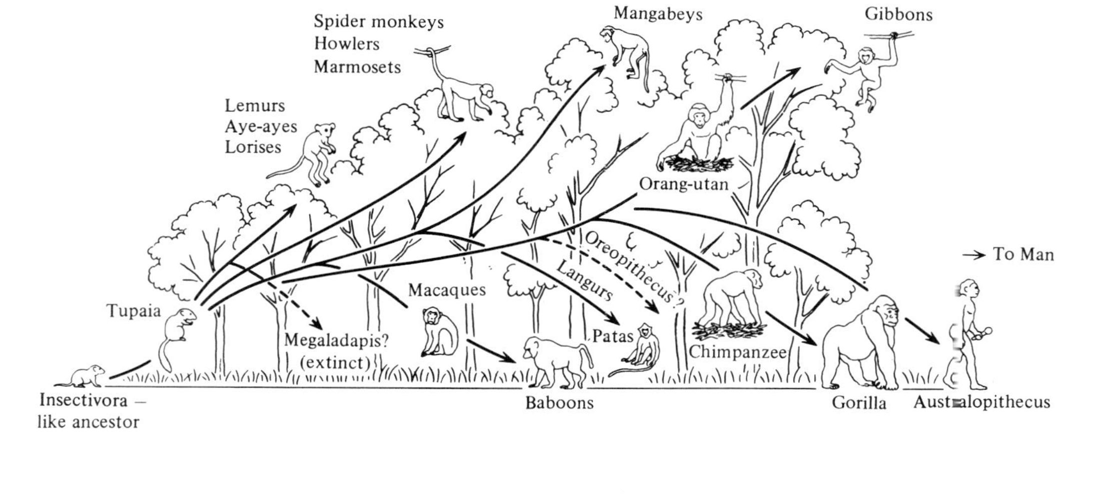

that vary so much; the mode of life is also diverse and various resources are exploited with the use of tools, e.g. 'fishing' for ants and termites (Goodall, 1963; Suzuki, 1966), and hunting other monkeys and antelopes (Goodall, 1965; Kawabe, 1966; Suzuki, 1971). Such unique flexibility of the mode of life might have been necessary for the evolution of mankind.

Comparing the gorilla and the chimpanzee, living in much the same environment, this flexibility seems to have existed as a species-specific life-style during the arboreal life. Kawai (1972) thinks that the carnivorous habit had already started when they came down from the trees. It is difficult to determine its origin but I would rather think that it was the invasion of the savannah that opened up the possibility for this new mode of life, for the only other species with carnivorous habits among the higher primates are the savannah baboons and the hamadryas. However, there is no reason to believe that the carnivorous habit of baboons developed in the same process as in the chimpanzee. In the baboons, perhaps the defence function of large males with sharp canine teeth has been converted to the predatory habit. The carnivorous habit of chimpanzees in the forest has been observed many times by Suzuki (1977) and this is the basis of Kawai's idea. Of course the preadaptation would have existed for this habit. However, how much survival value it had would be a different problem. My view is that it gave a large survival value for the first time in the savannah. Thus Crook & Gartlan (1966) and Jolly (1972) thought that the early human society, facing seasonal shortages of food in the savannah, at first had a social structure similar to that of the chimpanzee and then developed into a polygynous group or a multi-male group guarded by males with weapons.

When we consider the chimpanzee-type society, the question remains of how the family of man has come about from such a fluid and promiscuous group (as with Itani and others I would take a loosely organized group of a few dozen individuals as the basic unit). There is no human society, in all primitive races studied, that lacks a monogamous or polygynous matrimonial relation. The possibility that the social structure of *Australopithecus* was of the gelada-type cannot be denied. The question of whether or not it was possible to establish a polygynous group from the beginning of savannah life when early man had little sexual dimorphism must await future investigation.

Of the carnivorous habit, F. Engels wrote in 1876 in his book *Anteil der Arbeit an der Menschwerdung des Affen* that, along with the effect of a nutritious meat diet on the development of the brain and its possible consequences in the use of fire and domestication of animals, the habit of meat-eating shortened the time required for various vegetative processes of the body corresponding to the plant life (*Pflanzenleben*) and, as a result, more time and material and desire were given to the performance of the original animal life [*Tierische* (*Animalische*) *leben*]. We have seen the distribution of captured prey among the members of a group in wolves and lycaons, but such a habit has never been recorded for the Japanese macaque or the gorilla. Baboons eat meat but when they kill a large prey, such as the Thompson gazelle, the dominant male takes the best part (e.g. brain) and the low-ranking individuals eat the remainder (Eimerl & DeVore, 1965). In the chimpanzee, distribution of the kill has been confirmed. Goodall (1965) reported that a male carrying a large antelope tore a large lump of meat and gave it to a begging male who in turn tore a small piece from his meat and gave it to another male. This distribution of food has been confirmed a number of times by Japanese workers. Cooperative hunting has also been seen (Nishida, 1975; Suzuki, 1977; a review on meat eating of chimpanzees by G. Teleki, 1975, is cited by A. Suzuki, 1976). Other examples of food distribution by primates have been recorded for the brown-headed tamarin *Saguinus fascicollis* of the New World monkeys. They catch large grasshoppers (which are much bigger than the small palm of the tamarin) and give them to begging young (old enough to catch insects themselves) (Izawa, 1976b).

So long as the hominoid ape remained herbivorous in the savannah there would be no circumstance for the advancement and dissemination of food distribution as a culture. They would be better off eating on the spot rather than collecting and distributing the buds and fruit of trees which had low calorific value and were found everywhere (even in the best conditions of rain forest vegetarians would have to spend 40% of the daytime eating, see. p. 360). It is when they began eating high-calorie meat of large size, corresponding to a significant portion of their own weight, that 'distribution' as a basis of social relations appeared. Besides, meat-eating (though started during the arboreal life) became strongly advantageous in selection only after they had colonized the grassland – where the dry season brought food shortages. [Colonization of the savannah by chimpanzees and (perhaps) early man may be thought as the colonization of an environment in which the food supply changes as a result of the dry season or drought. However, they were too specialized to adopt the high-fecundity strategy. It would be reasonable to think of their flexible mode of life – the fluid strategy – as a factor for their success in the savannah.] It may have been the shortage of high-calorie food and the flexibility of their hands, permitting the production and use of tools on a permanent basis, that popularized the meat-eating habit in early man. In the sense that it marks the beginning of food distribution, the significance of the adoption of meat-eating is much greater than has been thought in the past. This is also considered to be a result of conversion from forest living to savannah living.

There have been many discussions as to the origin of man, but it is generally accepted that his evolution started with the adoption of bipedal locomotion in the savannah and the resultant production of tools, together leading to the development of brain. Of course there are still some uncertainties and discrepancies of details to be resolved, for example, the question of *Homo habilis*, differences in the estimates of the period of hominid differentiation between those based on fossil evidence and those based on the speed of molecular evolution (Zihlman *et al.*, 1978), but I support the view that man arrived a few million years ago as a species of *Australopithecus* type, which may be traced back to *Ramapithecus* found in East Africa and India of the mid- to late-Tertiary period (Simons, 1977). As a summary a schematic diagram of Simons & Ettel (1970) is presented in Fig. 6.16.

Early man probably lived in groups in which sexual relations were not

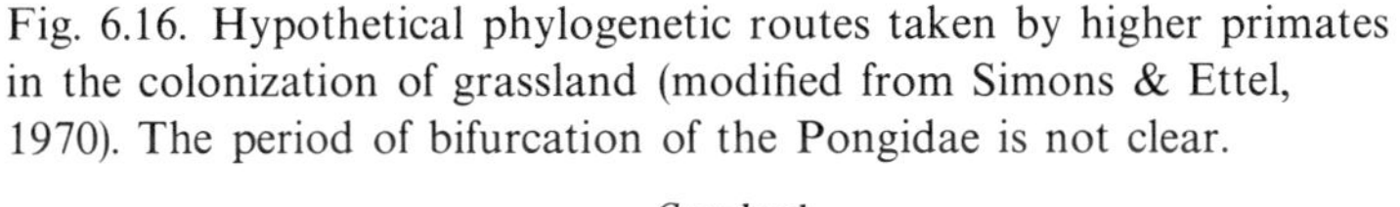

Fig. 6.16. Hypothetical phylogenetic routes taken by higher primates in the colonization of grassland (modified from Simons & Ettel, 1970). The period of bifurcation of the Pongidae is not clear.

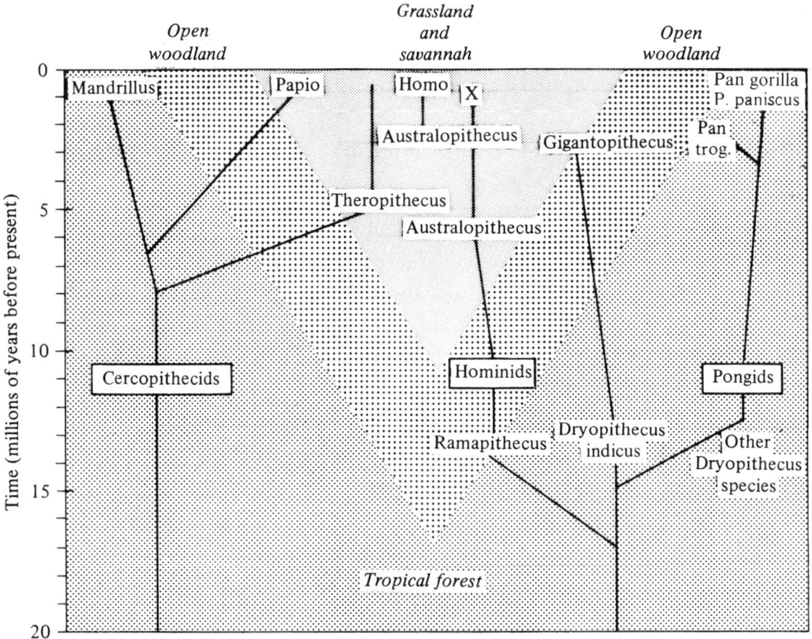

constant but, contrary to the idea of primitive promiscuity expressed by L. H. Morgan and F. Engels, the monogamous relation might have been established for a sufficiently long time subsequently to permit the development of prolonged parental care. This is supported by the investigation of present-day primitive societies of man. If this was the case when did the conflict between the family and the group explicit in the proposition of Espinas disappear? The gorilla with a strong polygynous tendency does not form a multi-male group while the chimpanzee that forms large groups shows no fixed pair relations. To provide for young for an extended period of parental care without the participation of the male, the female chimpanzees have had to develop post-parturition anoestrus of at least three years (Kano, 1979). This is accompanied by a lowering of the intrinsic rate of natural increase, *r*, of the population. The key to the success of man might have been the duality of group life and monogamous life which prevented the lowering of *r*. How this was made possible is still unknown, but it would be interesting in this connection to re-examine the society of the wolf which maintains pair relations within group life – including adoption and nursing of puppies from different families within the group.

The motif of this book was the theory that animals always faced the choice between two opposing strategies of high fecundity and low fecundity/parental protection. The former evolved in the environment in which it was relatively easy for the young to obtain food and the latter evolved in the opposite environmental conditions. Either strategy was equally necessary for survival; the latter was not particularly more advanced than the former in evolution. For example, pest insects and weeds evolved from species with the high-fecundity strategy. However, through the development of parental care the low-fecundity strategy was linked to the origin of the warm-blooded vertebrates, and indeed led to the origin of man. The significance of parental care associated with low fecundity is evident in the evolution of the survivorship curve from a concave shape to a convex shape. This was also the strategy to minimize the fluctuation of population density in a stable environment in which it was difficult for the young to obtain food, rather than the strategy of extinction/recolonization in the unstable environment.

The two strategies and the different environments in which they have evolved apply to both prey and predators. Thus from the prey's point of view, not only is the availability of food high but so is the risk of predation in the environment in which high fecundity developed. Conversely, the environment in which the low-fecundity/parental-protection strategy has evolved is an environment in which both young prey and young predators find it difficult to obtain food. We have seen that the typical environments for the high-fecundity strategy are the ocean, the lower reaches of large rivers and grassland whereas the typical environments for the low-fecundity/parental-protection strategy are mountain streams, the forest canopy and rocky mountains.

In fishes and birds parental care (by both parents) accompanied by territoriality has developed in the latter environments, and group life without internal family structure has developed in the former environments. Thus typical territories and parental care are found among stream-dwelling fishes and birds of the forest canopy. Among the rodents territoriality seems to be developed in arboreal species with low fecundity. At this stage territoriality may have evolved in association with the low-fecundity strategy. However, with the advent of large mammals the situation has changed. In the ungulates (except Suidae) the litter size is usually one or two regardless of whether the species is found in forest or in grassland. But they too have group life in grassland and family life with territories in the forest and high mountains. It is easy to understand that the group life in grassland is a product of an environment where young animals find it easy to obtain food but at the same time are exposed to a high risk of predation. On this point territoriality seems to be an extension of the low-fecundity/parental-protection strategy and living in large herds is an extension of the high-fecundity strategy.

Of course to substantiate such a claim we must demonstrate differences in the degree and period of dependence of young on parents between arboreal animals and grassland animals of the same litter size, or the period of post-parturition anoestrus in the females of such species. However, I am putting forward the above proposition as a working hypothesis concerning the factors that determine the differentiation of group life and family life.

Now, the early monkeys with arboreal habits not only became preadapted morphologically to erect locomotion but also developed a primordial family system. When the apes descended to the ground they would have had to integrate the inherited family life into a group life which was demanded by the new environment. How this has been achieved remains unknown.

The social structure found in the colony of eusocial insects is undoubtedly much advanced but is it comparable to the societies of vertebrates including man? As explained in Chapter 5, the evolution of insect societies, which once headed towards low fecundity and parental care (it is interesting to note that the male then participated in parental protection by defending the nest), later transmuted the early insect society into a supra-organismic organization, in which both the queen and the workers have lost the functions of independent individuals. The insect society is thus considered to be a closed one. The human society, on the other hand, seems to have arisen from the main stream of societal evolution among the vertebrate animals and emphasized individuality

REFERENCES

The Russian entries are romanized. The Japanese and Chinese titles are given in English in square brackets. An asterisk indicates that the English title is given by the author(s) and an English summary accompanies the paper. Unless otherwise stated all such entries refer to publications in Japanese.

Abildgård, F., Andersen, J. & Barndorft-Nielsen, O. (1972). The hare population (*Lepus europaeus* Pallas) of Illumo Island, Denmark. A report on the analysis of data from 1957–70. *Dan. Rev. Game Biol.*, **6(5)** 1–32.

Agwu, S. I. (1974). The population dynamics of *Leucoptera spartifoliella* (Hb.) (Lepidoptera: Lyonetiidae) in south-eastern England. *J. Anim. Ecol.*, **43**, 439–53.

Aikawa, H. (1949). [*Textbook of Fisheries Resources Study.*] Tokyo: Sangyō-Tosho.

Alexander, R. D. (1961). Aggressiveness, territoriality, and sexual behaviour in field crickets (Orthoptera: Gryllidae). *Behaviour*, **17**, 130–223.

Allee, W. C. (1938). *The Social Life of Animals.* New York: Norton.

Allee, W. C. (1951). *Co-operation among Animals with Human Implications.* London: Pitman.

Allee, W. C. (1952). Dominance and hierarchy in societies of vertebrates. In *Structure et Physiologie des Sociétés Animales: Colloques Internationaux, Centre National de la Recherche Scientifique*, **34**, pp. 157–81. Paris: Centre N.R.S.

Allee, W. C., Emerson, A. E., Park, O., Park, T. & Schmidt, K. P. (1949). *Principles of Animal Ecology.* Philadelphia: Saunders.

Andersen, J. (1953). Analysis of a Danish roe-deer population. *Dan. Rev. Game Biol.*, **2**, 127–55.

Anderson, J. F. (1971). Autogeny and mating and their relationship to biting in the saltmarsh deer fly, *Chrysops atlanticus* (Diptera: Tabanidae). *Ann. ent. Soc. Amer.*, **64**, 1421–4.

Andrewartha, H. G. & Birch, L. C. (1954). *The Distribution and Abundance of Animals.* Chicago: University of Chicago Press.

Anon. (1965). Studies on the control of pine moth *Dendrolimus spectabilis* Butler. *Ent. Res. Bull., Ent. Inst., Korea University* **1**, 1–109.

Anthony, A. (1955). Behavior patterns in a laboratory colony of prairie dog, *Cynomys ludovicianus. J. Mammal.*, **36**, 69–78.

Armitage, K. B. (1962). Social behaviour of a colony of the yellow-bellied marmot (*Marmota flaviventris*). *Anim. Behav.*, **10**, 319–31.

Armitage, K. B. & Downhower, J. F. (1974). Demography of yellow-bellied marmot populations. *Ecology*, **55**, 1233–45.

Armstrong, E. A. (1965). *Bird Display and Behaviour*, revised edition. New York: Dover.

Aronson, L. R. (1957). Reproductive and parental behavior. In *The Physiology of Fishes*, vol. 2, ed. M. E. Brown, pp. 271–304. New York: Academic Press.

Auer, C. (1968). Erste Ergebnisse einfacher stochastischer Modelluntersuchungen über die Ursachen der Populationsbewegung des grauen Larchenwicklers *Zeiraphera diniana* Gn. (= *Z. griseana* Hb.) im Oberengadin, 1949/66. *Z. angew. Ent.*, **62**, 202–35.

Azuma, S. & Toyoshima, A. (1963). Progress report of the survey of chimpanzees in their natural habitat, Kabogo Point Area, Tanganyika. *Primates*, **3**, 61–70.

Baker, R. R. (1972). Territorial behaviour of the nymphalid butterflies, *Aglais urticae* (L.) and *Inachis io* (L.). *J. Anim. Ecol.*, **41**, 453–69.

Bakker, R. T. (1971). Ecology of brontosaurs. *Nature, London*, **229**, 172–4.

Ballinger, R. E. (1973). Comparative demography of two viviparous iguanid lizards (*Sceloporus jarrovi* and *Sceloporus poinsetti*). *Ecology*, **54**, 269–83.

Baltensweiler, W. (1964). Zur Regelung von Insektenpopulationen. *Die Grüne, No.* **26**, 806–16.

Baltensweiler, W. (1971). The relevance of changes in the composition of larch bud moth populations for the dynamics of its numbers. In *Proceedings of the Advanced Study Institute on 'Dynamics of Numbers in Populations'*, ed. P. J. den Boer & G. R. Gradwell, pp. 208–19. Wageningen: P.U.D.O.C. (Centre for Agricultural Publishing and Documentation).

Baltensweiler, W. (1977). Natural control factors operating in some European forest insect populations. In *Proceedings of XV International Congress of Entomology*, ed. D. White, pp. 617–21. College Park, Maryland: Entomological Society of America.

Banfield, A. W. F. (1955). A provisional life table for the Barren Ground caribou. *Canad. J. Zool.*, **33**, 143–7.

Barker, A. N. (1942). The seasonal incidence, occurrence and distribution of protozoa in the bacteria bed process of sewage disposal. *Ann. app. Biol.*, **29**, 23–33.

Barker, A. N. (1946). The ecology and function of protozoa in sewage purification. *Ann. appl. Biol.*, **33**, 314–25.

Beaver, R. A. (1966). The development and expression of population tables for the bark beetle *Scolytus scolytus* (F.). *J. Anim. Ecol.*, **35**, 27–41.

Beer, J. R. (1955). Survival and movements of banded big brown bats. *J. Mammal.*, **36**, 242–8.

Benke, A. C. & Benke, S. S. (1975). Comparative dynamics and life histories of coexisting dragonfly populations. *Ecology*, **56**, 302–17.

Bennett, M. A. (1939). The social hierarchy in ring doves. *Ecology*, **20**, 337–57.

Bergerud, A. T. (1971). The population dynamics of Newfoundland caribou. *Wildl. Monogr.*, **25**, 1–57.

Bertram, B. C. R. (1975). The social system of lions. *Sci. Amer.*, **232**(**5**), 54–63.

Bess, H. A. (1961). Population ecology of the gypsy moth *Porthetria dispar* L. (Lepidoptera: Lymantridae). *Bull. Connect. agric. exp. Sta. No.* 646, 1–43.

Bingham, H. C. (1932). Gorillas in a native habitat. *Publ. Carnegie Inst. No.* 426, 1–66.

Birch, M. (ed.) (1974). *Pheromones.* New York: American Elsevier.

Blank, T. H. & Ash, J. S. (1956). The concept of territory in the partridge *Perdix perdix*. *Ibis*, **98**, 379–89.

Blickle, R. L. (1959). Observations on the hovering and mating of *Tabanus fishoppi* Stone (Diptera, Tabanidae). *Ann. end. Soc. Amer.*, **52**, 183–90.

Blum, M. S. (1974). Pheromonal bases of social manifestations in insects. In *Pheromones*, ed. M. C. Birch, pp. 190–9. New York: American Elsevier.

Bodenheimer, F. S. (1930). Studien zur Epidemiologie, und Physiologie der Afrikanischen Wanderheuschrecke (*Schistocerca gregaria* Fork). *Z. angew. Ent.*, **15**, 1–125; **16**, 433–50.

Borutzky, E. V. (1939). Dinamika biomassui *Ch. plumosus* profundali belogo ozera. *Trud. Limnol. Statz. v Kossine*, **22**, 156–95.
Bourlière, F. (1947). La longévité des petits mammifères sauvages. *Mammalia*, **11**, 111–15.
Bourlière, F. (1961). Patterns of social grouping among wild primates. In *Social Life of Early Man*, ed. S. L. Washburn, pp. 1–10. New York: Wenner-Gren Foundation.
Bourlière, F. (1964). *The Natural History of Mammals*, 3rd edn. New York: Knopf.
Bourlière, F., Bertrand, M. & Hunkeler, C. (1970). Ecology and behavior of Lowe's guenon (*Cercopithecus campbelli lowei*) in the Ivory Coast. In *Systematics and Behavior of the Old World Monkeys*, ed. J. R. Napier, pp. 297–350. New York: Academic Press.
Bovey, P. (1966). Le problème de la Tordeuse grise du mélèze (*Zeiraphera diniana* Gn.) dans les forêts alpine. *Bull. Murithienne*, **83**, 1–33.
Branch, G. M. (1975). Mechanisms reducing intraspecific competition in *Patella* spp.: migration, differentiation and territorial behaviour. *J. Anim. Ecol.*, **44**, 575–600.
Brattstrom, B. H. (1974). The evolution of reptilian social behavior. *Amer. Zool.*, **14**, 35–49.
Bremer, H. (1929). Grundsätzliches über den Massenwechsel von Insekten. *Z. angew. Ent.*, **14**, 254–72.
Brian, A. D. (1952). Division of labour and foraging in *Bombus agrorum* Fabricius. *J. Anim. Ecol.*, **21**, 223–40.
Brian, M. V. (1955). Food collection by a Scottish ant community. *J. Anim. Ecol.*, **24**, 336–51.
Brown, E. S., Betts, E. & Rainy, R. C. (1969). Seasonal changes in distribution of the African armyworm, *Spodoptera exempta* (Wlk.) (Lep., Noctuidae), with special reference to Eastern Africa. *Bull. ent. Res.*, **58**, 661–728.
Brown, J. L. (1974). Alternate routes to sociality in jays – with a theory for the evolution of altruism and communal breeding. *Amer. Zool.*, **14**, 63–80.
Bryant, E. H. (1971). Life history consequences of natural selection: Cole's result. *Amer. Natur.*, **105**, 75–7.
Buckner, C. H. (1966). The role of vertebrate predators in the biological control of forest insects. *Annu. Rev. Ent.*, **11**, 449–70.
Buechner, H. K. (1960). The bighorn sheep in the United States, its past, present, and future. *Wildl. Monogr.*, **4**, 1–174.
Buechner, H. K. (1961). Territorial behavior in Uganda kob. *Science*, **133**, 698–9.
Bunnell, P. (1973). Vocalizations in the territorial behavior of the frog *Dendrobates pumilio*. *Copeia* (1973), 277–84.
Burgess, J. W. (1976). Social spiders. *Sci. Amer.*, **234(3)**, 100–6.
Burt, W. H. (1940). Territorial behavior and populations of some small mammals in southern Michigan, *Misc. Publ. Mus. Zool., Univ. Michigan*, **45**, 1–56.
Burt, W. H. (1943). Territoriality and home range concepts as applied to mammals. *J. Mammal.*, **24**, 345–52.
Bustard, H. R. (1970). The role of behavior in the natural regulation of numbers in the gekkonid lizard *Gehyra variegata. Ecology*, **51**, 724–8.
Butler, C. G. (1954). The method and importance of the recognition by a colony of honeybees (*Apis mellifera*) of the presence of its queen. *Trans. roy. ent. Soc. Lond.*, **105**, 11–29.
Butler, C. G., Callow, R. K. & Johnston, N. C. (1959). Extraction and purification of 'queen substance' from queen bees. *Nature, London*, **184**, 1871.
Butler, L. (1951). Population cycles and color phase genetics of the colored fox in Quebec. *Canad. J. Zool.*, **29**, 24–41.
Buxton, A. P. (1952). Observations on the diurnal behaviour of the redtail monkey (*Cercopithecus ascanius schmidti* Matschie) in a small forest in Uganda. *J. Anim. Ecol.*, **21**, 25–58.

Calef, G. W. (1973). Natural mortality of tadpoles in a population of *Rana aurora*. *Ecology*, **54**, 741–58.

Carl, E. A. (1971). Population control in arctic ground squirrels. *Ecology*, **52**, 396–413.

Carpenter, C. R. (1934). A field study of the behavior and social relations of howling monkeys. *Comp. psychol. Monogr.*, **10(2)**, 1–168.

Carpenter, C. R. (1935). Behavior of red spider monkeys in Panama. *J. Mammal.*, **16**, 171–80.

Carpenter, C. R. (1938). A survey of wildlife conditions in Atjeh of North Sumatra, with special reference to the orang-utan. *In Netherland Committee for International Nature Protection, Communications No.* 12, pp. 1–34. Amsterdam: Netherland Committee for International Nature Protection.

Carpenter, C. R. (1940). A field study in Siam of the behavior and social relations of the gibbon (*Hylobates lar*). *Comp. psychol. Monogr.*, **16(5)**, 1–212.

Carpenter, C. R. (1952). Social behavior of non-human primates. In *Structure et Physiologie des Sociétés Animales*, pp. 227–46. *Coloq. Int. CNRS*, 34.

Carpenter, C. R. (1965). The howlers of Barro Colorado Island. In *Primate Behavior. Field Studies of Monkeys and Apes*, ed. I. DeVore, pp. 250–91. New York: Holt, Rinehart & Winston.

Caughley, G. (1964). Social organization and daily activity of the red kangaroo and the grey kangaroo. *J. Mammal.*, **45**, 429–36.

Caughley, G. (1966). Mortality patterns in mammals. *Ecology*, **47**, 906–18.

Chapman, F. M. (1935). The courtship of Gould's manakin (*Manacus vitellinus vitellinus*) on Barro Colorado Island, Canal Zone. *Bull. Amer. Mus. Nat. Hist.*, **68**, 471–525.

Chapman, J. A., Henny, C. J. & Wight, H. M. (1969). The status, population dynamics, and harvest of the dusky Canada geese. *Wildl. Monogr.*, **18**, 1–48.

Chitty, D. (1959). A note on shock disease. *Ecology*, **40**, 728–31.

Christian, J. J. (1950). The adreno-pituitary system and population cycles in mammals. *J. Mammal.*, **31**, 247–59.

Cleveland, L. R. (1934). The wood-feeding roach *Cryptocercus*, its protozoa, and the symbiosis between protozoa and roach. *Mem. Amer. Acad. Arts & Sci.*, **17(2)**, 185–342.

Clough, G. C. (1965). Lemmings and population problems. *Amer. Sci.*, **53**, 199–212.

Cockrum, L. (1956). Homing, movements, and longevity of bats. *J. Mammal.*, **37**, 48–57.

Cody, M. L. (1966). A general theory of clutch size. *Evolution*, **20**, 174–84.

Cody, M. L. (1971). Ecological aspects of reproduction. In *Avian Physiology* 1, ed. D. S. Ferner & J. R. King, pp. 462–512. New York: Acamedic Press.

Colbert, E. H. (1969). *Evolution of Vertebrates*, 2nd edn. New York: Wiley.

Cole, LaMont C. (1954). The population consequences of life history phenomena. *Quart. Rev. Biol.*, **29**, 103–37.

Collias, N. E. & Taber, R. D. (1951). A field study of some grouping and dominance relations in ring-necked pheasants. *Condor*, **53**, 265–75.

Connell, J. H. (1954). Home range and mobility of brush rabbits in California chaparral. *J. Mammal.*, **35**, 392–405.

Connell, J. H. (1963). Territorial behavior and dispersion in some marine invertebrates. *Res. Popul. Ecol.*, **5**, 87–101.

Corbet, P. S. (1962). *A Biology of Dragonflies*. London: Witherby.

Crane, J. (1941). Crabs of the genus *Uca* from the west coast of Central America, *Zoologica*, **26**, 145–208.

Crawley, M. C. (1973). A live-trapping study of Australian brush-tailed possums, *Trichosurus vulpecula* (Kerr), in the Orongorongo Valley, Wellington, New Zealand. *Aust. J. Zool.*, **21**, 75–90.

Crook, J. H. (1964). The evolution of social organization and visual communication in the weaver birds (Ploceinae). *Behaviour* (Suppl.), **10**, 1–178.

Crook, J. H. (1965). The adaptive significance of avian social organization. *Symp. zool. Soc. Lond.*, **14**, 181–218.
Crook, J. H. (1966). Gelada baboon herd structure and movement: A comparative report. *Symp. zool. Soc. London.*, **18**, 237–58.
Crook, J. H., Ellis, J. E. & Goss-Custard, J. D. (1976). Mammalian social systems: Structure and function. *Anim. Behav.*, **24**, 261–74.
Crook, J. H. & Gartlan, J. S. (1966). Evolution of primate societies. *Nature, London*, **210**, 1200–3.
Darling, F. (1937). *A Herd of Red Deer*. London: Oxford University Press.
Darwin, C. R. (1859). *The Origin of Species by Means of Natural Selection or the Preservation of Favoured Races in the Struggle for Life*. London: Murray (numerous editions).
Dasmann, R. F. & Taber, R. D. (1956). Behavior of Columbian blacktailed deer with reference to population ecology. *J. Mammal.*, **37**, 143–64.
Davis, D. E. (1940a). Social nesting habits of the smooth-billed ani. *Auk*, **57**, 179–218.
Davis, D. E. (1940b). Social nesting habits of *Guira guira*. *Auk*, **57**, 472–84.
Davis, D. E. (1942). The phylogeny of social nesting habits in the Crotophaginae. *Quart. Rev. Biol.*, **17**, 115–34.
Deevey, E. S. Jr. (1947). Life tables for natural populations of animals. *Quart. Rev. Biol.*, **22**, 283–314.
DeMars, C. J. Jr., Dahlsten, D. L. & Stark, R. W. (1970). Survivorship curves for eight generations of the western pine beetle in California, 1962–1965, and a preliminary life table. In *Studies on the Population Dynamics of the Western Pine Beetle*, Dendrolimus brevicomis *LeConte* (*Coleoptera: Scolytidae*), ed. R. W. Stark & D. L. Dahlsten, pp. 134–45. Berkeley: University of California Agricultural Publications.
Dementjeva, T. F. (1957). Researches in the U.S.S.R. on Baltic herring and cod. *J. du Conseil*, **22**, 309–21.
Dempster, J. P. (1957). The population dynamics of the Moroccan locust (*Dociostaurus maroccanus* Thunberg) in Cyprus. *Anti-Locust Bull.*, **27**, 1–60.
Dempster, J. P. (1971). The population ecology of the cinnabar moth, *Tyria jacobaeae* L. (Lepidoptera, Arctiidae). *Oecologia*, **7**, 26–67.
Dempster, J. P. (1975). *Animal Population Ecology*. London: Academic Press.
DeVore, I. (1963). A comparison of the ecology and behavior of monkeys and apes. In *Classification and Human Evolution*, ed. S. L. Washburn, pp. 301–19. New York: Wenner-Gren Foundation.
DeVore, I. & Hall, K. R. L. (1965). Baboon ecology. In *Primate Behavior. Field Studies of Monkeys and Apes*, ed. I. DeVore, pp. 20–52. New York: Holt, Rinehart & Winston.
Dingle, H. (1968). Life history and population consequences of density, photoperiod, and temperature in a migrant insect, the milkweed bug *Oncopeltus*. *Amer. Natur.*, **102**, 149–63.
Dingle, H. (1974). The experimental analysis of migration and life-history strategies in insects. In *Experimental Analysis of Insect Behaviour*, ed. L. B. Brown, pp. 329–42. Berlin: Springer-Verlag.
Duellman, W. E. (1966). Aggressive behavior in dendrobatid frogs. *Herpetologica*, **22**, 217–21.
Dunnet, G. M. (1964). A field study of local populations of the brush-tailed possum *Trichosurus vulpecula* in eastern Australia. *Proc. zool. Soc. Lond.*, **142**, 665–95.
Eaton, R. L. (1973). [The cheetah of the plain, its territory, dominance order and family life.] *Anima* (1973), (**9**), 29–33.
Eberhardt, M. J. W. (1969). The social biology of polistine wasps. *Misc. Publ. Mus. Zool., Univ. Michigan*, **140**, 1–101.

Eibl-Eibesfeldt, I. (1966). Das Verteidigen der Eiablageplätze bei der Hood-Meerechse (*Amblyrhynchus cristatus venustissimus*). *Z. Tierpsych.*, **23**, 627–31.
Eidmann, H. (1937). Zur Theorie der Bevölkerungsbewegung der Insekten. *Anz. Schädlingsk.*, **13**, 47–52.
Eimerl, S. & DeVore, I. (1965). *The Primates.* Chicago: Time-Life.
Eisenberg, J. F., Muckenhirn, N. A. & Budran, R. (1972). The relation between ecology and social structure in primates. *Science*, **176**, 863–74.
Elliott, J. M. (1973). The life cycle and production of the leech *Erpobdella octoculata* (L.) (Hirundinea, Erpobdellidae) in a Lake District Stream. *J. Anim. Ecol.*, **42**, 435–48.
Elliott, R. C. (1976). Observations on a small group of mountain gorillas (*G. g. beringei*). *Folia Primatol.*, **25**, 12–24.
Elton, C. (1924). Periodic fluctuations in the number of animals; their causes and effects. *Brit. J. exp. Biol.*, **2**, 119–63.
Elton, C. (1925). Plague and the regulation of numbers in wild mammals. *J. Hyg.*, **24**, 138–63.
Elton, C. (1927). *Animal Ecology.* London: Sidgwick & Jackson.
Elton, C. (1931). The study of epidemic diseases among wild animals. *J. Hyg.*, **31**, 435–56.
Elton, C. (1932). Territory among wood ants (*Formica rufa* L.) at Picket Hill. *J. Anim. Ecol.*, **1**, 69–76.
Elton, C. (1942). *Voles, Mice and Lemmings: Problems in Population Dynamics.* Oxford: Clarendon Press.
Elton, C. & Nicholson, M. (1942). The ten-year cycle in numbers of the lynx in Canada. *J. Anim. Ecol.*, **11**, 215–44.
Embree, D. G. (1965). The population dynamics of the winter moth in Nova Scotia, 1954–1962. *Mem. ent. Soc. Canad.*, **46**, 1–57.
Emerson, A. E. (1938). Termite nests – a study of the phylogeny of behavior. *Ecol. Monogr.*, **8**, 247–84.
Emlen, S. T. (1968). Territoriality in the bullfrog, *Rana catesbeiana. Copeia* (1968), 240–3.
Errickson, M. M. (1938). Territory, annual cycle, and numbers in a population of wren-tits (*Chamaea fasciata*). *Univ. Calif. Publ. Zool.*, **42**, 247–334.
Esch, H. (1967). The evolution of bee language. *Sci. Amer.*, **216(4)**, 96–104.
Escherich, K. (1914). *Die Forstinsekten Mitteleuropas, Bd.I. Allgemeiner Teil.* Berlin: Paul Parey.
Eschmeyer, R. W. (1939). Analysis of the complete fish population from How Lake, Crowford County, Michigan. *Pap. Mich. Acad. Sci.*, **24(2)**, 117–37.
Espinas, A. V. (1877). *Des Sociétés Animales.* Paris: Baillière.
Estes, R. D. (1969). Territorial behavior of the wildebeest (*Connodiaetes taurinus* Burchell, 1823). *Z. Tierpsychol.*, **26**, 283–370.
Estes, R. D. (1974). Social organization of the African Bovidae. In *The Behaviour of Ungulates and its Relation to Management*, ed. V. Geist & F. Walther, pp. 166–205. Morges: International Union for Conservation of Nature and Natural Resources.
Estes, R. D. & Goddard, J. (1967). Prey selection and hunting behavior of the African wild dog. *J. Wildl. Manage.*, **31**, 52–70.
Etkin, W. (1964). Types of social organization in birds and mammals. In *Social Behavior and Organization among Vertebrates*, ed. W. Etkin, pp. 256–97. Chicago: University of Chicago Press.
Evans, H. E. (1953). Comparative ethology and the systematics of spider wasps. *Syst. Zool.*, **2**, 155–72.
Evans, H. E. (1958). The evolution of social life in wasps. In *Proceedings of the Tenth International Congress of Entomology* (Montreal, 1956), vol. 2, ed. J. A. Downes, pp. 449–57.

Evans, H. E. (1964). Observations on the nesting behavior of *Moniaecera asperata* (Fox) (Hymenoptera; Sphecidae, Crabroninae) with comments on communal nesting in solitary wasps. *Insectes Sociaux*, **11**, 71–8.
Evans, L. T. (1936a). A study of a social hierarchy in the lizard, *Anolis carolinensis*. *J. genet. Psych.*, **48**, 88–111.
Evans, L. T. (1936b). Social behavior of the normal and castrated lizard, *Anolis carolinensis*. *Science*, **83**, 104.
Evans, L. T. (1936c). Behavior of castrated lizard. *J. genet. Psych.*, **48**, 217–21.
Evans, L. T. (1936d). Territorial behavior of normal and castrated females of *Anolis carolinensis*. *J. genet. Psych.*, **49**, 49–60.
Evans, L. T. (1938). Cuban field studies on territoriality of the lizard, *Anolis sagrei*. *J. comp. Psychol.*, **25**, 97–125.
Evans, S. M. (1973). A study of fighting reactions in some nereid polychaetes. *Anim. Behav.*, **21**, 138–46.
Farrow, R. A. (1975). The African migratory locust in its main outbreak area of the middle Niger: quantitative studies of solitary populations in relation to environmental factors. *Locusta* (Org. Int. Cent. Criquet Migr. Afr.), **11**, 1–198.
Fielder, D. R. (1965). A dominance order for shelter in the spiny lobster *Jasus lalandei* (H. Milne-Edwards). *Behaviour*, **24**, 236–45.
Fitch, H. S. (1940). A field study of the growth and behavior of the fence lizard. *Univ. Calif. Publ. Zool.*, **44(2)**, 51–72.
Fitch, H. S. (1960). Autecology of the copperhead. *Univ. Kansas Publ. Mus. Nat. Hist.*, **13(4)**, 85–288.
Foerster, R. E. (1934). An investigation of the life history and propagation of the sockeye salmon (*Oncorhynchus nerka*) at Cultus Lake, B.C. 4. The life history cycle of 1925 year class with natural propagation. *Contr. Canada Biol. Fish.*, **8(27)**, 345–55.
Force, D. C. (1970). Competition among four hymenopterous parasites of an endemic insect host. *Ann. ent. Soc. Amer.*, **63**, 1675–88.
Force, D. C. (1972). *r*- and *K*-strategists in endemic host-parasitoid communities. *Bull. ent. Soc. Amer.*, **18**, 135–7.
Force, D. C. (1975). Succession of *r* and *K* strategists in parasitoids. In *Evolutionary Strategies of Parasitic Insects and Mites*, ed. P. W. Price, pp. 112–29. New York: Plenum.
Fossey, D. (1974). Observation on the home range of mountain gorillas. *Anim. Behav.*, **22**, 568–81.
Frank, F. (1962). Zur Biologie des Bergelemmings, *Lemmus lemmus* (L.). Ein Beitrag zum Lemming-Problem. *Z. Morph. Oekol. Tiere*, **51**, 87–164.
Freeman, B. E. (1973). Preliminary studies on the population dynamics of *Sceliphron assimile* Dahlbom (Hymenoptera: Sphecidae) in Jamaica. *J. Anim. Ecol.*, **42**, 173–82.
Friedrichs, K. (1930). *Die Grundfragen und Gesetzmässigkeiten der land- und forstwirtschaftlichen Zoologie, insbesondere der Entomologie. I. Ökologischer Teil.* Berlin: Paul Parey.
Friedrichs, K. (1935). Folgerungen aus den neuen Untersuchungen über die Forleule. *Anz. Schädlingsk.*, **11(2)**, 19–23.
Friedrichs, K. (1937). *Ökologie aus Wissenschaft von der Natur oder biologische Raumforschung.* Bios 7. Leipzig: Johann Ambrosius Barth.
Frisch, K. von (1950). *Bees Their Vision, Chemical Senses, and Language.* New York: Cornell University Press.
Frost, W. E. & Brown, M. E. (1967). *The Trout.* London: Collins.
Fujisaki, K. (1977). [A pentatomid bug that forms a harem.] *Kagaku-Asahi* [*Science Asahi*] (1977), (**5**), 18–22.
Fukaya, M. & Nakatsuka, K. (1956). [*Forecasting abundance of the rice-stem borer.*]

Tokyo: Nippon Shokubutsu-Bōeki Kyōkai (Japanese Association of Plant Protection).
Furuya, Y. (1962). The social life of silvered leaf monkeys *Trachypithecus cristatus*. *Primates*, **3**, 41–60.
Furuya, Y. (1965). Social organization of the crab-eating monkey. *Primates*, **6**, 285–336.
Gadgil, M. & Solbrig, O. T. (1972). The concept of *r*- and *K*-selection: evidence from wild flowers and some theoretical considerations. *Amer. Natur.*, **106**, 14–31.
Gartlan, J. S. (1970). Preliminary notes on the ecology and behaviour of the drill, *Mandrillus leucophaeus* Ritgen, 1824. In *Old World Monkeys: Evolution, Systematics and Behaviour*, ed. J. R. Napier & P. H. Napier, pp. 445–80. New York: Academic Press.
Gautier, J. P. & Gautier-Hion, A. (1969). Association polyspecifique chez les *Cercopithecus* du Gabon. *Terre et Vie*, **23**, 164–201.
Gehrs, C. W. & Robertson, A. (1975). Use of life tables in analyzing the dynamics of copepod populations. *Ecology*, **56**, 665–72.
Geist, V. (1971). *Mountain Sheep: A Study in Behavior and Evolution*. Chicago: University of Chicago Press.
Geist, V. (1974). On the relationship of social evolution and ecology in ungulates. *Amer. Zool.*, **14**, 205–20.
Gibb, J. (1956). Territory in the genus *Parus*. *Ibis*, **98**, 420–9.
Ginn, H. B. (1969). The use of annual ringing and nest record card totals as indicators of bird population levels. *Bird Study*, **16**, 210–48.
Goetsch, W. (1941). Staatgründung und Kastenbildung bei Termiten. *Naturwiss.*, **29**, 1–13.
Goodall, J. (1963). Feeding behaviour of wild chimpanzees. *Symp. zool. Soc. Lond.*, **10**, 39–47.
Goodall, J. (1965). Chimpanzees of the Gombe Stream Reserve. In *Primate Behavior. Field Studies of Monkeys and Apes*, ed. I. DeVore, pp. 425–73. New York: Holt, Rinehart & Winston.
Graf, W. (1956). Territorialism in deer. *J. Mammal.*, **37**, 165–70.
Graham, S. A. (1939). *Principles of Forest Entomology*, 2nd edn. New York: McGraw-Hill.
Graunt, J. (1662). *Natural and Political Observations Mentioned in a Following Index, and Made upon the Bills of Mortality*. London: J. Martyn. (Japanese transl. (1968) by S. Kuruma. Tokyo: Kurita Publishing Association).
Green, R. G. & Larson, C. L. (1938a). A description of shock disease in the snowshoe hare. *Amer. J. Hyg.*, **28**, 190–212.
Green, R. G. & Larson, C. L. (1938b). Shock disease and the snowshoe hare cycle. *Science*, **87**, 298–9.
Greenberg, B. & Noble, G. K. (1944). Social behavior of the American chameleon (*Anolis carolinensis* Voigt.). *Physiol. Zoöl.*, **17**, 392–9.
Greer, A. E. Jr. (1971). Crocodilian nesting habits and evolution. *Fauna*, **2**, 20–8.
Gulland, J. A. (1971). The effect of exploitation on the number of marine animals. In *Proceedings of the Advanced Study Institute on 'Dynamics of Numbers in Populations'*, ed. P. J. den Boer & G. R. Gradwell, pp. 450–67. Wageningen: P.U.D.O.C. (Centre for Agricultural Publishing and Documentation).
Haddow, A. J. (1952). Field and laboratory studies on an African monkey, *Cercopithecus ascanius schmidti* Matchie. *Proc. zool. Soc. Lond.*, **122**, 297–394.
Hall, K. R. L. (1962). Variations in the ecology of the chacma baboon, *Papio ursinus*. *Symp. zool. Soc. Lond.*, **10**, 1–28.
Hall, K. R. L. (1964). Aggression in monkey and ape societies. In *The Natural History of Aggression*, ed J. D. Carthy & F. J. Ebling, pp. 51–64. New York: Academic Press.

Hall, K. R. L. (1965). Behavior and ecology of the wild patas monkey, *Erythrocebus patas* in Uganda. *J. Zool.*, **148**, 15–87.
Hall, K. R. L. & DeVore, I. (1965). Baboon social behavior. In *Primate Behavior. Field Studies of Monkeys and Apes*, ed. I. DeVore, pp. 292–319. New York: Holt, Rinehart & Winston.
Halley, E. (1693). An estimate of the degrees of the mortality of mankind drawn from curious tables of the birth and funerals at the city of Breslau. *Phil. Trans. roy. Soc. Lond.*, **17**, 596–610.
Hamilton, W. D. (1964). The genetical theory of social behaviour. *J. theor. Biol.*, **7**, 1–52.
Hancock, D. A. (1971). The role of predators and parasites in a fishery for the mollusc *Cardium edule* L. In *Proceedings of the Advanced Study Institute on 'Dynamics of Numbers in Populations'*, ed. P. J. den Boer & G. R. Gradwell, pp. 419–39. Wageningen: P.U.D.O.C. (Centre for Agricultural Publishing and Documentation).
Hancock, D. A. (1973). The relationship between stock and recruitment in exploited invertebrates. *Rapp. P.-v. Réun. Cons. perm. int. Explor. Mer.*, **164**, 113–31.
Haneda, K. (ed.) (1975). [*Life of Wild Birds.*] Tokyo: Tsukiji-Shokan.
Haneda, K. & Iida, Y. (1966). [Life history of eastern carrion-crow (*Corvus corone orientalis*). I. Breeding season (1).]* *Jap. J. Ecol.*, **16**, 97–105.
Hanson, W. C. & Eberhardt, L. L. (1971). A Columbia River Canada geese population, 1950–1970. *Wildl. Monogr.*, **28**, 15–56.
Harada, E. & Kawanabe, H. (1955). [The behavior of the sand-crab, *Scopimera globosa* de Haan, with special reference to the problem of coaction between individuals.]* *Jap. J. Ecol.*, **4**, 162–5.
Harcourt, A. H., Stewart, K. S. & Fossey, D. (1976). Male emigration and female transfer in wild mountain gorilla. *Nature, London*, **263**, 226–7.
Harcourt, D. G. (1971). Major mortality factors in the population dynamics of the diamondback moth, *Plutella maculipennis* (Curt.) (Lepidoptera: Plutellidae). *Mem. ent. Soc. Canada*, **32**, 55–66.
Harukawa, C., Takato, R. & Kumashiro, S. (1934). Studies on the rice-borer. III. On the population dynamics of the rice-borer. *Berichte Ōhara Inst. landwirt. Forsch.*, **7**, 1–97.
Haskins, C. P. (1970). Researches in the biology and social behavior of primitive ants. In *Development and Evolution of Behavior*, ed. L. R. Aronson, E. Tobach, D. S. Lehrman & J. S. Rosenblatt, pp. 355–88. San Francisco: Freeman.
Haskins, C. P. & Haskins, E. F. (1950). Notes on the biology and social behavior of the genera *Myrmica* and *Promyrmica*. *Ann. ent. Soc. Amer.*, **43**, 401–91.
Hatta, S. (Ha, Sa) (1897). [Nests and eggs of the stickleback and their protection.] *Dōbutsugaku-Zasshi* [*Zool. Mag., Tokyo*], **9**, 135–7.
Hayashi, I. & Numata, M. (1968). [A theoretical consideration on the succession of plant communities.] *Zassō-Kenkyū* [*Weed Research*], **7**, 1–11.
Heering, H. (1956). Zur Biologie, Oekologie und zum Massenwechsel des Buchenprachtkäfers (*Agrilus viridis* L.). *Z. angew. Ent.*, **39**, 76–114.
Higashi, K. (1969). Territoriality and dispersal in the population of dragonfly, *Crocothemis servilia* Drury (Odonata: Anisoptera). *Mem. Fac. Sci., Kyushu Univ. Ser. E*, **5**, 95–113.
Hinde, R. A. (1956). The biological significance of the territories of birds. *Ibis*, **98**, 340–69.
Hirabayashi, K. (1976). [Living in the Japan Alps.] (An article in a special issue on the rock ptarmigan.) *Anima* (1976), (7), 13–21.
Hirose, Y. (1977). Annual population fluctuation of *Platypleura kaempferi* (Fabricius) (Homoptera, Cicadidae). *Kontyū*, **45**, 314–19.

Hjort, J. (1926). Fluctuations in the year classes of important fishes. *J. du Conseil*, **1**, 1–38.
Holm, C. H. (1973). Breeding sex ratios, territoriality, and reproductive success in the red-winged blackbird (*Agelaius phoeniceus*). *Ecology*, **54**, 356–65.
Horikawa, Y. (1921). [Observations on the stickleback.] *Dōbutsugaku-zasshi* [*Zool. Mag., Tokyo*], **24**, 418–20.
Horn, H. S. (1968). The adaptive significance of colonial nesting in the Brewer's blackbird (*Euphagus cyanocephalus*). *Ecology*, **49**, 682–94.
Howard, L. O. & Fiske, W. F. (1911). The importation into the United States of the parasites of the gypsy moth and the brown-tail moth. *Bull. US. Bur. Ent.*, **91**, 1–344.
Howard, W. E. (1949). Dispersal, amount of inbreeding, and longevity in a local population of prairie deermice on the George Reserve, Southern Michigan. *Contr. Lab. Vertebr. Biol., Univ. Mich.*, **43**, 1–50.
Howard, W. E. & Childs, H. E. Jr. (1959). Ecology of pocket gophers with emphasis on *Thomomys bottae* Mewa. *Hilgardia*, **29**, 277.
Howard, W. E. & Emlen, J. T. Jr. (1942). Intercovey social relations in the valley quail. *Wilson Bull.*, **54**, 162–70.
Humphreys, W. F. (1976). The population dynamics of an Australian wolf spider. *Geolycosa godeffroyi* (L. Koch, 1865) (Araneae: Lycosidae). *J. Anim. Ecol.*, **45**, 59–80.
Ikan, R., Bergmann, E. D., Ishay. J & Gitter, S. (1968). Proteolytic enzyme activity in the various colony members of the oriental hornet, *Vespa orientalis* F. *Life Science*, **7(18)**, 929–34.
Imaizumi, Y. (1976). [Is the population size below 40? (study on the ecology of a newly discovered Japanese wild cat *Mayailurus iriomotensis*).] *Anima* (1976), (**5**), 13–20.
Imanishi, K. (1951). [*Infrahuman Societies.*] Tokyo: Iwanami.
Ishay, J. & Ikan, R. (1969). Gluconeogenesis in the oriental hornet, *Vespa orientalis* F. *Ecology*, **49**, 169–71.
Ishida, T. (1952). [Japanese herring fisheries and its biological consideration.] *Suisanchō Gyogyō-Kagaku Sōsho* [*Fisheries Agency Publications, Fisheries Science*], **4**, 1–57.
Ishii, S. (1971). [*Physiologically Active Substances of Insects*], 2nd edn. Tokyo: Nankōdo.
Ishikura, H. (1950). [*Abundance Forecasting of Insect Pests of Field Crops.*] Tokyo: Kawade-Shobō.
Ishikura, H. & Nakatsuka, K. (1955). [*Population Fluctuation of the Paddy Borer Moth and its Forecasting – a Review.*] Tokyo: Nippon Shokubutsu-Bōeki Kyōkai (Japanese Association of Plant Protection).
Itani, J. (1952). [Socio-ecological study on wild Japanese macaques.] *Jikken-dōbutsu Ihō* [*J. Exp. Animals*], **1(6)**, 49–53.
Itani, J. (1954). [Monkeys on Mt. Takasaki.] In [*Natural History of Japanese Animals*], vol. 2, ed. K. Imanishi pp. 1–284. Tokyo: Kōbunsha.
Itani, J. (1958). [Africa.] *Shizen* [*Nature, Tokyo*] (1959), (**1**), 58–65.
Itani, J. (1959). Parental care in the wild Japanese monkey, *Macaca fuscata*. *Primates*, **2**, 61–93.
Itani, J. (1961). The society of Japanese monkeys. *Japan Quarterly*, **8(4)**, 421–30.
Itani, J. (1966). [Social structure of wild chimpanzees.] *Shizen* [*Nature, Tokyo*] (1966), (**8**), 17–30.
Itani, J. (1967). [From the society of primates to the society of man.] *Kagaku* [*Science, Tokyo*], **37(4)**, 170–4.
Itani, J. (1972). [*The Social Structure of Primates.*] Tokyo: Kyōritsu.
Itani, J. (ed.) (1977). [*The Natural History of Chimpanzees.*] Tokyo: Kōdansha.
Itani, J. & Suzuki, A. (1967). The social unit of chimpanzees. *Primates*, **8**, 355–81.

Itani, J. & Tokuda, K. (1954). [The nomadism of the wild Japanese monkey, *Macaca fuscata fuscata*, in Takasaki Yama.]* *Jap. J. Ecol.*, **4**, 22–8.
Itô, Y. (1952a). [The growth form of populations in some aphids, with special reference to the relation between population density and the movements.]* *Res. Popul. Ecol.*, **1**, 36–48.
Itô, Y. (1952b). [On the population increase and migration in three species of barley aphids. Studies on the mechanisms of ecological segregatation in barley aphids.]* *Ōyō-Kontū* [*Applied Entomology*], **7**, 169–76.
Itô, Y. (1953). [Studies on the population increase and the movements of soy bean aphid, *Aphis glycines* Matsumura. I. On the two types of population increase; II. On the movements from plant to plant.]* *Ōyō-Kontū* [*Applied Entomology*], **8**, 141–8.
Itô, Y. (1958). [Changes in mortality and mortality factors during the insect outbreak.] *Seibutsu Kagaku* [*Biological Science, Tokyo*], **10**, 11–25.
Itô, Y. (1959). [*Comparative Ecology*.] Tokyo: Iwanami.
Itô, Y. (1960a). Ecological studies on population increase and habitat segregation among barley aphids. *Bull. National Inst. agric. Sci., Ser. C*, **11**, 45–130.
Itô, Y. (1960b). Territorialism and residentiality in a dragonfly, *Orthetrum albistylum speciosum* Uhler (Odonata, Anisoptera). *Ann. ent. Soc. Amer.*, **53**, 851–3.
Itô, Y. (1966). [Origin of Man.] Tokyo: Kinokuniya.
Itô, Y. (1970). Groups and family bonds in animals in relation to their habitat. In *Development and Evolution of Behavior*, ed. L. R. Aronson, E. Tobach, D. S. Lehrman & J. S. Rosenblatt, pp. 389–415. San Francisco: Freeman.
Itô, Y. (1975–76). [*Animal Ecology*], 2 vols. Tokyo: Kokin-shoin.
Itô, Y. (1977). Birth and death. In *Adaptation and Speciation in the Fall Webworm*, ed. T. Hidaka, pp. 101–28. Tokyo: Kōdansha.
Itô, Y. & Kiritani, K. (1971). [*How are the numbers of animals determined?*] Tokyo: Nippon Hōsō Shuppan Kyōkai (N.H.K. Publishers).
Itô, Y. & Miyashita, K. (1955). [Dispersion and mortality rate of fall brood population of cabbage armyworm in a Chinese cabbage field. Studies on the mortality rate and behaviour of field population of agricultural pest insects (2nd report).]* *Ōyō-Kontū* [*Applied Entomology*], **11**, 144–9.
Itô, Y. & Miyashita, K. (1956). Some considerations on the theory of population dynamics in insects. (Paper presented at the Section 24 of the VIII Congress of the International Union of Forestry Research Organizations Oxford, 1956.)
Itô, Y. & Miyashita, K. (1968). Biology of *Hyphantria cunea* Drury (Lepidoptera: Arctiidae) in Japan. V. Preliminary life tables and mortality data in urban areas. *Res. Popul. Ecol.*, **10**, 177–209.
Itô, Y. & Murai, M. (1977). [*Methods in Animal Ecology*], 2 vols. Tokyo: Kokin-shoin.
Itô, Y. & Nagamine, M. (1974). Distribution of infestations of a sugar cane cicada, *Mogannia iwasakii* Matsumura (Hemiptera: Cicadidae) in Okinawa Island with a discussion on the cause of outbreak. *Appl. Ent. Zool.*, **9**, 58–64.
Itô, Y. & Nagamine, M. (1976a). [Outbreaks of a cicada.] *Anima* (1976), (**2**), 82–9.
Itô, Y. & Nagamine, M. (1976b). Dynamics of epidemic sugarcane cicada (*Mogannia minuta*) population. (Paper presented to section 5 Ecology (moderator D. E. Leonard) at the International congress of Entomology, Washington, D.C., 19–27 August 1976.)
Itô, Y., Shibazaki, A. & Iwahashi, O. (1969). Biology of *Hyphantria cunea* Drury (Lepidoptera: Arctiidae) in Japan. IX. Population dynamics. *Res. Popul. Ecol.*, **11**, 211–28.
Iwata, K. (1940). [*Life of Wasps and Bees*.] Tokyo: Kōbundō.
Iwata, K. (1942). Comparative studies on the habits of solitary wasps. *Tenthredo*, **4**, 1–146.

Iwata, K. (1955a). [Evolution of insect habits.] *Shizen* [*Nature, Tokyo*] (1955), (**8**), 22–9; (**9**), 46–55.

Iwata, K. (1955b). The comparative anatomy of the ovary in Hymenoptera. I. Aculeata. *Mushi*, **29**, 17–34.

Iwata, K. (1956). [Eggs of Symphyta in Japan.]* *Botyu-Kagaku* [*Insect Control Science*], **22**, 13–9.

Iwata, K. (1958a). The comparative anatomy of the ovary in Hymenoptera. II. Symphyta. *Mushi*, **31**, 47–60.

Iwata, K. (1958b). Ovarian eggs of 233 species of the Japanese Ichneumonidae (Hymenoptera). *Acta Hymenoptologica*, **1**, 63–74.

Iwata, K. (1959a). The comparative anatomy of the ovary in Hymenoptera. III. Braconidae. *Kontyū*, **27**, 18–20.

Iwata, K. (1959b). The comparative anatomy of the ovary in Hymenoptera. IV. Proctorupoidea, etc. *Kontyū*, **27**, 231–8.

Iwata, K. (1960). The comparative anatomy of the ovary in Hymenoptera. V. Ichneumonidae. *Acta Hymenoptologica*, **1**, 115–69.

Iwata, K. (1964). Egg gigantism in subsocial Hymenoptera, with ethological discussion on tropical bamboo carpenter bees. *Nature & Life in Southeast Asia*, **3**, 399–434.

Iwata, K. (1966a). [Large-sized eggs in Curculionoidea (Coleoptera).]* *Res. Bull. Hyōgo Agric. College*, **7**, 43–5.

Iwata, K. (1966b). [Ovarian eggs in Scarabaeoidea (Coleoptera).]* *Seibutsu-Kenkyū* [Biological Studies, Fukui], **10**, 1–3.

Iwata, K. (1976). *Evolution of Instinct. Comparative Ethology of Hymenoptera.* New Delhi: Amerind Publishing Co. (Japanese edn in 1971).

Iwata, K. & Sakagami, F. S. (1966). Gigantism and dwarfism in bee eggs in relation to the modes of life, with notes on the number of ovaries. *Jap. J. Ecol.*, **16**, 4–16.

Izawa, K. (1970). Unit groups of chimpanzees and their nomadism in the savanna woodland. *Primates*, **11**, 1–45.

Izawa, K. (1976a). [New World monkeys.] [In a supplement to *Science* (Japan) published as *From Monkeys to Man*], pp. 106–15. Tokyo: Nippon Keizaishinbun.

Izawa, K. (1976b). [Monkeys living in the jungle.] *Anima* (1976), (**11**), 11–21.

Izawa, K. (1976c). Group sizes and compositions of monkeys in Upper Amazon Basin. *Primates*, **17**, 367–99.

Izawa, K. (1977). [Chimpanzees in Kasakati Basin.] In [*The Natural History of Chimpanzees*], ed. J. Itani, pp. 187–248. Tokyo: Kōdansha.

Izawa, K. (1978). A field study of ecology and behavior of the black mantle tamarin (*Saguinus nigricollis*). *Primates*, **19**, 241–74.

Izawa, K. & Nishida, T. (1963). Monkeys living in the northern limit of their distribution. *Primates*, **4**, 67–88.

Izumi, T. (1976). [From nursery to emergency exit – nest structure and society of the Norway rat.] *Anima* (1976), (**9**), 31–6.

Jacobs, M. E. (1955). Studies on territorialism and sexual selection in dragonflies. *Ecology*, **36**, 566–86.

Jalihal, D. R. & Sankolli, K. N. (1975). On the abbreviated metamorphosis of the fresh water prawn *Macrobrachium hendersodayanum* (Tiwari) in the laboratory. *Karnatak University Journal: Science*, **20**, 283–91.

Jameson, D. L. (1957). Life history and phylogeny in the Salientians. *Syst. Zool.*, **16**, 75–8.

Jarman, P. J. (1974). The social organization of antelope in relation to their ecology. *Behaviour*, **48**, 215–67.

Jay, P. (1963). The Indian langur monkey (*Presbytis entellus*). In *Primate Social Behavior*, ed. C. H. Southwick, pp. 114–23. Princeton: van Nostrand.

Jay, P. (1965). The common langur of north India. In *Primate Behavior. Field Studies of Monkeys and Apes*, ed. I. DeVore, pp. 197–249. New York: Holt, Rinehart & Winston.
Jeannel, R. (1956). [Subterranean evolution of animals.] *Kagaku* [*Science, Tokyo*], **26**, 237–43.
Jenkins, D., Watson, A. & Miller, G. R. (1963). Population studies on red grouse, *Lagopus lagopus scoticus* in northeast Scotland. *J. Anim. Ecol.*, **32**, 317–76.
Jenkins, D. W. (1944). Territory as a result of despotism and social organization in geese. *Auk*, **61**, 30–47.
Jensen, A. J. C. (1930). On the influence of hydrographical factors upon the yield of the mackerel fishery in the Sound. *Rept. Danish biol. Staat.*, **36**, 69–88.
Johnson, C. (1961). Breeding behavior and oviposition in *Hetaerina americana* (Fabricius) and *H. titia* (Drury) (Odonata: Agriidae). *Canad. Ent.*, **93**, 260–6.
Johnson, C. (1962). Breeding behavior and oviposition in *Calopteryx maculatum* (Beauvois) (Odonata: Calopterygidae). *Amer. Midl. Nat.*, **68**, 242–7.
Johnson, W. E. (1965). On mechanisms of self-regulation of population abundance in *Oncorhynchus nerka. Mitt. int. Verein. Limnol.*, **13**, 66–87.
Jolly, A. (1966). *Lemur Behavior*. Chicago: University of Chicago Press.
Jolly, A. (1972). *The Evolution of Primate Behavior*. New York: Macmillan.
Kagasaki, T. (1973). [Mating ceremony of the kestrel.] *Anima* (1973), (**7**), 5–25.
Kalmus, H. (1941). Defence of source of food by bees. *Nature, London*, **148**, 228.
Kamiya, K. (1938). [Relationship between environmental conditions and occurrences of parasitoids of the pine caterpillar.] *Oyo-Dobutsugaku-Zasshi* [*J. Appl. Zool.*], **10**, 85–9.
Kano, T. (1979). [Sex and society in the pygmy chimpanzee.] *Anima* (1979), (**1**), 58–66.
Kaufmann, J. H. (1974). Social ethology of whiptail wallaby, *Macropus parryi*, in northeastern New South Wales. *Anim. Behav.*, **22**, 281–369.
Kawabata, M. (1954). [Socio-ecological studies on the killi-fish, *Alocheilus latipes*. I. General remarks on the social behavior.]* *Jap. J. Ecol.*, **4**, 109–13.
Kawabe, M. (1966). One observed case of hunting behaviour among wild chimpanzees living in the savanna woodland of Western Tanzania. *Primates*, **7**, 393–6.
Kawaguchi, M. (1937). [*Ecological studies of Japanese birds*]. Tokyo: Sōrin Shobō.
Kawai, M. (1955). [Domestic rabbits]. In [*Natural History of Japanese Animals*], vol. 1, ed. K. Imanishi, pp. 141–283. Tokyo: Kōbunsha.
Kawai, M. (1972). [Forests and monkeys]. *Kikan Jinruigaku* [*Quarterly Anthropology*], **3**(**1**), 3–51.
Kawai, M. (1974). [The gelada baboon of Ethiopian highlands] *Anima* (1974), (**10**), 5–22.
Kawai, M. (1976). [Gelada baboons]. [In a supplement to *Science* (Japan) published as *From monkeys to Man*], pp. 63–77. Tokyo: Nippon Keizaishinbun.
Kawai, M. (ed.) (1979). *Ecological and Sociological Studies of Gelada Baboons*. Tokyo: Kōdansha.
Kawai, M., Iwamoto, M. & Yoshiba, K. (1968). [*Primates of the World*] Tokyo: Mainichi Shinbunsha.
Kawai, M. & Mizuhara, H. (1962). An ecological study on the wild mountain gorilla (*Gorilla gorilla beringei*). *Primates*, **2**, 1–42.
Kawamichi, T. (1976). [The society of tupaia testifying evolution]. Shizen [*Nature, Tokyo*] (1976), (**6**), 26–35; (**7**), 74–86; (**8**), 56–65.
Kawamichi, T. & Kawamichi, M. (1979). Spatial organization and territory of tree shrews (*Tupaia glis*). *Anim. Behav.*, **27**, 381–93.
Kawamura, T. (1950). [Social life of Nara deer].* *Seiri-Seitai* [*Physiology and Ecology, Kyoto*], **4**, 75–87.
Kawamura, T. (1952). [Social life of Japanese deer in Nara Park]. In [*Natural History of Japanese Animals*], vol. 4, ed. K. Imanishi, pp. 1–165. Tokyo: Kōbunsha.

Kawanabe, H. (1966). [Productivity of species populations in communities]. In [*Modern Biology* 9 – *Ecology and Evolution*], pp. 123–53. Tokyo: Iwanami.
Kawano, S. (1975). [Biology and adaptive strategy in weeds]. *Zasso Kenkyū* [*Weed Research, Tokyo*], **20**, 145–9.
Kawano, S. & Nagai, Y. (1975). The productive and reproductive biology of flowering plants. I. Life history strategies of three *Allium* species in Japan. *Bot. Mag., Tokyo*, **88**, 281–318.
Keith, L. B. (1963). *Wildlife's Ten-Year Cycle*. Madison: University of Wisconsin Press.
Kemp, G. A. & Keith, L. B. (1970). Dynamics and regulation of red squirrel (*Tamiasciurus hudsonicus*) populations. *Ecology*, **51**, 763–79.
Kennedy, J. S. (1961). Continuous polymorphism in locusts. In *Insect Polymorphism* ed. J. S. Kennedy, *Symposia of the Royal Entomological Society of London, No.* 1, pp. 80–90. London: Royal Entomological Society.
Kennedy, J. S. & Stroyan, H. L. G. (1959). Biology of aphids. *Annu. Rev. Ent.*, **4**, 139–60.
Kenyon, K. W., Scheffer, V. B. & Chapman, D. G. (1954). A population study of the Alaska fur-seal herd. *Special Sci. Rep. Wildl., U.S. Dept. Fish & Wildlife Service*, **12**, 1–75.
Kerr, W. E. (1950). Genetic determination of castes in the genus *Melipona*. *Genetics*, **35**, 143–52.
Kikuchi, T. (1958). [On the residentiality of a green frog, *Rana nigromaculata* (Hallowell) (1).]* *Jap. J. Ecol.*, **8**, 20–6.
Kilham, L. (1958). Territorial behavior in pikas. *J. Mammal.*, **39**, 307.
King, C. E. (1964). Relative abundance of species and MacArthur's model. *Ecology*, **45**, 716–27.
King, J. A. (1955). Social behavior, social organization, and population dynamics in a black-tailed prairie dog town in the Black Hills of South Dakota. *Contr. Lab. vert. Biol., Univ. Michigan*, **67**, 1–123.
Kipling, C. & Frost, W. E. (1970). A study of the mortality, population numbers, year class strengths, production and food consumption of pike, *Esox lucius* L., in Windermere from 1944 to 1962. *J. Anim. Ecol.*, **39**, 115–57.
Kiritani, K. (1964). Natural control of populations of the southern green stink bug, *Nezara viridula*. *Res. Popul. Ecol.*, **6**, 88–98.
Kiritani, K. (1971). Distribution and abundance of the southern green stink bug, *Nezara viridula*. *Proc. Symp. on Rice Insects* (Tokyo, 1971), 235–48.
Kiritani, K. & Hokyo, N. (1962). Studies on the life table of the southern green stink bug. *Jap. J. appl. Ent. Zool.*, **6**, 124–40.
Kiritani, K., Hokyo, N. & Kimura, K. (1963). Survival rate and reproductivity of the adult southern green stink bug, *Nezara viridula*, in the field cage. *Jap. J. appl. Ent. Zool.*, **7**, 113–24.
Kiritani, K., Hokyo, N. & Kimura, K. (1967). The study of the regulatory system of the population of the southern green stink bug, *Nezara viridula* L. (Heteroptera: Pentatomidae) under semi-natural conditions. *Appl. Ent. Zool.*, **2**, 39–50.
Kiritani, K., Hokyo, N., Sasaba, T. & Nakasuji, F. (1970). Studies on population dynamics of the green rice leafhopper, *Nephotettix cincticeps* Uhler: Regulatory mechanism of the population density. *Res. Popul. Ecol.*, **12**, 137–53.
Kiritani, K. & Nakasuji, F. (1967). Estimation of the stage-specific survival rate in the insect population with overlapping stages. *Res. Popul. Ecol.*, **9**, 143–52.
Kisimoto, R. (1965). [Studies on the polymorphism and its role playing in the population growth of the brown planthopper, *Nilaparvata lugens* Stål.]* *Bull. Shikoku Agric. Exp. Sta.*, **13**, 1–106.
Kisimoto, R. (1971). Long distance migration of planthoppers, *Sogatella furcifera* and *Nilaparvata lugens*. *Proc. Symp. tropical agric. Res.* (Tokyo, 1971), 201–16.

Kitahara, T. (Ki, Ta) (1897). [Oviposition of the Japanese sculpin.] *Dōbutsugaku-Zasshi* [*Zool. Mag., Tokyo*], **9**, 137–8.
Kleiman, D. G. & Eisenberg, J. F. (1973). Comparisons of canid and felid social systems from an evolutionary perspective. *Anim. Behav.*, **21**, 637–59.
Klomp, H. (1965). The dynamics of a field population of the pine looper, *Bupalus piniarius* L. (Lep. Geom.). *Adv. ecol. Res.*, **3**, 207–305.
Kluijver (Kluyver), H. N. (1951). The population ecology of the great tit, *Parus m. major*. *Ardea*, **39**, 1–135.
Kluijver (Kluyver), H. N. (1971). Regulation of numbers in populations of great tits (*Parus m. major*). In *Proceedings of the Advanced Study Institute on 'Dynamics of Numbers in Populations'*, ed. P. J. den Boer & G. R. Gradwell, pp. 507–23. Wageningen: P.U.D.O.C. (centre for Agricultural Publishing and Documentation).
Kobayashi, J. (1933). Ecology of a stickleback, *Pungitius sinensis* var. *kaibarae* (Tanaka). *Jour. Sci. Hiroshima Univ., Ser. B. Div* 1, **2**(**5**), 71–89.
Koford, C. B. (1957). The vicuña and the puma. *Ecol. Monogr.*, **27**, 153–219.
Komatsu, M. (1935). [Ecology of the black-tailed gull *Larus crassirostris* Vieillot on Kabushima, Hachinoe-shi, Aomori Prefecture (Summary report).] *Tori*, **8**(**40**), 446–61.
Konishi, M. & Itô, Y. (1973). Early entomology in east Asia. In *History of Entomology*, ed. R. F. Smith, T. E. Mittler & C. N. Smith), pp. 1–20. Palo Alto, Calif.: Annual Reviews Incorporated.
Kortlandt, A. (1962). [Chimpanzees in their natural habitats.] *Shizen* [*Nature, Tokyo*] (1962), (**8**), 28–37.
Koyama, N. (1976). [The gibbons.] [In a supplement to *Science* (Japan) published as *From Monkeys to Man*], pp. 78–91. Tokyo: Nippon Keizaishinbun.
Krebs C. J. (1964). The lemming cycle at Baker Lake, Northwest Territories, during 1959–62. *Arctic Inst. N. Amer., Technical Paper*, **15**, 1–104.
Krebs, C. J. (1966). Demographic changes in fluctuating populations of *Microtus californicus*. *Ecol. Monogr.*, **36**, 236–73.
Krebs, C. J. & DeLong, K. T. (1965). A *Microtus* population with supplementary food. *J. Mammal.*, **46**, 566–73.
Krebs, J. R. (1971). Territory and breeding density in the great tit, *Parus major* L. *Ecology*, **52**, 2–22.
Kruuk, H. (1972). *The Spotted Hyena: A Study of Predation and Social Behavior*. Chicago: University of Chicago Press.
Kubo, H. (1975). [The magpie – Our neighbour living in rice paddies with creeks.] *Anima* (1975), (**2**), 5–23.
Kubo, I. (1957). [On cycles seen in fisheries records.] *Seibutsu Kagaku* [*Biological Science, Tokyo*], **9**, 126–33.
Kuenzler, E. J. (1958). Niche relations of three species of lycosid spiders. *Ecology*, **39**, 494–500.
Kullmann, E. J. (1972). Evolution of social behavior in spiders (Araneae: Eresidae and Theridiidae). *Amer. Zool.*, **12**, 419–26.
Kumagai, S. (1952). *Nests and Eggs of Japanese Birds*. Sendai: Private Publication.
Kummer, H. (1968). *Social Organization of Hamadryas Baboons. A Field Study*. Chicago: University of Chicago Press.
Kummer, H. & Kurt, F. (1963). Social unit of a free-living population of hamadryas baboons. *Folia Primatol.*, **1**, 4–19.
Kuno, E. (1968). [Studies on the population dynamics of rice leafhoppers in a paddy field.]* *Bull. Kyushu agric. exp. Sta.*, **14**, 131–246.
Kuno, E. & Hokyo, N. (1970). Comparative analysis of the population dynamics of rice leafhoppers, *Nephotettix cincticeps* Uhler and *Nilaparvata lugens* Stål, with special reference to natural regulation of their numbers. *Res. Popul. Ecol.*, **12**, 154–84.

Kuroda, N. (1955). [Studies on the grey starling. The 1st report: from winter to breeding season.]* *Misc. Rept. Yamashina Inst. Ornithol.*, **7**, 277–89.
Kuroda, N. (1956a). [Studies on the grey starling. The 1st report (continued).]* *Misc. Rept. Yamashina Inst. Ornithol.*, **8**, 318–28.
Kuroda, N. (1956b). [Studies on the grey starling. The 2nd report: breeding (1).]* *Misc. Rept. Yamashina Inst. Ornithol.*, **9**, 375–86.
Kuroda, N. (1957). [Studies on the grey starling. The 2nd report: breeding (2).* *Misc. Rept. Yamashina Inst. Ornithol.*, **10**, 413–26.
Kuroda, N. (1962). [Aves.] In [*Animal Systematics and Taxonomy*], vol. 10A, ed. T. Uchida, pp. 1–341. Tokyo: Nakayama.
Kuroda, N. (1967). [*Studies on Birds.*] Tokyo: Shinshichō sha.
Lack, D. (1946a). Competition for food by birds of prey. *J. Anim. Ecol*, **15**, 123–9.
Lack, D. (1946b). *The Life of the Robin.* London: Penguin Books.
Lack, D. (1949). Family size in certain thrushes (Turdidae). *Evolution*, **3**, 57–66.
Lack, D. (1954a). The evolution of reproductive rates. In *Evolution as a Process*, ed. J. Huxley, A. C. Hardy & E. B. Ford, pp. 143–56. London: Allen & Unwin.
Lack, D. (1954b). *The Natural Regulation of Animal Numbers.* Oxford: Clarendon.
Lack, D. (1966). *Population Studies of Birds.* Oxford: Clarendon.
Lack, D. (1968). *Ecological Adaptations for Breeding in Birds.* London: Chapman & Hall.
Larkin, P. A. & McDonald, J. G. (1968). Factors in the population biology of the sockeye salmon of the Skeena River. *J. Anim. Ecol.*, **37**, 229–58.
Lawick, H. van & Lawick-Goodall, J. (1971). *Innocent Killers.* Boston: Houghton Mifflin.
Laws, R. M. (1962). Some effects of whaling on the southern stocks of baleen whales. In *The Exploitation of Natural Animal Populations*, ed. E. D. LeCren & M. W. Holgate, pp. 137–58. Oxford: Blackwell.
LeRoux, E. J., Paradis, R. O. & Hudon, M. (1963). Major mortality factors in the population dynamics of the eye-spotted bud moth, the pistol casebearer, the fruit-tree leaf roller, and the European corn borer in Quebec. *Mem. ent. Soc. Can.*, **32**, 67–82.
Levins, R. (1968). *Evolution in Changing Environments. Some Theoretical Explorations.* Princeton: Princeton University Press.
Lewontin, R. C. (1965). Selection for colonizing ability. In *The Genetics of Colonizing Species*, ed. H. G. Baker & G. L. Stebbins, pp. 79–94. New York: Academic Press.
Li, K. P. Wong, H. H. & Woo, W. S. (1964). [Route of the seasonal migration of the oriental armyworm moth in the eastern part of China as indicated by a three-year result of releasing and recapturing of marked moths.]* [*Acta Phyto-phylacica Sinica*], **3**, 101–10. (In Chinese.)
Lidicker, W. Z. Jr. (1973). Regulation of numbers in an island population of the California vole, a problem in community dynamics. *Ecol. Monogr.*, **43**, 271–302.
Lin, N. (1963). Territorial behavior in the cicada killer wasp *Sphecius speciosus* (Drury) (Hymenoptera: Sphecidae). *Behaviour*, **20**, 115–33.
Linsdale, J. & Tomich, Q. P. (1953). *A Herd of Mule Deer.* Berkeley: University of California Press.
Linsenmair, K. E. & Linsenmair, C. (1971). Paarbildung und Paarzusammenhalt bei der monogamen Wustenassel *Hemilepistus reaumuri* (Crustacea, Isopoda, Oniscoidea). *Z. Tierpsychol.*, **29**, 134–55.
Lloyd, M. & Dybas, H. S. (1966). The periodical cicada problem. *Evolution* **20**, 133–49, 466–505.
Lotka, A. J. (1956) *Elements of Mathematical Biology.* New York: Dover, (reprint).
Lowe, V. P. W. (1969). Population dynamics of the red deer (*Cervus elaphus* L.) on Rhum. *J. Anim. Ecol.*, **38**, 425–58.
Lumsden, W. H. R. (1951). The night resting habits of monkeys in a small area on the

edge of the Semliki forest, Uganda. A study in relation to the epidemiology of sylvan yellow fever. *J. Anim. Ecol.*, **20**, 11–30.

Lüscher, M. (1961). Social control of polymorphism in termites. In *Insect Polymorphism*, ed. J. S. Kennedy, *Symposia of the Royal Entomological Society of London, No.* 1, pp. 57–67. London: Royal Entomological Society.

Ma, S. C. (1958). [The population dynamics of the oriental migratory locust (*Locusta migratoria manilensis* Meyen) in China.]* [*Acta entomol. sinica*], **8**, 1–40. (In Chinese.)

Ma, S. C., Ting, Y. C., & Li, D. M. (1965). [Study on long-term prediction of locust population fluctuations.]* [*Acta entomol. sinica*], **14**, 319–38. (In Chinese.)

Macan, T. T. (1974). Twenty generations of *Pyrrhosoma nymphula* (Sulzer) and *Enallagma cyathigerum* (Charpentier) (Zygoptera: Coenagrionidae). *Odonatologica*, **3**, 107–19.

MacArthur, R. H. (1962). Some generalized theorems of natural selection. *Proc. natl. Acad. Sci.*, **48**, 1893–7.

MacArthur, R. H. & Wilson, E. O. (1967). *The Theory of Island Biogeography*. Princeton: Princeton University Press.

Mackenzie, J. M. D. (1952). Fluctuations in the numbers of British tetraonids. *J. Anim. Ecol.* **21**, 128–53.

MacKinnon, J. (1974). The behavior and ecology of wild orang-utans (*Pongo pygmaeus*). *Anim. Behav.*, **22**, 3–74.

McLean, D. (1940). The deer of California with particular reference to the Rocky Mountain mule deer. *Calif. Fish & Game*, **26**, 139–66.

MacLulich, D. A. (1937). Fluctuations in the numbers of the varying hare (*Lepus americanus*). *Univ. Toronto Stud. Biol. Ser.*, **43**, 1–136.

Macpherson, A. H. (1969). The dynamics of Canadian arctic fox populations. *Canad. Wildl. Service Report Ser.*, **8**, 1–52.

Mann, K. H. (1953). The history of *Erpobdella octoculata* (Linnaeus, 1758). *J. Anim. Ecol.*, **22**, 199–207.

Mann, K. H. (1957). A study of a population of the leech *Glossiphonia complanata* (L.). *J. Anim. Ecol.*, **26**, 99–111.

Marshall, N. B. (1953). Egg size in arctic, antarctic and deep-sea fishes. *Evolution*, **7**, 328–41.

Martof, B. S. (1953a). Territoriality in the green frog, *Rana clamitans*. *Ecology*, **34**, 165–74.

Martof, B. S. (1953b). Home range and movements of the green frog, *Rana clamitans*. *Ecology*, **34**, 529–43.

Martynov, A. V. (1938). Wings of termites and phylogeny of Isoptera and of allied groups of insects. In *A l'Academician N. V. Massonov*, pp. 83–150. Moscow: Academy of Sciences U.S.S.R.

Marukawa, H. (1933). Biological and fishery research on Japanese king-crab *Paralithodes camtschatica* (Pilesius.). *J. Imp. Fish. Exp. Sta.*, **4**, 1–159.

Masure, R. & Allee, W. C. (1934). The social order in flocks of the common chicken and the pigeon. *Auk*, **51**, 306–27.

Matsui, M. (1975). A new type of Japanese toad larvae living in mountain torrents. *Dōbutsugaku-Zasshi* [*Zool. Mag., Tokyo*], **84**, 196–204.

Matsui, M. (1976). A new toad from Japan. *Contr. biol. Lab., Kyoto Univ.*, **25**, 1–10.

Matsura, T. (1976). [Ecological studies of a coccinellid, *Aiolocaria hexaspilota* Hope. I. Interaction between field populations of *A. hexaspilota* and its prey, the walnut leaf beetle (*Gastrolina depressa* Baly).]* *Jap. J. Ecol.*, **26**, 147–56.

Matsuura, M. (1977). [Life of paper wasps.] *Shizen* [*Nature, Tokyo*] (1977), (**2**), 26–36.

Matthews, R. W. (1968). *Microstigmus comes*: sociality in a sphecid wasp. *Science*, **160**, 787–8.

Mayr, E. (1935). Bernard Altum and the territory theory. *Proc. Linn. Soc. N. Y. Nos.*, **45/46**, 24–38.

Michener, C. D. (1958). The evolution of social behavior in bees. In *Proceedings of the Tenth International Congress of Entomology* (Montreal, 1956), vol. 2, ed. J. A. Downes, pp. 441–7.

Michener, C. D. (1962). Biological observations on the primitively social bees of the genus *Allodapula* in the Australian region (Hymenoptera, Xylocopinae). *Insectes Sociaux*, **9**, 355–73.

Michener, C. D. (1969). Comparative social behavior of bees. *Annu. Rev. Entomol.*, **14**, 299–342.

Michener, C. D. (1974). *The Social Behavior of the Bees: A comparative Study.* Cambridge, Mass.: Harvard University Press (Belknap Press).

Michener, C. D. & Lange, R. B. (1958). Distinctive type of primitive social behavior among bees. *Science*, **127**, 1046–7.

Michener, C. D. & Lange, R. B. (1959). Observations on the behavior of Brazilian halictid bees (Hymenoptera, Apoidea). IV. *Augochloropsis*, with notes on extralimital forms. *Amer. Mus. Novitates*, **1924**, 1–41.

Middleton, A. D. (1934). Periodic fluctuations in British game populations. *J. Anim. Ecol.*, **3**, 231–49.

Miller, C. A. (1966). The black-headed budworm in eastern Canada. *Can. Ent.*, **98**, 592–613.

Miller, E. M. (1969). Caste differentiation in the lower termites. In *Biology of Termites*, vol. 1, ed. K. Krishna & F. M. Weesner, pp. 283–310. New York: Academic Press.

Milne, A. (1957a). The natural control of insect populations, *Can. Ent.*, **89**, 193–213.

Milne, A. (1957b). Theories of natural control of insect populations. *Cold Spring Harbor Symp. quant Biol.*, **22**, 253–67.

Milne, L. J. & Milne, M. (1976). The social behavior of burying beetles. *Sci. Amer.*, **235(2)**, 84–9.

Milstead, W. W. (ed.) (1967). *Lizard Ecology; a Symposium.* Columbia: University of Missouri Press.

Miyadi, D., Kawabata, M. & Ueda, K. (1952). [Standard density of 'ayu' on the basis of its behavior and grazing unit area.] [*Kyoto Univ. Fac. Sci. Physiol. Ecol. Res. Rept.*], **75**, 1–25.

Miyadi, D., Kawanabe, H. & Mizuno, N. (1976). [*Freshwater Fishes in Japan, Illustrated in Colour*], new edn. Ōsaka: Hoikusha.

Miyashita, K. (1954). [An analysis on the process of increase of green peach aphid population under closed and open conditions.]* *Jap. J. Ecol.*, **4**, 16–20.

Miyashita, K. (1955). [Some consideration on the population fluctuation of the rice stem borer.]* *Bull. natl. Inst. agric. Sci. Japan, Ser. C.*, **5**, 99–106.

Miyashita, K. (1963). Outbreaks and population fluctuations of insects, with special reference to agricultural insect pests in Japan. *Bull. natl. Inst. agric. Sci. Japan, Ser. C.*, **15**, 99–170.

Miyashita, K., Itô, Y. & Gotoh, A. (1956). [Study on the fluctuation of egg and larval populations of common cabbage butterfly and the factors affecting it. Studies on the mortality rate and behavior of field population of agricultural pest insect (3rd report).]* *Ōyō-Kontū* [*Appl. Entomol.*], **12**, 50–5.

Miyashita, K., Itô, Y., Nakamura, K., Nakamura, M. & Kondo, M. (1965). Population dynamics of the chestnut gall wasp, *Dryocosmus kuriphilus* Yasumatsu (Hymenoptera: Cynipidae). III. Five year observation on population fluctuations. *Jap. J. appl. Ent. Zool.*, **9**, 42–52.

Miyazaki, M. (1974). [The Japanese serow in central Japan Alps – its winter life.] *Anima* (1974), (**1**), 5–24.

Mizuno, N. (1961). [Study on the gobioid fish, 'yoshinobori' *Rhinogobius similis* Gill.]* *Bull. Jap. Soc. Sci. Fish.*, **27**, 6–11, 307–12.
Mizuno, N. & Kawanabe, H. (1957). [Behaviour of salmon-like fish 'ayu' in an area with closely established territories.]* *Jap. J. Ecol.*, **7**, 26–30.
Mizuno, N. & Niwa, H. (1961). [Two ecological types of the freshwater scorpion, *Cottus pollux* Gunther.]* *Dōbutsugaku-Zasshi* [*Zool. Mag., Tokyo*], **70**, 267–75.
Mizuno, T. & Funakawa, T. (1956). [Social behavior of the jumping spider (*Menemerus confluses*).]* *Jap. J. Ecol.*, **6**, 93–6.
Moore, N. W. (1952). On the so-called 'territories' of dragonflies (Odonata–Anisoptera). *Behaviour*, **4**, 85–99.
Moore, N. W. (1953). Population density in adult dragonflies (Odonata–Anisoptera). *J. Anim. Ecol.*, **22**, 344–59.
Moore, N. W. (1964). Intra- and interspecific competition among dragonflies (Odonata). An account of observations and field experiments on population density control in Dorset, 1954—60. *J. Anim. Ecol.*, **33**, 49–71.
Moran, P. A. P. (1949). The statistical analysis of the sunspot and lynx cycles. *J. Anim. Ecol.*, **18**, 115–16.
Moreau, R. E. (1944). Clutch size: A comparative study, with special reference to African birds. *Ibis*, **86**, 286–347.
Mori, S. (1956). [Social organization of a group of young girelloid fish, *Girella punctata* Gray, confined in a tide-pool, with special reference to the relation between social hierarchy and territorial system.]* *Jap. J. Ecol.*, **5**, 145–50.
Morisita, M. (1939). [On the interrelationships among several species of ants on trees.] *Kansai Konchū-gakkai Kaiho* [*Proc. Kansai entomol. Soc.*], **9(2)**, 22–42.
Morisita, M. (1950). [Population density and movement of the water strider *Gerris lacustris* – observation and discussion on animal aggregations.] *Kyoto University Fac. Sci. Physiol. Ecol. Res. Rept.*, **65**, 1–149.
Morisita, M. (1975). [Developmental stages of survivorship curve studies.] [*Proc. Soc. Popul. Ecol.*], **26/27**, 12–7.
Morisita, M. (1976). [*Societies of Animals.*] Tokyo: Kyōritsu.
Morris, R. F. (ed.) (1963). On the dynamics of epidemic spruce budworm populations. *Mem. ent. Soc. Can.*, **31**, 1–332.
Morris, R. F. & Miller, C. A. (1954). The development of life tables for the spruce budworm. *Canad. J. Zool.*, **32**, 283–301.
Morris, R. F., Miller, C. A., Greenbank, D. O. & Mott, D. G. (1958). The population dynamics of the spruce budworm in eastern Canada. In *Proceedings of the Tenth International Congress of Entomology* (Montreal, 1956), vol. 4, ed. E. B. Watson, B. M. McGugan & M. L. Prebble, pp. 137–49.
Murie, A. (1944). *The wolves of Mt. McKinley.* Washington, D.C.: U.S. National Parks Service.
Murton, R. K., Westwood, N. J. & Isaacson, A. J. (1974). A study of wood-pigeon shooting: the exploitation of a natural animal population. *J. appl. Ecol.*, **11**, 61–81.
Myers, J. H. & Krebs, C. J. (1974). Population cycles in rodents. *Sci. Amer.*, **230(6)**, 38–46.
Myers, K. (1971). The rabbits in Australia. In *Proceedings of the Advanced Study Institute on 'Dynamics of Numbers in Populations'*, ed. P. J. den Boer & G. R. Gradwell, pp. 478–506. Wageningen: P.U.D.O.C. (Centre for Agricultural Publishing and Documentation).
Nagamine, M., Teruya, R. & Itô, Y. (1975). A life table of *Mogannia iwasakii* (Homoptera: Cicadiidae) in sugarcane field of Okinawa. *Res. Popul. Ecol.*, **17**, 39–50.
Nakai, Z. (1962). Studies relevant to mechanisms underlying the fluctuation in the catch

of the Japanese sardine, *Sardinops melanosticta* (Temminck & Schlegel). *Jap. J. Ichthyol.*, **9**, 1–115.
Nakai, Z., Usami, S., Hattori, S., Honjō, K. & Hayashi, S. (1955). *Progress Report of the Cooperative Iwashi Resources Investigations.* Tokyo: Fisheries Agency.
Nakamura, K., Itô, Y., Nakamura, M., Matsumoto, T. & Hayakawa, K. (1971). Estimation of population productivity of *Parapleurus alliaceus* Germer (Orthoptera, Acridiidae) on a *Miscanthus sinensis* grassland. I. Estimation of population parameters. *Oecologia*, **7**, 1–15.
Nakamura, K., Nakamura, M., Matsumoto, T. & Itô, Y. (1975). Fluctuations of grasshopper populations on a *Miscanthus sinensis* grassland. *Res. Popul. Ecol.*, **16**, 198–206.
Nakamura, T. (1969). [Structure of flock home range in the long-tailed tit. I. Winter flock, its home range and territory.]* *Misc. Rept. Yamashina Inst. Ornithol.*, **5(31)**, 433–61.
Nakamura, T. (1976a). [*The Social Life of Birds.*] Tokyo: Shisakusha.
Nakamura, T. (1976b). [Dances of ptarmigan and grouse.] *Anima* (1976), **(7)**, 22–8.
Napier, J. R. & Napier, P. H. (1967). *A Handbook of Living Primates.* London: Academic Press.
Neill, S. R. St. J. & Cullen, J. M. (1974). Experiments on whether schooling by their prey affects the hunting behaviour of cephalopods and fish predators. *J. Zool., Lond.*, **172**, 549–69.
Neilson, M. M. & Morris, R. F. (1964). The regulation of European spruce sawfly numbers in the maritime Provinces of Canada from 1937 to 1963. *Can. Ent.*, **96**, 773–84.
Nice, M. M. (1941). The role of territory in bird life. *Amer. Midl. Nat.*, **26**, 441–87.
Nicholson, A. J. (1933). The balance of animal populations. *J. Anim. Ecol.* (*Suppl.*), **2**, 132–78.
Nicholson, A. J. (1957). The self-adjustment of populations to change. *Cold Spring Harbor Symp. quant. Biol.*, **22**, 153–73.
Nishida, T. (1968). The social group of wild chimpanzees in the Mahali Mountains. *Primates*, **9**, 167–227.
Nishida, T. (1975). ['Sex' and 'society' of wild chimpanzees.] *Anima* (1975), **(3)**, 5–20.
Nishimura, A. & Izawa, K. (1975). The group characteristics of woolly monkeys (*Lagothrix lagothrix*) in the upper Amazonian Basin. In *Contemporary Primatology*, ed. S. Kondo, M. Kawai & A. Ehara, pp. 351–7, Basel: S. Karger.
Nissen, H. W. (1931). A field study of the chimpanzee. Observations of chimpanzee behavior and environment in Western French Guinea. *Comp. psych. Monogr.*, **8(36)**, 1–122.
Noble, G. K. (1938). Sexual selections among fishes. *Biol. Rev.*, **13**, 133–58.
Noble, G. K. (1939a). The role of dominance in the social life of birds. *Auk*, **56**, 264–73.
Noble, G. K. (1939b). The experimental animal from the naturalist's point of view. *Amer. Natur.*, **73**, 113–26.
Noble, G. K. & Curtis, B. (1939). The social behavior of the jewel fish, *Hemichromis bimaculatus* Gill. *Bull. Amer. Mus. Nat. Hist.*, **76**, 1–46.
Noirot, C. (1961). Formation of castes in the higher termites. In *Biology of Termites*, vol. 1, ed. K. Krishna & F. M. Weesner, pp. 311–50. New York: Academic Press.
Odum, E. P. (1971). *Fundamentals of Ecology*, 3rd edn. Philadelphia: W. B. Saunders.
Ohgushi, R. (1953). On the environmental resistances of *Ectemnius* (*Hypocrabro*) *rubicola* (Dufour & Perris) (Hymenoptera: Sphecidae). *Trans. Shikoku ent. Soc.*, **4**, 83–6.
Ohgushi, R. (1955). [Ethological studies on the intertidal limpets. II. Analytical studies on the homing behavior of two species of limpets.]* *Jap. J. Ecol.*, **5**, 31–5.
Okada, Y. (1955). *Fishes of Japan.* Tokyo: Maruzen.
Oku, T. & Kobayashi, T. (1973). Some dynamic aspects of field populations of the cab-

bage armyworm, *Mamestra brassiae* Linne, in Tohoku District. II. Mortalities during the progress and breakdown of an outbreak on the sugarbeet. *Kontyū*, **41**, 267–79.
Oliver, A. (1956). Reproduction in the king cobra, *Ophiophagus hannah* Cantor. *Zoologica*, **41**, 145–57.
Ono, Y. (1957). [The interrelation among individuals of a fiddler crab, *Ilyoplax pusillus* De Haan.]* *Jap. J. Ecol.*, **7**, 45–51.
Organ, J. A. (1961). Studies of the population dynamics of the salamander genus *Desmognathus* in Virginia. *Ecol. Monogr.*, **31**, 189–220.
Orians, G. H. (1961). The ecology of blackbird (*Agelaius*) social systems. *Ecol. Monogr.*, **31**, 285–312.
Orians, G. H. (1969). On the evolution of mating systems in birds and mammals. *Amer. Natur.*, **103**, 589–603.
Otte, D. & Joern, A. (1975). Insect territoriality and its evolution: population studies of desert grasshoppers on creosote bushes. *J. Anim. Ecol.*, **44**, 29–54.
Owen, D. F. (1977). Latitudinal gradients in clutch size: an extension of David Lack's theory. In *Evolutionary Ecology*, ed. B. Stonehouse & C. Perrins, pp. 171–9. London: Macmillan.
Palmer, R. S. (1951). The whitetail deer of Tombegan Camps, Main, with added notes on fecundity. *J. Mammal.*, **32**, 267–81.
Paradis, R. O. & LeRoux, E. J. (1965). Recherches sur la biologie et la dynamique des populations naturelles d'*Archips argyrospilus* (Wlk.) (Lépidoptères: Tortricidae) dans le sud-ouest du Québec. *Mem. Soc. ent. Can.*, **43**, 1–77.
Pardi, L. (1948). Dominance order in *Polistes* wasps. *Physiol. Zoöl.*, **21**, 1–13.
Pardi, L. (1952). Dominazione e gerarchia in alcuni invertebrati. In *Structure et Physiologie des Sociétés Animales: Colloques Internationaux, Centre National de la Recherche Scientifique*, **34**, pp. 183–97. Paris: Centre N.R.S.
Pearl, R. (1922). *The Biology of Death.* Philadelphia: Lippincott.
Pearl, R. & Miner, J. R. (1935). Experimental studies on the duration of life. XIV. The comparative mortality of certain lower organisms. *Quart. Rev. Biol.*, **10**, 60–79.
Pearson, P. G. (1955). Population ecology of the spade foot toad, *Scaphiopus h. holbrooki* (Harlan). *Ecol. Monogr.*, **25**, 233–69.
Peckham, D. J., Kurczewski, F. E. & Peckham, D. B. (1973). Nesting behavior of nearctic species of *Oxybelus* (Hymenoptera: Sphecidae). *Ann. ent. Soc. Amer.*, **66**, 647–61.
Peek, J. M., Lovaas, A. L. & Rouse, R. A. (1967). Population changes within the Gallatin elk herd, 1932–65. *J. Wildl. Manage.*, **31**, 304–16.
Petrides, G. A. & Swank, W. G. (1966). Estimating the productivity and energy relations of an African elephant population. In *Proceedings of the Ninth International Grassland Congress*, pp. 831–42. São Paulo: State of São Paulo (Department of Animal Production, Secretary of Agriculture).
Petter, J. (1962). Recherches sur l'ecologie et l'ethologie des Lemuriens malagaches. *Mém. Mus. Hist. Nat., Paris Ser. A* (*Zool.*), **27**, 1–146.
Petter, J. (1965). The lemurs of Madagascar. In *Primate Behavior. Field Studies of Monkeys and Apes*, ed. I. DeVore, pp. 292–319. New York: Holt, Rinehart & Winston.
Pianka, E. R. (1970). On *r*- and *K*-selection. *Amer. Natur.*, **104**, 592–7.
Pianka, E. R. (1974). *Evolutionary Ecology.* New York: Harper.
Piccioli, M. T. M. & Pardi, L. (1970). Studi sulla biologia di *Belonogaster* (Hymenoptera, Vespidae). I. Sull'ectogramma di *Belonogaster griseus* (Fab.). *Monitore zool. Ital.* (N.S.), Suppl. III, 197–225.
Pickles, W. (1940). Fluctuations in the populations, weights and biomasses of ants at Thornhill, Yorkshire, from 1935 to 1939. *Trans. ent. Soc. Lond.*, **90(17)**, 467–85.
Pitelka, F. A. (1957a). Some characteristics of microtine cycles in the Arctic. In *Arctic*

Biology, ed. H. P. Hansen, *18th Biology Colloqium*, pp. 73–88. Corvallis: Oregon University Press.

Pitelka, F. A. (1957b). Some aspects of population structure in the short-term cycle of the brown lemming in Northern Alaska. *Cold Spring Harbor Symp. quant. Biol.*, **22**, 237–51.

Pitelka, F. A. (1973). Cyclic pattern in lemming populations near Barrow, Alaska. In *Alaskan Arctic Tundra*, ed. M. E. Britton, *Arctic Institute of North America, Technical Paper No. 25*, pp. 199–215. Montreal: Arctic Institute of North America.

Plateaux-Quénu, C. (1959). Un nouveau type de société d'insectes: *Halictus marginatus* Brullé (Hymenoptera, Apidae). *Année biol.*, **35**, 325–444.

Plateaux-Quénu, C. (1962). Biology of *Halictus marginatus* Brullé. *J. apicultural Res.*, **1**, 41–51.

Pooley, A. C. & Gans, C. (1976). The Nile crocodile. *Sci. Amer.*, **234(4)**, 114–24.

Porter, K. R. (1972). *Herpetology*. Philadelphia: Saunders.

Price, P. W. (1973). Parasitoid strategies and community organization. *Environmental Entomol.*, **2**, 623–6.

Price, P. W. (1975). Reproductive strategies of parasitoids. In *Evolutionary Strategies of Parasitic Insects and Mites*, ed. P. W. Price, pp. 87–111. New York: Plenum.

Primate Research Group (Japan) (1957). [The status of natural troops of Japanese macaques in various parts of the country.] Jikken-Dōbutsu [*Experimental Animals*], **6(4)**, 97–101.

Prokopy, R. J. (1972). Evidence for a marking pheromone deterring repeated oviposition in apple maggot flies. *Environmental Entomol.*, **1**, 326–32.

Pukowski, E. (1933). Ökologische Untersuchungen an *Necrophorus* F. *Z. Morphol. Ökol. Tiere*, **27**, 518–86.

Raatikanen, M. (1967). Bionomics, enemies and population dynamics of *Javesella pellucida* (F.) (Hom., Delphacidae). *Ann. Agric. Fenniae*, **6**, 1–149.

Rabb, G. B. (1973). Evolutionary aspects of the reproductive behavior of frogs. In *Evolutionary Biology of the Anurans*, ed. J. L. Vial, pp. 213–27. Columbia: University of Missouri Press.

Radovich, J. (1962). Effects of sardine spawning stock size and environment on year-class production. *Calif. Dept. Fish. Game Bull.*, **48**, 123–40.

Rand, A. S. (1967). The adaptive significance of territoriality in iguanid lizards. In *Lizard Ecology; a Symposium*, ed. W. W. Milstead, pp. 106–15. Columbia: University of Missouri Press.

Raw, A. L. (1975). Territoriality and scent marking by *Centris* males (Hymenoptera: Anthophoridae) in Jamaica. *Behaviour*, **54**, 311–21.

Reynolds, V. & Reynolds, F. (1965). Chimpanzees of the budongo forest. In *Primate Behavior. Field Studies of Monkeys and Apes*, ed. I DeVore, pp. 368–424. New York: Holt, Rinehart & Winston.

Reynoldson, T. B. (1955). Factors influencing populations of *Urceolaria mitra* (Peritricha) epizoic on freshwater triclads. *J. Anim. Ecol.*, **24**, 57–83.

Reynoldson, T. B. (1964). Evidence for intra-specific competition in field populations of triclads. *J. Anim. Ecol.*, **33** (suppl.), 187–201.

Richards, O. W. & Southwood, T. R. E. (1968). The abundance of insects: Introduction. In *Insect Abundance*, ed. T. R. E. Southwood, pp. 1–7. Oxford: Blackwell.

Richards, O. W. & Waloff, N. (1954). Studies on the biology and population dynamics of British grasshoppers. *Anti-Locust Bull.*, **17**, 1–182.

Richards, O. W. & Waloff, N. (1961). A study of natural population of *Phytodecta olivacea* (Forster) (Coleoptera; Chrysomelidae). *Phil. Trans. roy. Soc. Lond.*, **B244**, 205–57.

Richdale, L. E. (1957). *A Population Study of Penguins*. Oxford: Clarendon.

Ricker, W. E. (1954). Stock and recruitment. *J. Fish. Res. Bd. Can.*, **11**, 559–623.

Rodman, P. S. (1973). Population composition and adaptive organization among orang-utans of the Kutai Reserve. In *Comparative Ecology and Behaviour of Primates*, ed. R. P. Michael & J. H. Crook, pp. 171–209. New York & London: Academic Press.

Rowell, T. E. (1966). Forest-living baboons in Uganda. *J. Zool., Lond.*, **149**, 344–64.

Royama, T. (1969). A model for the global variation of clutch size in birds. *Oikos*, **20**, 562–7.

Russell, F. S. (1936). The seasonal abundance of the pelagic young of teleostean fishes in the Plymouth area. III. *J. Marine biol. Assoc.*, N.S., **20(2)**, 595–604.

Safriel, U. N. (1975). On the significance of clutch size in nidifugous birds. *Ecology*, **56**, 703–8.

Sakagami, S. F. (1958). [Forest protection with an ant, *Formica rufa*, in Germany.] *Shinrin-Bōeki Nyūsu* [*Forest Protection News*], **7(5)**, 13–17.

Sakagami, S. F. (1970). [*The Path Taken by the Honeybee.*] Tokyo: Shisakusha.

Sakagami, S. F. (1975). [*Brazil and Her Bees in My Life.*] Tokyo: Shisakusha.

Sakagami, S. F. & Fukuda, H. (1968). Life tables for worker honeybees. *Res. Popul. Ecol.*, **10**, 127–39.

Sakagami, S. F. & Hayashida, K. (1958). Biology of the primitive social bee, *Halictus duplex* Dalla Torre. I. Preliminary report on the general life history. *Annot. Zool. Japon.*, **31**, 151–5.

Sakagami, S. F. & Hayashida, K. (1968). Bionomics and sociology of the summer matrifilial phase in the social halctine bee, *Lasioglossum duplex*. *J. Fac. Sci., Hokkaido Univ., Ser. VI (Zool.)*, **16**, 413–513.

Sakagami, S. F., Hirashima, Y. & Ohé, Y. (1966). Bionomics of two new Japanese halictine bees (Hymenoptera, Apoidea). *J. Fac. Agric., Kyushu Univ.*, **13**, 674–703.

Sakagami, S. F., Laroca, S. & Moure, J. S. (1967). Two Brazilian apid nests worth recording in reference to comparative bee sociology, with description of *Euglossa melanotricha* Moure sp. n. (Hymenoptera, Apidae). *Annot. Zool. Japan.*, **40**, 45–54.

Sakagami, S. F., Montenegro, M. J. & Kerr, W. E. (1965). Behavior studies of the stingless bees, with special reference to the oviposition process. I. *Melipona compressipes manaosensis* Schwarz. *J. Fac. Sci., Hokkaido Univ., Ser. VI (Zool.)*, **15**, 300–18.

Sakagami, S. F., Mori, H. & Kikuchi, H. (1956). [Miscellaneous observations on a pika, *Ochotona* sp., inhabiting near Lake Shikaribetsu, Taisetsuzan National Park, Hokkaido.]* *Jap. J. appl. Zool.*, **21**, 1–8.

Sakagami, S. F., Ubukata, H., Iga, M. & Toda, M. J. (1974). Observations on the behavior of some Odonata in the Bonin Islands, with considerations on the evolution of reproductive behavior in Libellulidae. *J. Fac. Sci., Hokkaido Univ., Ser. VI (Zool.)*, **19**, 722–57.

Sakagami, S. F. & Zucchi, R. (1967). Behavior studies of the stingless bees, with special reference to the oviposition process. VI. *Trigona (Tetragona) clavipes*. *J. Fac. Sci., Hokkaido Univ., Ser. VI (Zool.)*, **16**, 292–313.

Sakagami, S. F. & Zucchi, R. (1968). Oviposition behavior of an Amazonic stingless bee, *Trigona (Duckeola) ghilianii*. *J. Fac. Sci., Hokkaido Univ., Ser. VI (Zool.)*, **16**, 564–81.

Sakaguchi, S. (1922). [On the fighting fish, *Macropodus opercularis*.] *Dōbutsugaku-Zasshi* [*Zool. Mag., Tokyo*], **34**, 915–20.

Salisbury, E. J. (1942). *The Reproductive Capacity of Plants – Studies in Quantitative Biology*. London: G. Bell & Sons.

Sano, M. (1973). [Structure of population range in *Passer montanus*.]* *Misc. Rept. Yamashina Inst. Ornithol.*, **7(39)**, 73–86.

Schaller, G. B. (1961). The orang-utan in Sarawak. *Zoologica*, **46**, 73–82.

Schaller, G. B. (1963). *The Mountain Gorilla. Ecology and Behavior*. Chicago: University of Chicago Press.

Schaller, G. B. (1972). *The Serengeti Lion. A Study of Predator–Prey Relations.* Chicago: University of Chicago Press.
Schjelderup-Ebbe, T. (1922). Beiträge zur Sozialpsychologie des Haushuhns. *Z. Psychol.*, **88**, 225–52.
Schneirla, T. C. (1933). Studies on army-ants in Panama. *J. comp. Psych.*, **15**, 267–99.
Schneirla, T. C. (1938). A theory of army-ant behavior based upon the analysis of activities in a representative species. *J. comp. Psych.*, **25**, 51–90.
Schneirla, T. C. (1940). Further studies on the army-ant behavior pattern. Mass organization in the swarm-riders. *J. comp. Psych.*, **29**, 401–60.
Schneirla, T. C. (1949). Army-ant life and behavior under dry-season conditions. 3. The course of reproduction and colony behavior. *Bull. Amer. Mus. Nat. Hist.*, **94(1)**, 3–81.
Schneirla, T. C. (1952). Basic correlations and coordinations in insect societies with special reference to ants. In *Structure et Physiologie des Sociétés Animales: Colloques Internationaux, Centre National de la Recherche Scientifique*, **34**, pp. 247–69. Paris: Centre N.R.S.
Schneirla, T. C., Brown, R. Z. & Brown, F. C. (1954). The bivouac or temporary nests as an adaptive factor in certain terrestrial species of army ants. *Ecol. Monogr.*, **24**, 269–96.
Schneirla, T. C. & Piel, G. (1948). The army ant. *Sci. Amer.*, **178(6)**, 16–23.
Schoener, T. W. (1968). Sizes of feeding territories among birds. *Ecology*, **49**, 123–44.
Schultz, A. M. (1964). The nutrient recovery hypothesis for arctic microtine cycles. II. Ecosystem variables in relation to arctic microtine cycles. In *Grazing in Terrestrial and Marine Environments*, ed. D. J. Crisp, pp. 57–68. Oxford: Blackwell.
Schultz, A. M. (1969). A study of an ecosystem: The arctic tundra. In *The Ecosystem Concept in Natural Resource Management*, ed. G. M. van Dyne, pp. 77–93. New York: Academic Press.
Schwerdtfeger, F. (1932). Betrachtungen zur Epidemiologie des Kiefernspanners. *Z. angew. Ent.*, **19**, 104–29.
Schwerdtfeger, F. (1935a). Studien der Massenwechsel einiger Forstschädlinge. I. Das Klima der Schadgebiete von *Bupalus piniarius*, *Panolis flammea* und *Dendrolimus pini* in Deutschland, *Z. Forst.- und Jagdw.*, **67**, 15–38, 85–104.
Schwerdtfeger, F. (1935b). Studien der Massenwechsel einiger Forstchädlinge. II. Über die Populationsdichte von *Bupalus piniarius*, *Panolis flammea*, *Dendrolimus pini*, *Sphinx pinastri* und ihrer zeitlichen Wechsel. *Z. Forst.-und Jagdw.*, **67**, 449–82, 513–40.
Schwerdtfeger, F. (1935c). Studien der Massenwechsel einiger Forstschädlinge. III. Untersuchungen über die Mortalität der Forleule (*Panolis flammea* Schiff.) in Krisenjahr einer Epidemie. *Mitt. Forstwirts. Forstwiss.*, **1934**, 417–74.
Schwerdtfeger, F. (1942). Über die Ursachen des Massenwechsels der Insekten. *Z. angew. Ent.*, **28**, 254–304.
Scott, J. W. (1942). Mating behavior of the sage grouse. *Auk*, **59**, 477–98.
Scott, J. W. (1944). Additional observations on the mating behavior of the sage grouse. *Anat. Rec., Suppl.*, **89**, 24 (abstract).
Seton, E. T. (1911). *The Arctic Prairies.* New York: International University Press.
Sette, O. E. (1943). Biology of Atlantic mackerel (*Scomber scombrus*) of the North America. *U.S. Fish & Wildl. Serv. Bull.*, **50**, 147–237.
Sexton, O. J. (1960). Some aspects of the behavior and of the territory of a dendrobatid frog, *Prostherapis trinitatis. Ecology*, **41**, 107–15.
Sexton, O. J. (1962). Apparent territorialism in *Leptodactylus insularum* Barbour. *Herpetologica*, **18**, 212–14.
Shelford, V. E. (1927). An experimental investigation of the relations of the codling moth to weather and climate. *Bull. Ill. nat. Hist. Survey*, **16**, 307–440.
Shelford, V. E. (1929). *Laboratory and Field Ecology.* Baltimore: Williams & Wilkins.

Shelford, V. E. (1932). An experimental and observational study of the chinch bug in relation to climate and weather. *Bull. Ill. nat. Hist. Survey*, **19**, 487–547.
Shimomura, K. (1938). [*Handbook of Wild Birds, with Colour Illustrations.*] Tokyo: Sanseido.
Shokita, S. (1973a). Abbreviated larval development of fresh-water atyid shrimp, *Caridina brevirostris* Stimpson from Iriomote Island of the Ryukyus. *Bull. Sci. Eng. Div., Univ. Ryukyus, Math. & Nat. Sci.*, **16**, 222–31.
Shokita, S. (1973b). Abbreviated larval development of the fresh-water prawn, *Macrobrachium shokitai* Fujino & Baba (Decapoda, Palaemonidae) from Iriomote Island of the Ryukyus. *Annot. Zool. Japon.*, **46**, 111–26.
Shokita, S. (1975). [The distribution and speciation of the inland water shrimps and prawns from the Ryukyu Islands.]* *Bull. Sci. Eng. Div., Univ. Ryukyus, Math. & Nat. Sci.*, **18**, 115–36.
Shokita, S. (1976). Early life-history of the land-locked atyid shrimp, *Caridina denticulata ishigakiensis* Fujino & Shokita, from the Ryukyu Islands. *Researches on Crustacea (Tokyo)*, **7**, 1–10.
Silverstone, P. A. (1975). A revision of the poison-arrow frogs of the genus *Dendrobates* Wagler, *Nat. Hist. Mus. Los Angeles County Sci. Bull.*, **21**, 1–55.
Silverstone, P. A. (1976). A revision of the poison-arrow frogs of the genus *Phyllobates* Bibron *in* Sagra (Family Dendrobatidae). *Nat. Hist. Mus. Los Angeles County Sci. Bull.*, **27**, 1–53.
Simonds, P. E. (1965). The bonnet macaque in south India. In *Primate Behavior. Field Studies of Monkeys and Apes*, ed. I. DeVore, pp. 175–96. New York: Holt, Rinehart & Winston.
Simons, E. L. (1977). Ramapithecus *Sci. Amer.*, **236(5)**, 28–35.
Simons, E. L. & Ettel, P. C. (1970). Gigantopithecus. *Sci. Amer.*, **222(1)**, 76–85.
Slade, N. A. & Balph, D. F. (1974). Population ecology of Uinta ground squirrels. *Ecology*, **55**, 989–1003.
Smith, C. C. (1968). The adaptive nature of social organization in the genus of tree squirrels *Tamiasciurus*. *Ecol. Monogr.*, **38**, 31–63.
Smith, H. S. (1935). The role of biotic factors in the determination of population densities. *J. econ. Ent.*, **28**, 873–98.
Smith, W. J., Smith, S. L., Oppenheimer, E. C., Villa, J. G. de & Ulmer, F. A. (1973). Behavior of a captive population of black-tailed prairie dogs. Annual cycle of social behavior. *Behaviour*, **46**, 189–220.
Snow, D. W. & Lill, A. (1974). Longevity records for some neotropical land birds. *Condor*, **76**, 262–7.
Snyder, T. H. (1949). Catalog of the termites (Isoptera) of the world. *Smithsonian Miscel. Collect.*, **112**, 1–490.
Solomon, M. E. (1949). The natural control of animal populations. *J. Anim. Ecol.*, **18**, 1–35.
Solomon, M. E. (1953). The population dynamics of storage pests. *Trans. IXth int. Congr. Ent.*, **2**, 235–48.
Solomon, M. E. (1957). Dynamics of insect populations. *Annu. Rev. Ent.*, **2**, 121–42.
Solomon, M. E. (1969). *Population Dynamics*. London: Arnold.
Southern H. N. (1970). The natural control of a population of tawny owls (*Strix aluco*). *J. Zool., Lond.*, **162**, 197–285.
Southward, A. J. & Crisp, D. J. (1956). Fluctuations in the distribution and abundance of intertidal barnacles. *J. marine biol. Assoc.*, N.S., **35**, 211–29.
Southwick, C. H., Beg, M. A. & Siddiqi, M. R. (1965). Rhesus monkeys in north India. In *Primate Behavior. Field Studies of Monkeys and Apes*, ed. I. DeVore, pp. 111–59. New York: Holt, Rinehart & Winston.

Southwood, T. R. E. (1977). Habitat, the templet for ecological strategies? *J. Anim. Ecol.*, **46**, 337–65.
Southwood, T. R. E. & Reader, P. M. (1976). Population census data and key-factor analysis for the viburnum whitefly, *Aleurotrachelus jelinekii* (Franenf.), on three bushes. *J. Anim. Ecol.*, **45**, 313–25.
Spinage, C. A. (1972). African ungulate life tables. *Ecology*, **53**, 645–52.
Stebbins, R. C. & Robinson, H. B. (1946). Further analysis of a population of a lizard, *Sceloporus graciosus gracilis. Univ. Calif. Publ. Zool.*, **48**, 149–68.
Stimson, J. (1970). Territorial behavior of the owl limpet, *Lottia gigantea. Ecology*, **51**, 113–18.
Stoddart; M. (1976). *Mammalian Odours and Pheromones.* London: Arnold.
Stortenbeker, C. W. (1967). Observations on the population dynamics of the red locust, *Nomadacris septemfasciata* (Serville) in its outbreak areas. *Mededeling Nr.* 84 (Inst. Toegepast Biol. Onderzoek Natuur), 1–118.
Struhsaker, T. T. (1969). Correlates of ecology and social organization among African cerocopithecines. *Folia Primatol.*, **11**, 80–118.
Suehiro, Y. (1942). [*The Life of Fishes.*] Tokyo: Iwanami.
Sugiyama, Y. (1961). [The social structure of a sand-crab, *Scopimera globosa* De Haan, with special reference to its population.]* *Seiri-Seitai* [*Physiology and Ecology*, Kyoto], **10**, 10–17.
Sugiyama, Y. (1964). Group composition, population density and some sociological observations of hunuman langurs (Presbytis entellus). *Primates*, **5**, 7–37.
Sugiyama, Y. (1968). Social organization of chimpanzees at the Budongo Forest, Uganda. *Primates*, **9**, 225–58.
Sugiyama, Y. (1969). Social behavior of chimpanzees in the Budongo Forest, Uganda. *Primates*, **10**, 197–225.
Sugiyama, Y. (1973). The social structure of wild chimpanzees. A review of field studies. In *Comparative Ecology and Behaviour of Primates*, ed. R. P. Michael & J. H. Crook, pp. 376–410. New York & London: Academic Press.
Sugiyama, Y. (1976). Life history of male Japanese monkeys. *Adv. Study of Behavior*, **7**, 255–84.
Suzuki, A. (1965). An ecological study of wild Japanese monkeys in snowy areas – focused on their food habits. *Primates*, **6**, 31–72.
Suzuki, A. (1966). On the insect-eating habits among wild chimpanzees living in the savanna woodland of western Tanzania. *Primates*, **7**, 481–7.
Suzuki, A. (1969). An ecological study of chimpanzees in a savanna woodland. *Primates*, **10**, 103–48.
Suzuki, A. (1971). Carnivory and cannibalism observed among forest-living chimpanzees. *J. anthrop. Soc. Nippon*, **79**, 30–48.
Suzuki, A. (1975). The origin of hominid hunting: a primatological perspective. In *Socioecology and Psychology of Primates*, ed. R. H. Tuttle, pp. 259–78. The Hague: Mouton.
Suzuki, A. (1976). [Ecology of primates.] In [*Treatise on Anthropology*. 2. *Primates*], ed. J. Itani, pp. 147–94. Tokyo: Yūzankaku.
Suzuki, A. (1977). [Society and adaptation of the chimpanzee.] In [*The Natural History of Chimpanzees*], ed. J. Itani, pp. 251–336. Tokyo: Kōdansha.
Suzuki, A., Wada, K., Yoshihiro, S., Tokita, E., Hara, S. & Aburada, Y. (1975). [Population dynamics and group movement of Japanese monkeys in Yokoyugawa Valley, Shiga Heights.]* *Seiri-Seitai* [*Physiology and Ecology, Kyoto*], **16**, 15–23.
Suzuki, Y. (1976). So-called territorial behaviour of the small copper, *Lycaena phlaeas daimio* Seits (Lepidoptera: Lycaenidae). *Kontyū*, **44**, 194–204.
Symmons, P. M. (1966). Assessing the size of populations of adults of the red locust,

Nomadacris septemfasciata (Serv.) in their outbreak areas by means of a helicopter. *Bull. ent. Res.*, **56**, 715–24.

Taber, R. D. (1949). Observations on the breeding behavior of the ring-necked pheasant. *Condor*, **51**, 153–75.

Taber, R. D. & Dasmann, R. F. (1957). The dynamics of three natural populations of the deer *Odocoileus hemionus columbianus*. *Ecology*, **38**, 233–46.

Tachikawa, M. (1957). [Ecology of a small dragonfly, *Nannophya pygmaea*.] *Saishū to Shiiku* [*Collecting & Breeding*], **20(6)**, 166–70.

Taffe, C. A. & Ittyeipe, K. (1976). Effect of nest substrata on the mortality of *Eumenes colona* Saussure (Hymenoptera) and its inquilines. *J. Anim. Ecol.*, **45**, 303–11.

Takai, A., Itô, Y., Miyashita, K. & Nakamura, K. (1963). [Population dynamics of *Mecostethus magister* (Orthoptera: Acrididae).]* *Jap. J. Ecol.*, **13**, 196–204.

Takedatsu, M. (1973). [*Vulpes schrencki*; its wild life. III.] *Anima* (1973), **(11)**, 5–23.

Taketō, A. (1958). [Some ecological observations on *Oligoaeschna pryeri* Martin (Aeschnidae).]* *Tombo* [*Dragonfly*], **1(2/3)**, 12–17.

Tanai, M. (1940). [On the stock of the yellowtail.] *Bull. Jap. Soc. Sci. Fish.*, **9**, 91–3.

Tanaka, J. (1965). Social structure of Nilgiri langurs. *Primates*, **6**, 107–22.

Tanaka, R. (1953). Home ranges and territories in a *Clethrionomys*-population on a peat-bog grassland in Hokkaido. *Bull. Kōchi Women's College*, **2**, 10–20.

Tanaka, R. (1954). [The Norway rat as a pest of crop plants.] [*Kusunoki Agric. News*] (Kōchi Pref. Agr. Exp. Sta.), **8(8)**, 1–14.

Tanaka, S. (1973). Significance of egg and larval surveys in the studies of population dynamics of fish. In *The Early History of Fish*, ed. J. H. S. Blaxter, pp. 151–7. Berlin: Springer-Verlag.

Tanaka, T. (1967). [Evolution of life history patterns and wing-morphs in aphids.] *Shokubutsu-Bōeki* [*Plant Protection*], **21(6)**, 249–54.

Taylor, R. W. (1978). *Nothomyrmecia macrops*: a living-fossil ant rediscovered. *Science*, **201**, 979–85.

Test, F. H. (1954). Social aggressiveness in an amphibian. *Science*, **120**, 140–1.

Thalenhorst, W. (1953). Vergleichende Betrachtungen über den Massenwechsel der Kiefernbuschhornblattwespen. *Z. angew. Ent.*, **35**, 168–82.

Thalenhorst, W. (1958). *Grundzüge der Populationsdynamik des grossen Fichtenborkenkäfers* Ips typographus *L.* Frankfurt: Sauerländers Verlag.

Thompson, W. R. (1928). A consideration to the study of biological control and parasite introduction in continental areas. *Parasitology*, **20**, 90–112.

Thompson, W. R. (1939). Biological control and the theories of the interactions of populations. *Parasitology*, **31**, 299–388.

Thompson, W. R. (1956). The fundamental theory of natural and biological control. *Annu. Rev. Ent.*, **1**, 379–402.

Thomson, J. A. & Owen, W. H. (1964). A field study of the Australian ringtail possum *Pseudocheirus peregrinus* (Marsupialia: Phalangeridae). *Ecol. Monogr.*, **34**, 27–52.

Thorington, R. W. (1968). Observations of squirrel monkeys in a Columbian forest. In *The Squirrel Monkey*, ed. L. A. Rosenblum & B. W. Cooper, pp. 69–87. New York: Academic Press.

Thorson, G. (1950). Reproductive and larval ecology of marine bottom invertebrates. *Biol. Rev.*, **25**, 1–45.

Tinbergen, N. (1953). *Social Behaviour in Animals, with Special Reference to Vertebrates.* London: Methuen.

Tinbergen, N. (1956). On the functions of territory in gulls. *Ibis*, **98**, 401–11.

Tinbergern, N. (1957). The functions of territory. *Bird Study*, **4**, 14–27.

Tinkle, D. W. (1969). The concept of reproductive effort and its relation to the evolution of life histories in lizards. *Amer. Natur.*, **103**, 501–16.

Tinkle, D. W. (1972). The dynamics of a Utah population of *Sceloporus undulatus*. *Herpetologica*, **28**, 351–9.

Tokuda, K. (1956). [Dominance–subordination relations in natural troops of Japanese macaques.] [*Biol. Sci., Tokyo, Supplement: Interrelations between Organisms and Environment*], 48–53.

Tokuda, M. (1957). [*Evolution*.], revised edn. Tokyo: Iwanami.

Tompa, F. S. (1964). Factors determining the numbers of song sparrows, *Melospiza melodia* (Wilson), on Mandarte Island, B.C., Canada. *Acta zool. Fennica*, **109**, 1–73.

Toyoshima, H. (1955). [An observation on the territorial behaviour of syrphid flies (Diptera).] *Dobutsu-Shinrigaku Nenpo* [*Ann. Anim. Psych.*], **5**, 63–5.

Tschirkova, A. F. (1955). Opuit massovoi glazomernoi otsenki chislennosti i prognozui 'urozhaya' pestsov (1944–1949 g.). *Voprosui Biologii Pushnuikh Zverei*, **14**, 1–74. English transl. in *Translations of Russian Game Reports* (1958) **3**, 101–65. Canadian Department of Northern Affairs & National Resources.

Tsuneki, K. & Adachi, Y. (1957). [The intra- and interspecific influence relations among nest populations of four species of ants.] *Jap. J. Ecol.*, **7**, 166–71.

Turner, F. B., Hoddenbach, G. A., Medica, P. A. & Lannom, J. R. (1970). The demography of the lizard, *Uta stansburiana* Baird & Girard, in southern Nevada. *J. Anim. Ecol.*, **39**, 505–19.

Tyndale-Biscoe, H. (1973). *Life of Marsupials*. London: Edward-Arnold.

Ubukata, H. (1975). [Territorial behaviour of a dragonfly, *Cordulia aenea*.] [*Insectarium*] (1975), (**12**) 196–99.

Uchida, T. (ed.) (1976). [*Yazu-Uchida Dictionary of Animal Nomenclature*.] Tokyo: Nakayama.

Ueda, T. (1970). [*Studies on Freshwater Prawns in Japan*.], revised edn. Matsue: Sonoyama.

Uramoto, M. (1966). [Life of Birds.] Tokyo: Kinokuniya.

Uramoto, M. (1967). [Crook's paper on the evolution of social organization and intraspecific communication in weaver birds.] [*Biol. Sci., Tokyo*], **19**, 49–55.

Uvarov, B. P. (1921). A review of the genus *Locusta* L. (= *Pachytulus* Fieb.), with a new theory as to the periodicity and migration of locusts, *Bull. ent. Res.*, **12**, 13–163.

Uvarov, B. P. (1931). Insect and climate. *Trans. ent. Soc. Lond.*, **79**, 1–247.

Varley, G. C. (1949). Population changes in German forest pests (special review). *J. Anim. Ecol.*, **18**, 117–22.

Varley, G. C. & Gradwell, G. R. (1958). Oak defoliators in England. In *Proceedings of the Tenth International Congress of Entomology* (Montreal, 1956), vol. 4, ed. E. B. Watson, B. M. McGugar & M. L. Prebble, pp. 133–6.

Varley, G. C. & Gradwell, G. R. (1960). Key factors in population studies. *J. Anim. Ecol.*, **29**, 399–401.

Varley, G. C. & Gradwell, G. R. (1963). The interpretation of insect population changes. *Proc. Ceylon Ass. Adv. Sci.*, **18(D)**, 142–56.

Varley, G. C. & Gradwell, G. R. (1968). Population models for the winter moth. In *Insect Abundance*, ed. T. R. E. Southwood, pp. 132–42. Oxford: Blackwell.

Varley, G. C., Gradwell, G. R. & Hassell, M. P. (1973). *Insect Population Ecology. An Analytical Approach*. Oxford: Blackwell.

Verner, J. & Wilson, M. F. (1966). The influence of habitats on mating systems of North American passerine birds. *Ecology*, **47**, 143–7.

Vial, J. L. (ed.) (1973). *Evolutionary Biology of the Anurans*. Columbia: University of Missouri Press.

Vial, J. L., Berger, T. L. & McWilliams, W. T. Jr. (1977). Quantitative demography of

copperheads, *Agkistrodon contortrix* (Serpentes: Viperidae). *Res. Popul. Ecol.*, **18**, 223–34.
Viktorov, G. A. (1955). K. voprosu o prichinakh massovuikh razmnozhenii nasekomuikh. *Zool. Zhurn.*, **34**, 259–66.
Vinegar, M. B. (1975). Demography of the striped plateau lizard *Sceloporus virgatus*. *Ecology* **56**, 172–82.
Volterra, V. (1926), Variazioni e fluctuazioni del numero d'individui in specie animali conviventi. *Mem. Acad. Lincei*, **602**, 31–113. English transl. in R. N. Chapman (1931) *Animal Ecology*. New York: McGraw-Hill.
Voûte, A. D. (1943). Classification of factors influencing the natural growth of a population of insects. *Acta Biotheor.*, **7**, 99–116.
Voûte, A. D. (1946). Regulation of the density of the insect-populations in virgin-forest and cultivated woods. *Arch. neérland. Zool.*, **7**, 435–70.
Waksman, S. A. (1927). *Principles of Soil Microbiology*. Baltimore: Williams & Wilkins.
Waloff, Z. (1966). The upsurges and recession of the desert locust plague: an historical survey, *Anti-Locust Memoir*, **8**, 1–111.
Waloff, Z. (1976). Some temporal characteristics of desert locust plagues. *Anti-Locust Memoir*, **13**, 1–36.
Wang, Y. K. (1937). [On the stock of the crab, *Paralithodes camtschatica*, in the seas of Hokkaido and Saghalien.]* *Bull. Jap. Soc. Sci. Fish.*, **5**, 291–4.
Washburn, S. L. (1950). The analysis of primate evolution with particular reference to the origin of man. *Cold Spring Harbor Symp. quantit. Biol.*, **15**, 67–78.
Washburn, S. L. & DeVore, I. (1961). Social behavior of baboons and early man. In *Social Life of Early Man*, ed. S. L. Washburn, pp. 91–105. New York: Wenner-Gren Foundation.
Watanabe, M. (1976). A preliminary study on population dynamics of the swallowtail butterfly, *Papilio xuthus* L. in a deforested area. *Res. Popul. Ecol.*, **17**, 200–10.
Watanabe, T. (1970). [Morphology and ecology of early stages of life in Japanese common mackerel, *Scomber japonicus* Houttuyn, with special reference to fluctuation of population.]* *Bull. Tokai reg. Fish Res. Lab.*, **62**, 1–283.
Watanabe, T. (1972). [The recent trend in the stock size of the Pacific population of the common mackerel off Honshu, Japan, as reviewed from egg abundance.]* *Bull. Jap. Soc. Sci. Fish.*, **38**, 439–44.
Watmough, R. H. (1968). Population studies on two species of Psyllidae (Homoptera, Sternorhyncha) on broom (*Sarothamus scoparius* (L.) Wimmer). *J. Anim. Ecol.*. **37**, 283–314.
Watson, A. (1971). Key factor analysis, density dependence and population limitation in red grouse. In *Proceedings of the Advanced Study Institute on 'Dynamics of Numbers in Populations'*, ed. P. J. den Boer & G. R. Gradwell, pp. 548–64. Wageningen: P.U.D.O.C. (Centre for Agricultural Publishing and Documentation).
Watt, K. E. F. (1968). *Ecology and Resource Management*. New York: McGraw-Hill.
Weatherly, A. H. (1972). *Growth and Ecology of Fish Populations*. London: Academic Press.
Weeden, R. B. & Theberge, J. B. (1972). The dynamics of a fluctuating population of rock ptarmigan in Alaska. In *Proceedings of the XV International Ornithological Congress*, ed. K. H. Voous, pp. 90–106. Leiden: E. J. Brill.
Wellington, W. G. (1957). Individual difference as a factor in population dynamics: The development of a problem. *Can. J. Zool.*, **35**, 293–323.
Welsh, B. L. (1975). The role of grass shrimp, *Palaemonetes pugio*, in a tidal marsh ecosystem. *Ecology*, **56**, 513–30.
Wheeler, W. M. (1923). *Social Life among the Insects*. New York: Harcourt, Brace.

Wheeler, W. M. (1928). *The Social Insects. Their Origin and Evolution.* New York: Knopf.

Whittaker, J. B. (1967). Estimation of production in grassland froghoppers and leafhoppers. In *Secondary Productivity of Terrestrial Ecosystems*, vol. 2, ed. K. Petrusewicz, pp. 779–89. Warszawa: Państwowe Wydawnictwo Naukowe.

Whittaker, J. B. (1973). Density regulation in a population of *Philaenus spumarius* (L.) (Homoptera: Cercopidae). *J. Anim. Ecol.*, **42**, 163–72.

Wilbur, H. M. (1975). The evolutionary and mathematical demography of the turtle *Crysemys picta. Ecology*, **56**, 64–77.

Wiley, H. (1973). Territoriality and non-random mating in sage grouse, *Centrocercus urophasianus. Anim. Behav. Monogr.*, **6(2)**, 85–167.

Wiley, H. (1974). Evolution of social organization and life-history patterns among grouse. *Quart. Rev. Biol.*, **49**, 201–27.

Williams, G. C. (1959). Ovary weights of darters: A test of the alleged association of parental care with reduced fecundity in fishes. *Copeia* (1959), (**1**), 18–24.

Williams, G. R. (1954). Population fluctuations in some Northern Hemisphere game birds (Tetraonidae). *J. Anim. Ecol.*, **23**, 1–34.

Williamson, M. H. (1961). An ecological survey of a Scottish herring fishery. Part IV: Changes in the plankton during the period 1949 to 1959. *Bull. Mar. Ecol.*, **5**, 207–29.

Wilson, E. O. (1968). The ergonomics of castes in the social insects. *Amer. Natur.*, **102**, 41–66.

Wilson, E. O. (1971). *The Insect Societies.* Cambridge, Mass.: Harvard University Press (Belknap Press).

Wilson, E. O. (1975). *Sociobiology. The New Synthesis.* Cambridge, Mass.: Harvard University Press (Belknap Press).

Wilson, E. O. & Eisner, T. (1957). Quantitative studies of liquid food transmission in ants. *Insectes Sociaux*, **4**, 157–66.

Winn, H. E. (1958). Comparative reproductive behaviour and ecology of fourteen species of darters (Pisces-Percidae). *Ecol. Monogr.*, **28**, 155–91.

Witter, J. A., Kulman, H. M. & Hodson, A. C. (1972). Life tables for the forest tent caterpillar. *Ann. ent. Soc. Amer.*, **65**, 25–31.

Wolf, L. L. & Stiles, F. G. (1970). Evolution of pair co-operation in a tropical hummingbird. *Evolution*, **24**, 759–73.

Woodgerd, W. (1964). Population dynamics of bighorn sheep on Wildhorse Island. *J. Wildl. Manage.*, **28**, 381–91.

Woolfenden, G. E. (1975). Florida scrub jay helpers at the nest. *Auk*, **92**, 1–15.

Wunder, W. (1934). Nestbau und Brutpflege bei Reptilien. *Erg. Bio.*, **10**, 1–36.

Yamagishi, H. & Fukuhara, H. (1971). Ecological studies on chironomids in Lake Suwa. *Oecologia*, **7**, 309–27.

Yamagishi, S. (1970). [Observations on the breeding biology of *Emberiza cioides*.]* *Misc. Rept. Yamashina Inst. Ornith.*, **6(33/34)**, 103–30.

Yamagishi, S. (1971). A study of the home range and the territory in meadow bunting (*Emberiza cioides*). I. Internal structure of home range under a high density in breeding season. *Misc. Rept. Yamashina Inst. Ornithol.*, **6(36)**, 356–88.

Yamagishi, S. (1973). [The world symbolized by songs – territory of the meadow bunting.] *Anima* (1973), (**7**), 65–73.

Yamagishi, S. (1976). [Three thousand crows in a roost – how does the roost change seasonally?] *Anima* (1976), (**2**), 13–20.

Yamaguchi, T. & Tanaka, M. (1974). [Studies on the ecology of sand bubbler crab, *Scopimera globosa* De Haan (Decapoda, Ocypodidae). I. Seasonal variation of population structure.]* *Jap. J. Ecol.*, **24**, 165–74.

Yano, S. (1922). [Some considerations on the causes of outbreaks of insect pests.] *Dōbutsugaku-Zasshi* [*Zool. Mag., Tokyo*], **34**, 359–65.

Yasuno, M. (1965). Territory of ants in the Kayano grassland at Mt. Hakkōda. *Sci. Rept. Tohoku Univ. Ser. 4* (*Biol.*), **31**, 195–206.

Yoshida, N. (1975). [Streaked shearwaters on Kammuri Island.] *Anima* (1975), (**12**), 5–24.

Yoshikawa, K. (1973). [*Social Insects, with Special Reference to Hunting Wasps.*] Tokyo: Kyōritsu.

Yoshikawa, K., Ohgushi, R. & Sakagami, S. F. (1969). Preliminary report on entomology of the Osaka City University 5th Scientific Expedition to Southeast Asia 1966, with descriptions of two new genera of stenogastrine wasps by J. van der Vecht. *Nature and Life in Southeast Asia*, **6**, 153–82.

Yushima, T. (1976). [*Insect Pheromones.*] Tokyo: Tokyo University Shuppan-Kai.

Zihlman, A. L., Cronin, J. E., Cramer, D. L. & Sarich, V. M. (1978). Pygmy chimpanzee as a possible prototype for the common ancestor of humans, chimpanzees and gorillas. *Nature, London*, **275**, 744–6.

Zucchi, R., Sakagami, S. F. & Camargo, J. M. F. de (1969). Biological observation on a neotropical parasocial bee, *Eulaema nigrita*, with a review on the biology of Euglossinae (Hymenoptera, Apidae). A comparative study. *J. Fac. Sci., Hokkaido Univ. Ser. VI* (*Zool.*), **17**, 271–380.

Zweifel, R. G., & Lowe, C. H. (1966). The ecology of a population of *Xantusia vigilis*, the desert night lizard. *Amer. Mus. Novit.*, **2247**, 1–57.

TAXONOMIC INDEX

AUTHOR INDEX

SUBJECT INDEX